THE SHORTER
SCIENCE AND CIVILISATION IN CHINA

COLIN A. RONAN

The Shorter Science and Civilisation in China

AN ABRIDGEMENT OF
JOSEPH NEEDHAM'S ORIGINAL TEXT

Volume 2

VOLUME III AND A SECTION OF
VOLUME IV, PART 1
OF THE MAJOR SERIES

MATHEMATICS
ASTRONOMY
METEOROLOGY
GEOGRAPHY & MAP-MAKING
GEOLOGY & RELATED SCIENCES
PHYSICS (excluding electricity & magnetism)

CAMBRIDGE UNIVERSITY PRESS
Cambridge, New York, Melbourne, Madrid, Cape Town,
Singapore, São Paulo, Delhi, Tokyo, Mexico City

Cambridge University Press
The Edinburgh Building, Cambridge CB2 8RU, UK

Published in the United States of America by
Cambridge University Press, New York

www.cambridge.org
Information on this title: www.cambridge.org/9780521315364

First published 1981
First paperback edition 1985
Reprinted 1992, 1995

A catalogue record for this publication is available from the British Library

Library of Congress Cataloguing in Publication data

ISBN 978-0-521-23582-2 Hardback
ISBN 978-0-521-31536-4 Paperback

CONTENTS

ILLUSTRATIONS

TABLES

PREFACE

With the second volume of the abridgement of Dr Joseph Needham's *Science and Civilisation in China* we start to look in some detail at the Chinese contributions to various sciences. Beginning with mathematics, written with the non-mathematical reader in mind, the text is next concerned with the sciences of the heavens – astronomy and meteorology (which was also considered a celestial subject in ancient times). The sciences of the earth follow next – geography and map-making, with geology and its allies, seismology and mineralogy. Finally there is a description of some, though not of all, Chinese physics – their predilection for wave theory as opposed to particles, their work in measurement, their studies of statics and hydrostatics, of motion, surface phenomena, heat, light and sound – and a note on the invention of the tempered scale, almost contemporary in China and Europe. This volume therefore covers what is in volume 3 of *Science and Civilisation in China* and about half the content of volume 4, part 1.

Once again I have been encouraged and helped by Joseph Needham, who has always been very ready with advice and has generously given of his valuable time to read critically through these pages. For this I am most grateful. As before this is no new edition; it is an abridgement of the original text. Nevertheless we have made a few minor changes here and there because time has brought further insight into some questions. On magic squares, however, we do not find much need for revision in spite of the criticisms of Schuyler Cammann, but anyone wishing to go further into the matter should certainly read what he has written on the subject. Again, Geneviève Guitel's discussion of our views on the Shang numerical system does not call for any major alterations, though interested readers may well wish to refer to her careful discussion of number systems in general. Nor have we altered the original translations of the Mohist canon, even though specialists will assuredly need to follow Angus Graham's new translations. Lastly, it is necessary to draw the attention of the reader to the work of

Yabuuchi Kiyoshi on the Chinese calendar-systems or astronomical tables and of Nathan Sivin on the history of Chinese mathematical astronomy. A note of all these works will be found in the bibliography, which has been generously compiled by Joseph Needham and Nathan Sivin.

The romanisation adopted is the same as that in volume 1; i.e. the system of Wade-Giles, with the substitution of *h* for the aspirate apostrophe (see Table 1 in volume 1 of this abridgement).

My warm thanks are also due to the Press editor, Dr Simon Mitton, for his patience, and to their copy editor for her immense care. I owe a debt of thanks, too, to Mrs Shirley Barry who has typed the manuscript so ably, and to my wife, Penny, for reading carefully through all my text and making some very helpful comments. I am indebted, too, to Mr Storm Dunlop for reading proofs, and to Miss Muriel Moyle for compiling the index.

Bar Hill, Cambridge Colin A. Ronan
2 February 1980

1

Mathematics

With this volume we enter on the second part of our study of science and civilisation in early China. Since mathematics and a mathematical approach to theories have been the backbone of modern science, it seems proper that this subject should come first in any description of Chinese science and technology. Western opinions of Chinese mathematics have often oscillated between two extremes – either extolling Chinese achievements to the point of exaggeration, or denigrating them by saying that mathematically the Chinese never did anything worth while, such knowledge as they had being transmitted from the Greeks. Yet, as will now become evident, the second opinion is certainly far from the truth.

THE WRITING OF NUMERALS, PLACE-VALUE AND ZERO

Table 19 shows the various forms which Chinese written numerals can take, or have taken in the past. Of these, the 'accountants' forms' came into use gradually during and after the Han (first century B.C.), being considered more elegant and less liable to falsification. In column A the lower digits are pictographs, but from the number 4 onwards there seems to have been a borrowing of similar sounding words (homophones) from botanical and zoological nomenclature. The oracle-bone inscriptions (fourteenth to eleventh centuries B.C.) and inscriptions on coins and bronze vessels (tenth to third centuries B.C.), given in columns C and D, are in some cases similar to those related to the 'counting-rod' characters, columns F and G, which have been supposed to originate from actual counting-rods laid out on a flat board. All later notations follow the counting-board system.

It has been said that the earliest book in which the rod-numerals appear is the *Wu Tshao Suan Ching* (Mathematical Manual of the Five Government Departments), written in the fifth, or perhaps the fourth century A.D., but no editions of the book seen by us show rod-numerals; the calculations are merely written out in the standard way. But the question

Table 19. *Ancient and medieval Chinese numeral signs*

	A Standard modern forms		B Accountants' forms	C Shang oracle-bone forms (14th to 11th centuries B.C.)	D Bronze and coin forms (10th to 3rd centuries B.C.)	E Other forms found on coins of Chou period (6th to 3rd centuries B.C.)	F Counting-rod forms (2nd century B.C. to 4th century A.D.) units	F tens	G Late counting-rod forms (13th century A.D.) units	G tens	H Commercial forms (from 16th century A.D.)
1	一	*i*	弌 or 壹	一	一	一 ｜	𝍩	𝍠	𝍠	𝍩	〡
2	二	*erh*	弍 or 貳	二	二	二 ‖	𝍪	𝍡	𝍡	𝍪	〢
3	三	*san*	叁	三	三	三 ‖｜	𝍫	𝍢	𝍢	𝍫	〣
4	四	*ssu*	肆	亖	亖	亖 ☒ 𝍣 ☐ ☒	𝍬	𝍣	𝍣 ✕	𝍬 ✕	〤
5	五	*wu*	伍	☒	☒	𝍭 ✕ ✕ ☒	𝍭	𝍤	𝍤 ○	𝍭 ○	〥
6	六	*liu*	陸	∧ 介	介	介 介 ⊥ ⊤	𝍮	𝍥	𝍥	𝍮	〦
7	七	*chhi*	柒	十	十	𝍯 𝍦	𝍯	𝍦	𝍦	𝍯	〧
8	八	*pa*	捌	)(	)(	✕ 𝍰	𝍰	𝍧	𝍧	𝍰	〨
9	九	*chiu*	玖	⺄	九	𝍱	𝍱	𝍨	𝍨 ✕	𝍱 ✕	〩
10	十	*shih*	拾	See Table 20	See Table 20	十 ϕ	Indicated by place		Indicated by place		十
100	佰	*pai*	佰			𠔁					百
1,000	仟	*chhien*	仟			千					千
10,000	萬	*wan*	萬			万					万
0	零	*ling*	零			Blank space until 8th century A.D.			○		○

turns out to be of little significance, for the printing of mathematical works began in the eleventh century A.D., and since there is ample evidence that rod-numerals had been in use more than a thousand years earlier, it must have depended on individual editors whether or not rod-numerals were used in any particular printed book. Moreover, Han mathematical texts certainly use expressions which imply the use of counting-rods.

On the other hand, an enigmatic representation in the *Tso Chuan* (Master Tsochhiu's Enlargement of the Spring and Summer Annals) under the date 542 B.C. has often been cited to show that rod-numerals go back to the middle of the Chou. The passage certainly shows an understanding of the place-value of digits,* but in view of the subsequent remodelling of the *Tso Chuan* text it would perhaps be unsafe to accept this as evidence for a period earlier than that of the Warring States for which, in any case, coins bear witness. If, as there is reason to think, the Chinese character for calculation, *suan* (筭), is a truly ancient pictograph of counting-rods, the numerals (as well as the counting-board) may go back to the first millennium B.C. Some of the oracle-bone numerals (column C), especially 5, 6, 7 and 10, certainly look like arranged counting-rods.

During the Chin and the Han, the functions of the two kinds of numerals such as ㅠ and ⊥ were stabilised. The former were used for units, the latter for tens, the former for hundreds, the latter for thousands, and so on. By the end of the third century A.D. at least, they were termed respectively *tsung* and *hêng* numerals. The *Sun Tzu Suan Ching* (Master Sun's Mathematical Manual) of this period says:

> In making calculations we must first know the positions (and structure) (*wei*) (of numerals). The units are vertical and the tens horizontal, the hundreds stand while the thousands lie down; thousands and tens therefore look the same, as also the ten thousands and the hundreds ... When we come to 6, we no longer pile up (strokes), and the five has not got a one (a ligature).

The system was thus stabilised as follows:

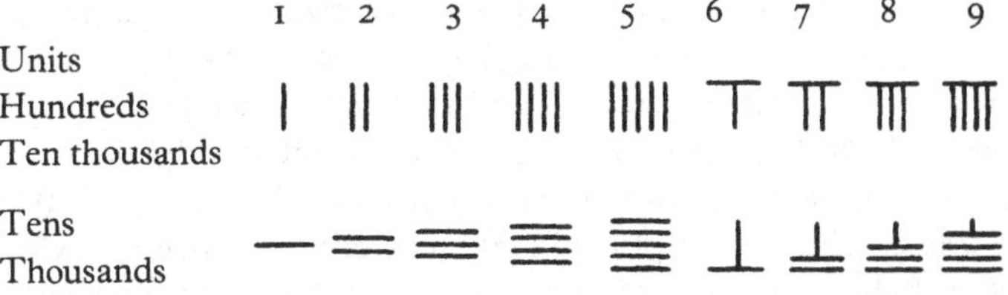

* The place-value of digits will be something familiar to every reader, if not by name at least from experience. Thus 1 is one, 10 (i.e. with the digit 1 moved one place to the left) indicates ten, 100 (the 1 two places to the left) one hundred, and so on. Decimals, of course, use place values to the right: thus 0·1 is one-tenth, 0·01 one-hundredth, etc.

Thus, for example, the number 4716 appeared as ≣ ╥ — ⊤. The separation of neighbouring powers of ten in this way facilitated the use of counting-boards without marked vertical columns. But it is worth noting the striking fact that, as far back as the oracle-bone numerals of the thirteenth century B.C., the symbols for 1 and 10 were both straight lines, the former horizontal and the latter vertical. This was just the opposite of the convention recorded by Master Sun, but the principle was identical.

The word *wei* in the quotation from the *Sun Tzu* referred essentially to the positions of the rods in the columns on a counting-board, in other words to place-value. Another word was 'rank' (*têng*, 等). Before the eighth century A.D., the place where a zero would now be written was always left vacant. For example, in a Thang manuscript from the Tunhuang cave-temples, one roll contains multiplication tables; here the results appear as rod-numerals, and 405 is shown as ≣ |||||.

The circular symbol for zero is first found in A.D. 1247 in print in the *Shu Shu Chiu Chang* (Mathematical Treatise in Nine Sections) of Chhin Chiu-Shao, but many have believed that it was in use during the preceding century at least. The usual view is that it was derived from India, where it first appears on the Bhojadeva inscriptions at Gwalior dated A.D. 870. But there is no positive evidence for this transmission, and the form could perhaps have been borrowed from the philosophical diagrams of which the twelfth-century A.D. Neo-Confucians were so fond. In any case, the Sung mathematicians had at their disposal a fully developed notation, as in this example from the work of Chhin Chiu-Shao, where the subtraction

$$1{,}405{,}536 = 1{,}470{,}000 - 64{,}464$$

appears as follows:

| ≣ O ≣ ||||| ═ ⊤ | ≣ ╥ O O O O
⊤ X |||| ⊥ X

However, while the first written evidence for the zero in India is of the late ninth century A.D., it has been discovered in use about two hundred years earlier in Indo-China and other parts of south-east Asia. This fact may be of much significance. The literary and written evidence for place-value in India has been conflicting. A dating of the eighth century A.D. seems the earliest that can be determined with any certainty. But Indo-Chinese inscriptions use place-values much earlier – A.D. 609 in Champa in eastern India, and A.D. 605 in Cambodia. And it was soon after this, in 683, that the first inscriptions showing zero appear simultaneously in Cambodia and Sumatra. The Indian numerals, with separate signs for 10 and its multiples, were no improvement on the Greek and Hebrew alphabetical numbers, so Indo-China seems at first sight rather an unlikely place for such a revolutionary discovery. It is possible, though, that the written zero symbol, and the more reliable calculations which it permitted, really originated in the eastern zone of Hindu culture where it met the southern zone of the culture of the Chinese.

What stimulus could have been received at this interface? Could the culture there have adopted a zero sign from the empty blanks left for zero on the Chinese counting-boards? The essential point is that the Chinese had possessed, long before the time of the *Sun Tzu Suan Ching* (late third century A.D.), a place-value system which was fundamentally decimal. It may then be that the 'emptiness' of Taoist mysticism, no less than the 'void' of Indian philosophy, contributed to the invention of a symbol for zero. The first appearance of zero in dated inscriptions on the borderline of the Indian and Chinese culture areas would then seem hardly likely to be a coincidence.

As far as place-value is concerned, the Chinese appear always to have used it. The Shang numeral system (thirteenth century B.C.) is the earliest of which we have evidence; its symbols are given in Table 20, and the significance of place-value is obvious. Clearly, it was more advanced than contemporary scripts in Babylonia and Egypt. Admittedly, all three systems started a new cycle of signs at 10 and multiples of 10, and each had a place-value component, but only the Shang Chinese were able to express any number, however large, using no more than 9 numerals and a counting-board. This was a great stride forward. What is more, they never used the somewhat cumbersome Roman subtractive place-value system of forming numerals (where, for instance, 9 is written IX). The Shang Chinese system seems, then, the simplest of the ancient methods, and appeared two thousand years before the West inherited what are usually called the 'Arabic' numerals.

The other question we must glance at concerns the origin and history of the Chinese written form for the word zero, *ling* (零). The ancient meaning of this word was the last small raindrops of a storm, or drops of rain remaining afterwards on things. Later it came to be applied to any remainder, especially when coupled with another word, and would appear in a phrase like 'five over the hundred'. From this the transition of the use of *ling* for expressing the zero in a number such as 105 may readily be understood. Nevertheless, this use seems to have risen late, and though to determine precisely when would require a special investigation, we have never met with its use as a word meaning zero in a mathematical text until the Ming (fourteenth century A.D.). On the other hand, the algebraists in Sung times (thirteenth century A.D.) used the symbol o extensively, as we have seen (page 4), and it is easy to find examples of numbers written out in which the term *ling* could have been employed. It is, then, a little difficult to explain why the use of *ling* came in during the Ming and not at any other

Table 20. *Notations for numerals higher than 10 on the Shang oracle-bones and on Chou bronze inscriptions*

	Bone forms	Coin forms	Bronze forms
11	11th month perhaps also		
12			
13			
14	presumably analogous but no example known		
15			
20			
30			
40			
50			
56	(i.e. *wu shih yu liu* 五十又六; five tens plus six)		
60			
88			
90			
100			
162			
200			
209	(i.e. *erh pai yu chiu* 二百又九; two hundreds and nine)		
300			
500			
600			
656	(i.e. *liu pai wu shih liu* 六百五十六; six hundreds, five tens, six.) This is the form which continued unchanged through the next three thousand years)		
1000			
3000			
4000			
5000			

time. However, we have suggested that perhaps the zero symbol had been pronounced 'ling' from the time of its first general use in the Sung, and it may well be that the use of the old character arose not only because it had long meant 'remainder', but also because the o symbol was shaped like a spherical raindrop.

SURVEY OF PRINCIPAL LANDMARKS IN CHINESE MATHEMATICAL LITERATURE

We give here a very brief summary of the most important books on mathematics produced by the Chinese over the centuries. Much Chinese mathematical literature was devoted to commentaries on them, and these often included original material.

From antiquity to the San Kuo period (third century A.D.)

Tradition has it that the *Chou Pei Suan Ching* (The Arithmetical Classic of the Gnomon and the Circular Paths of Heaven) is the earliest mathematical classic, but the first firm dates connected with it are some two centuries later than those associated with the *Chiu Chang Suan Shu* (Nine Chapters on the Mathematical Art). However, the question is a difficult one, and it will be convenient if we accept the traditional view here, especially since most of the book is so archaic that it is difficult not to believe that it goes back to the period of the Warring States (see chronological table, page 388).

The *Chou Pei* mentions Pythagoras' Theorem about right-angled triangles (the theorem that the square on the hypotenuse of the triangle is equal to the sum of the squares on the other two sides); this is at least as old as the sixth century B.C., but there is no proof of the kind later offered by Euclid (third century B.C.). It mentions the gnomon, a vertical stake that casts shadows the length and direction of which give information about the sun's altitude, and which was much used in early astronomy and calendar-making. There are also diagrams of stars close to the Pole Star and some other astronomical matters. Yet although concerned so much with astronomy, the book is interesting mathematically since there is some use of fractions, and discussion of their multiplication and division and the finding of common denominators. And even if the process of working out square roots is not given, the text makes it clear that square roots were certainly used.

The discussion of the right-angled triangle (Fig. 30), which occurs at the beginning of the book, is the oldest part of the text. A section of this runs as follows:

(1) Of old, Chou Kung addressed Shang Kao, saying, 'I have heard that the Grand Prefect (Shang Kao) is versed in the art of numbering. May I venture to enquire how Fu-Hsi anciently established the degrees of the celestial sphere? There are no steps by which one may ascend the heavens, and the earth is not measurable with a foot-rule. I should like to ask you what was the origin of these numbers?'

(2) Shang Kao replied, 'The art of numbering proceeds from the circle (*yuan*) and the square (*fang*). The circle is derived from the square and the square from the rectangle (lit. T-square or carpenter's square; *chü*).

(3) The rectangle originates from (the fact that) $9 \times 9 = 81$ (i.e. the multiplication table or the properties of numbers as such).

(4) Thus, let us cut a rectangle (diagonally), and make the width (*kou*) 3 (units) wide, and the length (*ku*) 4 (units) long. The diagonal (*ching*) between the (two) corners will then be 5 (units) long. Now after drawing a square on this diagonal, circumscribe it by half-rectangles like that which has been left outside, so as to form a (square) plate. Thus the (four) outer half-rectangles of width 3, length 4, and diagonal 5, together make (*tê chhêng*) two rectangles (of area 24); then (when this is subtracted from the square plate of area 49) the remainder (*chang*) is of area 25. This (process) is called 'piling up the rectangles' (*chi chü*[2]).

(5) The methods used by Yü the Great in governing the world were derived from these numbers.'

It will be remembered that the legendary Yü was the patron saint of hydraulic engineers and all those concerned with water-control, irrigation and conservancy. Epigraphic evidence from the Later Han, when the *Chou Pei* had taken its present form, shows us, in reliefs on the walls of the Wu Liang tomb-shrines (*c.* + 140), the legendary culture-heroes Fu-Hsi and Nü-Kua holding a carpenter's square and *quipu* (see Fig. 17 in volume 1 of this abridgement, page 62). The reference to Yü here undoubtedly indicates the ancient need for mensuration and applied mathematics.

Although no further commentary is needed, it is worth noting what seems to be one deeply significant point, namely the statement in paragraph

弦圖

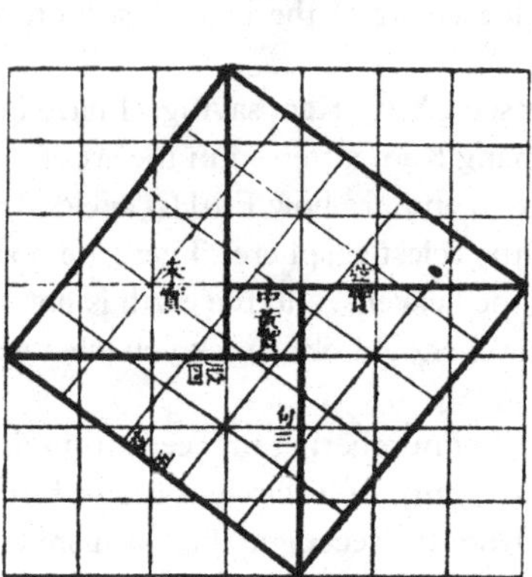

Fig. 30. The proof of Pythagoras' Theorem in the *Chou Pei Suan Ching*.

(3) that geometry arises from measurement. This seems to show the Chinese arithmetical-algebraic mind at work from the earliest times; they had little interest in abstract geometry, independent of concrete numbers, and consisting of theorems and propositions capable of proof, given only certain fundamental postulates to begin with. The precise numbers might be unknown, but numbers there had to be. And it is worth remembering that although the book was written at a time when astrology and divination were universally dominant, it speaks of the phenomena of the heavens and the earth without the slightest admixture of superstition.

The *Chiu Chang Suan Shu* represents a much more advanced state of mathematical knowledge than the *Chou Pei*. Dating the *Chiu Chang* is also difficult, and it is perhaps safest to regard it as a Chhin and Former Han book with Later Han accretions, bringing it down to some time in the second or early third century A.D. Perhaps the most influential of all Chinese mathematical books, it contains nine chapters with a total of 246 problems. These concern land surveying, engineering, the fair distribution of taxation, and other subjects, all of which bring various mathematical operations into play. The handling of fractions, the use of 'arithmetical' and 'geometrical' progressions of numbers, and the solution of simple simultaneous equations are all here, while the book also introduces the Rule of Three (a method of finding a fourth number from three given numbers where the ratio between two of them is the same as that between the third and the unknown fourth), and the Chinese invention, the Rule of False Position, for solving simple equations (page 49). Moreover, in the discussion on simultaneous equations, the problem of finding five unknowns with only four equations is discussed – a foreshadowing of indeterminate equations (page 49). In addition to the algebra contained in it, the *Chiu Chang* deals with geometrical questions – the areas of various figures and the volume of different solids, as well as the right-angled triangle and the famous problems associated with it, one of which is illustrated in Fig. 31. Some of these problems later found their way to Europe by way of India.

Apart from the two works described, there were many other mathematical books current during the Han dynasty; indeed the titles of some of them are known. Unfortunately all were afterwards lost. However, one Han book of considerable importance was the *Shu Shu Chi I* (Memoir on Some Traditions of Mathematical Art) by Hsü Yo, who lived and worked around A.D. 190. We know of it owing to a commentary written some four centuries later by Chen Luan, and it was clearly a very different kind of book from those already mentioned, being much nearer to Taoism and divination. Nevertheless it contains one of the earliest literary references to a magic square – a discovery in the theory of numbers which we shall consider on page 18 – and the earliest mention of the abacus.

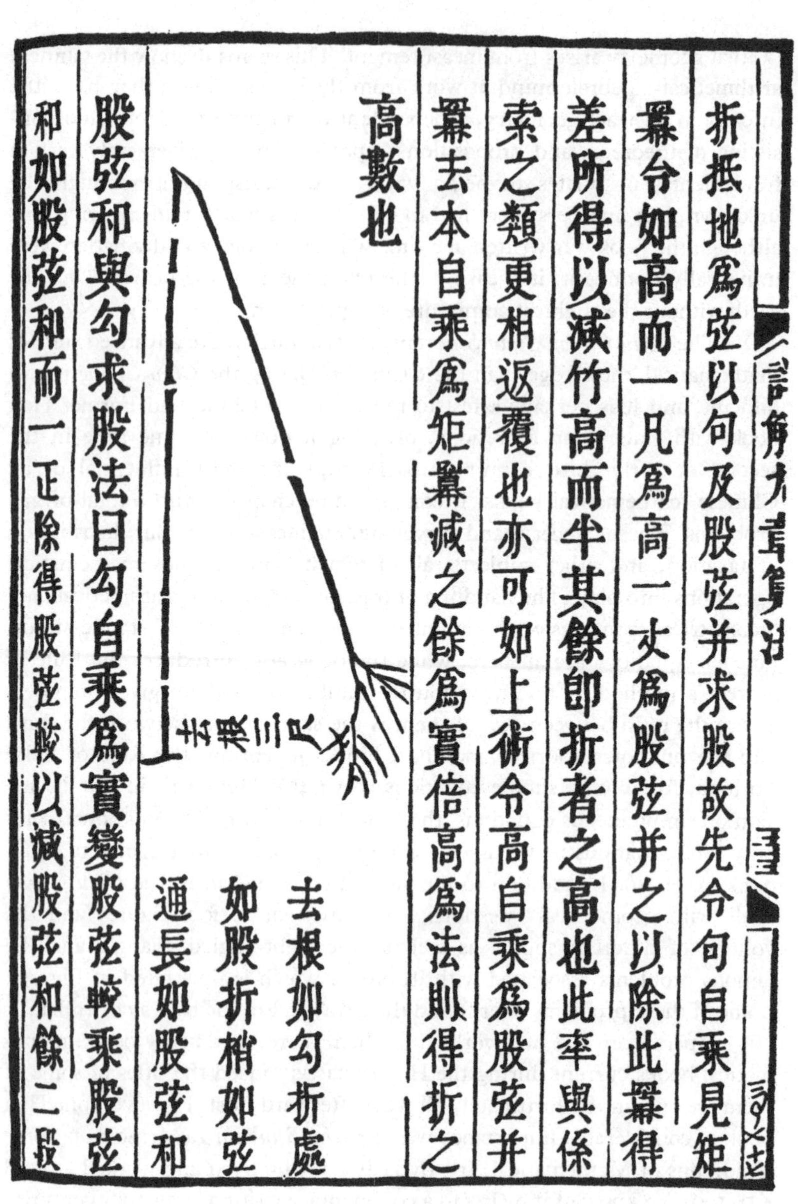
折抵地爲茲以句及股茲并求股故先令句自乘見矩
冪令如高而一凡爲高一丈爲股茲并之以除此冪得
差所得以減竹高而半其餘卽折者之高也此率與係
索之類更相返覆也亦可如上術令高自乘爲股茲并
冪去本自乘爲矩冪減之餘爲實倍高爲法則得折之
高數也

去根三尺

去根如勾折處
如股折梢如弦
通長如股茲和

股弦和與勾求股法曰勾自乘爲實變股茲較乘股茲
和如股茲和而一 正除得股茲較以減股茲和餘二段

Fig. 31. The problem of the Broken Bamboo (from Yang Hui's *Hsiang Chieh Chiu Chang Suan Fa*, A.D. 1261).

From the San Kuo to the beginning of the Sung (tenth century A.D.)

During the following centuries there appeared a dozen books which acquired renown. The earliest is the *Sun Tzu Suan Ching* (Master Sun's Arithmetical Manual), a straightforward book having as its chief mathematical advance its example of a worked-out problem in indeterminate analysis (see page 49). Probably the most important book of this period is the *Chui Shu* (Book of Connections) of Tsu Chhung-Chih (A.D. 430 to 501). It had a very great reputation, and was supposed to require far more study than any of the other mathematical books; few could understand it. The probability is that it was concerned with Tsu Chhung-Chih's remarkably accurate determination of π (Fig. 32; see also page 42), and with the determination of values of a quantity or 'function' at those points where it is unknown, or what we call the Method of Finite Differences – a technique important in astronomy and in the theory of the calendar (see page 50).

Tsu Chhung-Chih's book seems to be the chief exception to the rather slow development of Chinese mathematical literature after the *Chiu Chang Suan Shu*, since most of the books between the third and sixth centuries A.D. repeated the mistakes of the *Chiu Chang* while adding little that was new to its solid achievements. Nevertheless, in almost every case, each author made some special contribution of his own. This was especially true of Chang Chhiu-Chien, who worked between A.D. 560 and 580, who in his *Chang Chhiu-Chien Suan Ching* (Chang Chhiu-Chien's Mathematical Manual) was the first to deal with true indeterminate equations (page 49), devised new uses for quadratic equations (i.e. equations involving x^2), and expressed such equations in algebraic form, using words instead of numbers.

Equations of the third degree (i.e. involving x^3) appear first in 625, during the Thang dynasty (618 to 906), when mathematics made progress after the static period just mentioned. The equations arose out of the practical needs of engineers, architects and surveyors (Fig. 33). This was also the period of Li Shun-Fêng (late seventh century), probably the greatest commentator on mathematical books in all Chinese history.

The Sung, Yuan (Mongol) and Ming periods

The great period of Sung and Yuan algebra lay in the thirteenth and fourteenth centuries, but between this and the times of Li Shun-Fêng there stand a few mathematicians of importance. The most interesting is Shen Kua, whose many-sided genius was mentioned in the first volume of this abridgement. His book *Mêng Chhi Pi Than* (Dream Pool Essays) of 1086 is not a formal mathematical treatise, for it contains notes on almost every science of his time, but there is much of algebraic and geometric interest to be found in it. His need to survey led him to geometry and, in particular, to

Fig. 32. One of Tai Chen's illustrations of the *Chiu Chang Suan Shu*, explaining Liu Hui's method (A.D. 264) of finding the approximate value of π.

find the lengths of circular arcs, and he devised a method which formed the basis of the spherical trigonometry (trigonometry applied to figures not on a plane surface but on the surface of a sphere) evolved by Kuo Shou-Ching in the thirteenth century. His book also gives the first instance in Chinese mathematics of summing a series of numbers: it relates to the number of

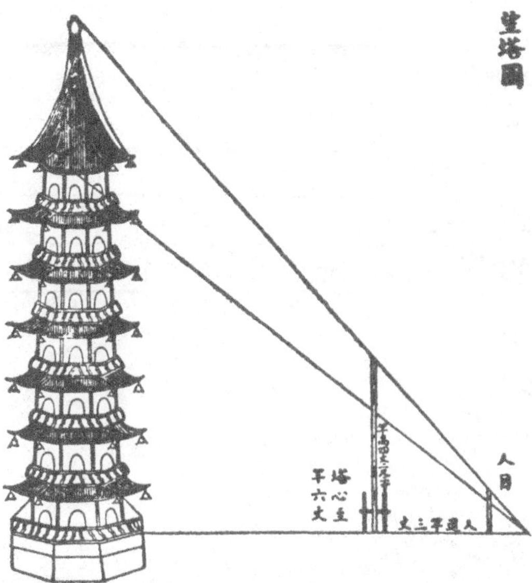

Fig. 33. Practical geometry; the measurement of the height of a pagoda as explained in Liu Hui's third-century A.D. *Hai Tao Ching* (illustration from the *Shu Shu Chiu Chang* of Chhin Chiu-Shao).

kegs which can be piled up in layers in a space shaped like a decapitated pyramid. It was in the eleventh century, too, that we find the first printed Chinese mathematical books, although it seems possible that the first could have appeared two hundred years earlier. However, the first such book of which we have knowledge was the *Hai Tao Suan Ching* (Sea Island Mathematical Manual) of Liu Hui.

It was in the later phase of Sung science (i.e. after 1200) that the greatest mathematical contributions were made, and mostly during the latter half of the thirteenth century. Those concerned were Chhin Chiu-Shao (thirteenth century), Li Yeh (1178 to 1265), Yang Hui (thirteenth century), and Chu Shih-Chieh (thirteenth century), but they seem to have worked in isolation – at least our knowledge of this medieval Chinese group is too fragmentary to allow us to trace any links. Socially there is a difference between the mathematicians of the thirteenth century and those in the Thang; in the Thang they were high officials but in the Sung mostly minor functionaries or men of the people whose attention was directed to practical problems of interest to the populace or to technicians. Thus in Chhin Chiu-Shao's *Shu Shu Chiu Chang* (Mathematical Treatise in Nine Sections), while the first section is devoted to indeterminate analysis (page 49), the rest is concerned with calculating complex areas and volumes –

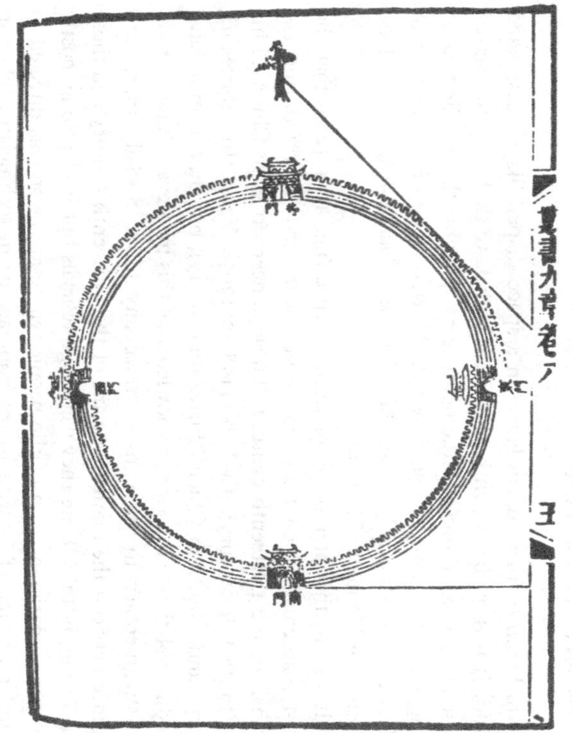

Fig. 34. Determination of the diameter and circumference of a circular walled city from distant observations (*Shu Shu Chiu Chang*, A.D. 1247).

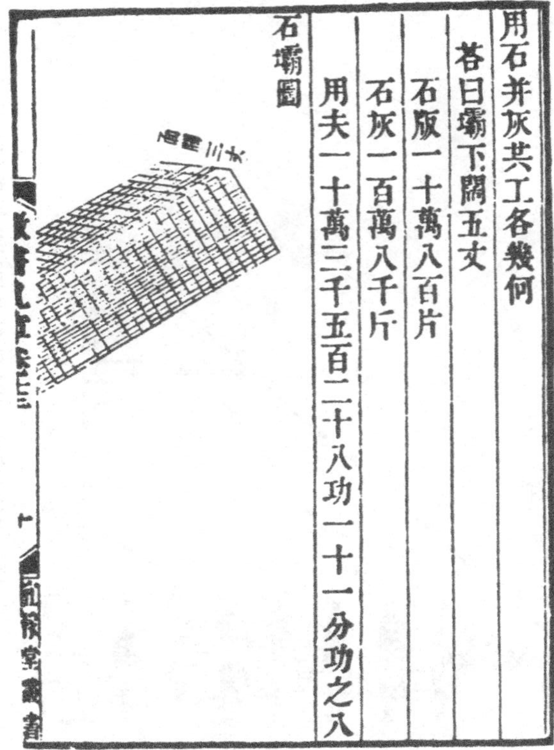

用石并灰共工各幾何
荅曰墻下闊五丈
石版一十萬八百片
石灰一百萬八千斤
用夫一十萬三千五百二十八功一十一分功之八
石墻圖

Fig. 35. A problem in dyke construction (*Shu Shu Chiu Chang*, A.D. 1247).

problems like determining the diameter and circumference of a city wall from a point outside (Fig. 34), or those concerned in dyke construction (Fig. 35), allocation of water for irrigation, and problems connected with financial affairs. However, in spite of the practical bias, Chhin Chiu-Shao's book made some novel introductions. It was here that a famous Chinese phrase *thien yuan* (literally 'heaven origin unit') appeared for the symbol of unity (1) placed at the top left-hand corner of the counting-board before beginning one of the most important parts of indeterminate analysis, while the symbol zero appears throughout the book. Negative and positive numbers were distinguished respectively by black and red characters, and Chhin Chiu-Shao, and all the Sung and Yuan algebraists after him, always arranged that in any equation the absolute term (i.e. the term without x) should be negative; this was directly equivalent to putting an equation equal to zero, a technique which did not arrive in Europe until the beginning of the seventeenth century.

Turning now to Li Yeh and Yang Hui, we find considerable differences between them. Li Yeh's *Tshê Yuan Hai Ching* (Sea Mirror of Circle Measurements), while it deals with circles inscribed in triangles, is mainly concerned with the solution of equations. His treatment is entirely algebraic, as it is also in his *I Ku Yen Tuan* (New Steps in Computation), in which books he introduced the use of a diagonal line across a digit to indicate that it was negative. Yang Hui stands a little apart from Li Yeh and, indeed, from the rest of the group, being interested in arithmetical progressions and the 'Rule of Mixtures' (the solution of problems concerning the mixing of articles of different qualities or values). Yang Hui was an expert in decimal fractions, and used what is the equivalent of a decimal point. It is also in Yang Hui's work that we first find quadratic equations with negative coefficients of x, although he says that Liu I had considered them before him.

With Chu Shih-Chieh, the fourth of our Sung mathematicians, the high-water mark of Chinese algebra is reached. In the first of his books, the *Suan Hsüeh Chhi Mêng* (Introduction to Mathematical Studies) of 1299, the rule for signs in algebraic addition and multiplication is given, and it forms, as a whole, an introduction to algebra. It was, however, in the *Ssu Yuan Yü Chien* (Precious Mirror of the Four Elements) of 1303 that Chu published his really important discoveries. The book contains an interesting preface written by a certain Tsu I-Chi, part of which runs as follows:

> People come like clouds from the four quarters to meet at his gate in order to learn from him ... By the aid of geometrical figures he explains the relations of heaven, earth, men and things (technical terms for the algebraic notation). Heaven corresponds to the base of

> the right-angled triangle, earth to the height, man to the hypotenuse, and things to the diameter of a circle inscribed in the triangle (*huang fang*), as may be seen from his diagrams. By moving the expressions upwards and downwards, and from side to side, by advancing and retiring, alternating and connecting, by changing, dividing and multiplying, by assuming the unreal for the real and using the imaginary for the true, by employing different signs for positive and negative, by keeping some and eliminating others and then changing the positions of the counting-rods, by attacking from the front or from one side, as shown in the four examples – he finally succeeds in working out the equations and roots in a profound yet natural manner ... By not using (a thing) yet it is used; by not using a number the number required is obtained. Mathematicians aforetime could not attain to the mysterious principles contained in the present profound book.

The interest of this quotation lies in the Taoist paradoxes which it contains, suggesting that they may have afforded some inspiration to the Chinese algebraists.

The work itself opens with a diagram identical to that which later on in the West came to be known as Pascal's triangle (page 54 and Fig. 53). Chu entitles it 'Diagram of the Old Method for finding Eighth and Lower Powers', suggesting that it had been known at least for some time. He also gives the 'four-element process' which was essentially to take accessory unknowns beside the one which is sought for and then, from the known relations given by the data of a problem, to get rid of the accessory unknowns. Hence the remark of Tsu I-Chi quoted above. Thus Chu's procedure, which was used for solving simultaneous equations of less than five variables, is practically identical to that of the nineteenth-century Western mathematician James Sylvester, except that Chu did not use the technique of determinants (page 48).

During the Thang, higher mathematics had been applied to calendar-making by I-Hsing and now, under the Yuan (Mongol) dynasty, this need led to the work of a very notable mathematician and astronomer, Kuo Shou-Ching (1231 to 1316). Though none of his original writings has survived, his methods can be studied in another work.

Being concerned with the calendar, Kuo Shou-Ching had to consider the spherical figures made on the sphere of the heavens by the intersections of the ecliptic (the sun's apparent path), the celestial equator and the path of the moon. This led him to discuss geometrical figures on a spherical surface and thus to lay the foundations of spherical trigonometry in China, although in saying this we must realise that he does not seem to have known the basic trigonometrical quantities such as the sine and cosine: it was not, then, a fully developed spherical trigonometry as we now understand it. Kuo also employed biquadratic equations and a method devised originally by Li Shun-Fêng in the Thang and equivalent to our Method of Finite Differences (page 50), which was a satisfactory way of calculating the speed across the sky of the sun's apparent motion.

Arab influence on Kuo Shou-Ching and his contemporaries is still an unknown quantity. During the Yuan dynasty the Arabs (largely Persians and Central Asians) certainly played a role in Chinese science and technology, rather as the Indians did during the Thang, and there is no doubt that there was every opportunity for Arabic and Persian mathematical influences to enter Chinese traditions from their observatories at Marāghah and Samarkand. But whether any important effect was actually exerted, we do not know.

During the first century and a half of the Ming dynasty, little of any interest occurred in mathematics, but after 1500 things began to change. Thang Shun-Chih (1507 to 1560), a military engineer as well as a mathematician, achieved renown for his work on circle measurements, while his contemporary Ku Ying-Hsiang, Governor of Yünan, systematised the formulae developed up to that time for dealing with the arcs and segments of a circle. Ku also wrote on equations. But neither of these two Ming mathematicians, however, was a master of the old Sung and Yuan algebra, which fell completely out of use. Even Chhêng Ta-Wei, one of the most interesting of the Ming mathematicians, did not adopt it. His *Suan Fa Thung Tsung* (Systematic Treatise on Arithmetic) of 1593 was primarily a practical treatise, which dealt with the determination of areas of peculiar shape and the mixing of alloys, and also contained a considerable number of magic squares. It was, too, the first book to give an illustration of a Chinese with an abacus, with instructions for its use.

With the coming of the Jesuits to Peking at the beginning of the seventeenth century, the period of what we may call 'indigenous mathematics' comes to an end. Chinese scholars and Jesuit mathematicians began to collaborate; the Jesuits introduced algebra from Europe and there was a flush of enthusiasm for it. It was not until Mei Ku-Chhêng (1681 to 1763), pointed out in his *Chhih Shui I Chen* (Pearls recovered from the Red River) that Chinese mathematics had gone very far before the seventeenth century that the revival of Chinese algebra began.

ARITHMETICA AND COMBINATORIAL ANALYSIS

Elementary theory of numbers

In ancient times the term *arithmetica* did not mean the simple computations which go by the name of arithmetic today, but concerned rather the elementary aspects of the theory of numbers. From the atmosphere of number-mysticism and numerology (see volume 1 of this abridgement, pages 157 ff) common to both Greek and Chinese beginnings, there emerged an appreciation of the existence of prime and composite numbers, figurate numbers, amicable numbers, and the like (Table 21). In Greece, certain 'books' in Euclid's *Elements* (third century B.C.) systematise such knowledge, and the 'sieve of Eratosthenes', of the same date, is an excellent example. Here prime numbers are found by sifting out all composite real whole numbers in the natural series, 1, 2, 3, 4, 5, . . . etc.

But the distinction between odd and even numbers must have been the first to arouse interest. In the West the odd numbers were known as gnomonic numbers (the reason is given in Table 21). Determining numbers this way necessarily involves counting rows of dots or rows of squares, a practice which was also used in contemporary China, though for other purposes, and it leads to the realisation that the sum of the first 1, 3, 5, 7, . . . n odd numbers is always a square. It was natural that the odd and even numbers should be associated with the two sexes, and this is found in ancient Chinese discussions as with the Pythagoreans of ancient Greece. The Chinese also shared the widespread superstition that odd numbers were lucky and even ones unlucky.

The Greeks were interested in all these aspects of number, as well as others such as perfect numbers and figurate numbers; Arab mathematicians in the ninth century A.D. were also intrigued by them. Yet all this was rather foreign to Chinese mathematics, where the preference was for concrete number, not numbers as such. They did, however, know of figurate numbers; Chhin Chiu-Shao had discussed 'square and round arrays' in 1247, but the origin of this goes back to the first century A.D. and the *Chiu Chang*'s expression of a series as 'progressive rows in a pyramidal array'.

Magic squares

Ancient Chinese interest was in combinatorial analysis, the construction of magic figures, i.e. arrangements of tables of numbers in various geometrical forms such that when a simple operation like addition is performed on them the sum is the same whichever way the addition is made (Fig. 36). Owing to acceptance by Western historians of quite untenable datings of Chinese classics, this branch of mathematics has been given a much higher antiquity than it deserves, but even when we take a justifiably

Table 21. *Types of numbers*

Amicable numbers	Numbers which are mutually equal to the whole sum of each other's aliquot parts, i.e. the whole sum of those numbers which will divide into the amicable number without a remainder. For example, 284 and 220 are amicable numbers, since 284 may be divided without remainder by 1, 2, 4, 71 and 142 and 1 + 2 + 4 + 71 + 142 = 220; while 220 may be divided without remainder by 1, 2, 4, 5, 10, 11, 20, 22, 44, 55 and 110, the sum of which is 284.
Complex numbers	A number composed of real and imaginary parts.
Composite numbers	Numbers made up of two or more whole number factors.
Fibonacci numbers	A sequence of numbers each one of which (except for the first and second) is the sum of the two preceding numbers. The sequence is 1, 1, 2, 3, 5, 8, 13, 21, ...
Figurate numbers	Numbers which are in a progression, beginning with 1, and so-called because they can be arranged in patterns or figures. Thus there are the triangular numbers 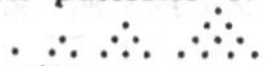 . ∴ ∴ ∴ or 1, 3, 6, 10, etc.; the square numbers  . :: ::: :::: or 1, 4, 9, 16, etc; and so on.
Gnomonic numbers	These are, simply, the odd numbers. The name arose because of the Greek development of the word 'gnomon' from meaning a stick placed vertically in the ground, to one line drawn perpendicular to another, and then to an instrument like a carpenter's square, ⌜. A natural transition from this was to a geometrical figure of this shape, i.e. the figure which remains over when a square is cut from a larger square. Applying this to square numbers we get  . :: ::: :::: where each number outside a gnomon is larger than the previous number by 2. Thus, starting with 1, we have 3, 5, 7, ..., or the series of odd whole numbers. For the algebraist each number may be expressed as $(2n + 1)$.
Imaginary numbers	Real numbers multiplied by the 'imaginary' quantity $\sqrt{-1}$.
Irrational numbers	Numbers which cannot be expressed as the ratio of two whole numbers (integers). Examples are $\sqrt{2}$ and π.
Perfect numbers	Numbers which are equal to the sum of their aliquot parts. Thus 6, whose aliquot parts are 1, 2 and 3, is a perfect number because 1 + 2 + 3 = 6. 28 is another perfect number.
Prime numbers	Those numbers which cannot be divided by any other number besides themselves and 1.
Rational numbers	Numbers which can be expressed as the ratio of two integers; e.g. $\frac{1}{4}$, $\frac{1}{2}$, $\frac{3}{4}$, etc.
Real numbers	Numbers which can be represented by a series, finite or infinite, of decimal places.
Square numbers	See 'Figurate numbers' above.
Triangular numbers	See 'Figurate numbers' above.

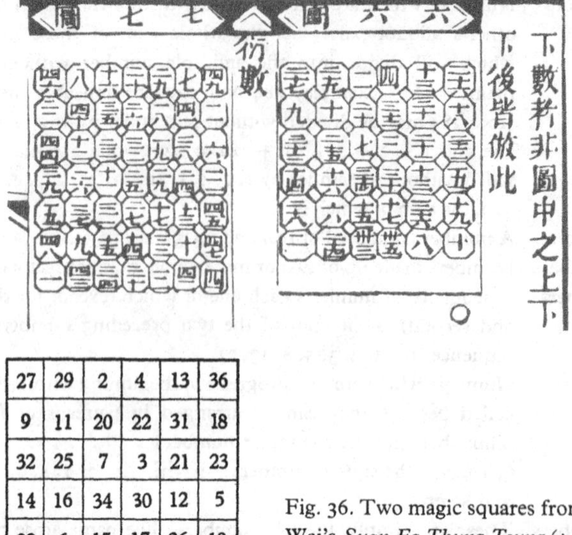

27	29	2	4	13	36
9	11	20	22	31	18
32	25	7	3	21	23
14	16	34	30	12	5
28	6	15	17	26	19
1	24	33	35	8	10

Fig. 36. Two magic squares from Chhêng Ta-Wei's *Suan Fa Thung Tsung* (A.D. 1593). The one on the right is shown below in Arabic numerals corrected by Li Nien.

conservative view of the sources the priority still seems to be Chinese. Since what has been aptly called 'the undatable Chinese tradition' goes back to the legendary period of history, the facts are rather difficult to ascertain, but they are approximately as follows.

One of the components of the corpus of legend was the story that to help him in governing the empire, the engineer-emperor Yü the Great was presented with two charts or diagrams by miraculous animals which emerged from the waters which he alone had been able to control. Thus the Lo Shu (Fig. 37) was the gift of a turtle from the River Lo, and the Ho Thu (Fig. 38) was the gift of a dragon-horse which came out of the Yellow River. There can be no doubt that this story is of great antiquity, and from internal evidence it can hardly be later than the fifth century B.C. There were then a couple of references in the fourth century, but it was only in the second century B.C. that the picture really began to develop. In the Great Appendix to the *I Ching* (Book of Changes) there are two passages. One mentions the diagrams in the usual way as having emerged from the rivers, but the other refers to them as ways of arranging the first series of nine numbers used in counting by tens – at least that is the traditional explanation given by the commentators of Sung times.

Late in the first century B.C. Ssuma Chhien recorded a curious story which is probably connected with the Ho Thu and Lo Shu, namely that Master Lu, one of Chhin Shih Huang Ti's magicians, presented him with a

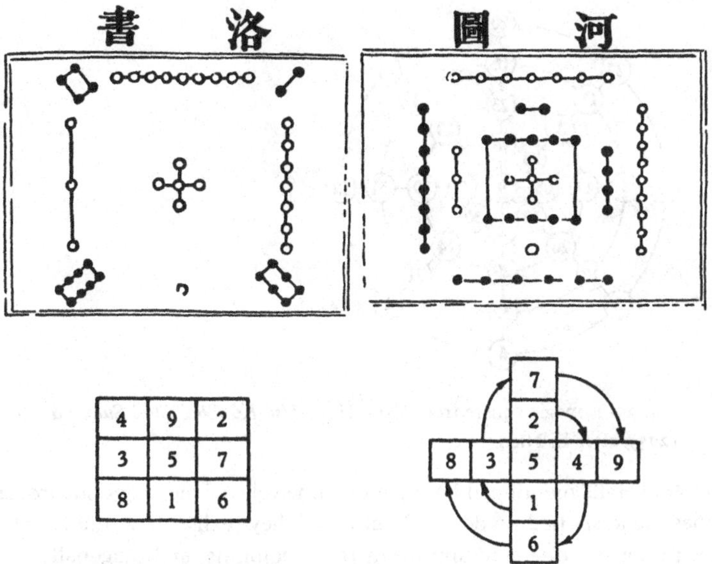

Fig. 37. The Lo Shu diagram. Fig. 38. The Ho Thu diagram.

book by a certain Lu Thu, containing prophecies. The account, which would date from about 100 B.C., reporting events of about 230 B.C., refers in all probability to the beginning of the process whereby the ancient diagrams became a nucleus for magical divination, being incorporated in the first and second centuries A.D. into new apocryphal treatises which were considered to have great authority. Before the first century A.D. there were only hints that numbers were involved but once these apocryphal treatises arrived, the situation changed. For instance, the *Ta Tai Li Chi* (Record of Rites compiled by Tai the Elder) of about A.D. 80 gives a clue to the simple arrangements concerned, while the *I Wei Chhien Tso Tu* (Apocryphal Treatise on the *Book of Changes:* A Penetration of the Regularities of the *kua* Chhien), which is either first or second century A.D., is specific and worth quoting:

> The Yang in operating advances, changing from 7 to 9 and thus symbolising the waxing of its *chhi.* The Yin in operating withdraws, changing from 8 to 6 and thus symbolising the waning of its *chhi.* Thus the Supreme Unity takes these numbers and circulates among the Nine Halls. (Whether they be added according to the direction of) the four compass points, or (according to) the four intermediary compass points, they always add up to 15.

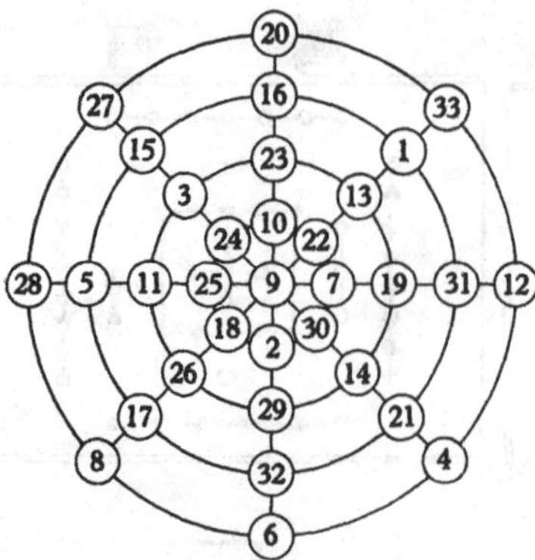

Fig. 39. A magic square from Yang Hui's *Hsü Ku Chai Chhi Suan Fa* (A.D. 1275), after Li Nien.

The Nine Halls mentioned here are the nine cells of the magic square; later Chinese scholars had no doubt about this. They realised that the Lo Shu is indeed a magic square adding up in rows, columns, and diagonally to 15, while the Ho Thu is so arranged that, disregarding the central 5 and 10 (in the upper part of Fig. 38), both odd and even sets add up to 20. After this, more mentions of the Lo Shu and Ho Thu continue, but there are also descriptions which make it clear that magic squares existed. Indeed, it is now clear that the true inventor of magic squares lived long before the identification in about A.D. 940 of the Lo Shu and Ho Thu as diagrams of magic square type.

However, before the thirteenth century the development of magic squares was absent from the main current of Chinese mathematical thought. Not until 1275 were these *tsung hêng thu* (vertical horizontal diagrams), as they were called, first studied as a mathematical problem by Yang Hui in his *Hsü Ku Chai Chhi Suan Fa* (Continuation of Ancient Mathematical Methods for Elucidating the Strange (Properties of Numbers)). Some of his magic squares were very complicated (Fig. 39), nevertheless he always gave rules for their construction. For example, a simple magic square may be constructed if the numbers from 1 to 16 are placed in an array of four columns and four horizontal rows (Fig. 40*a* and *c*), and the numbers at the corners of both inner and outer squares are transposed (Fig. 40*b* and *d*); we shall then have a magic square in which all columns, lines, and diagonals add up to 34.

Yang Hui's work was continued in 1593 by Chhêng Ta-Wei in his *Suan Fa Thung Tsung* (Systematic Treatise on Arithmetic) which gave 14 diagrams, one of which we saw in Fig. 36. Others were added by Fang Chung-Thung in 1661, and many more by Chang Chhao in 1670. At this time, and also during the eighteenth century, Japanese as well as Chinese

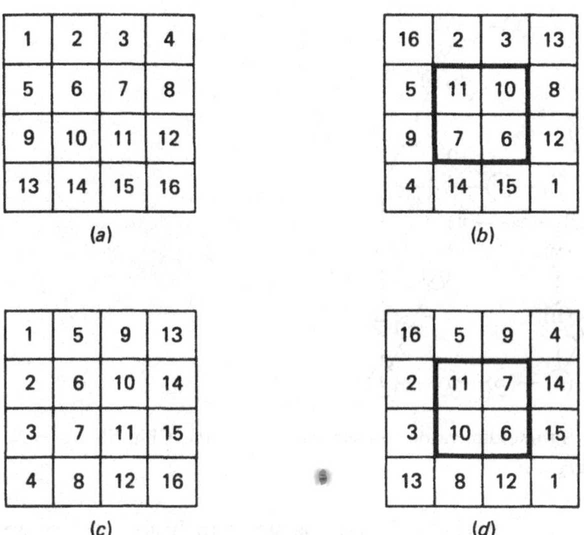

Fig. 40. The construction of a magic square.

mathematicians were particularly interested in these aspects of mathematics, with the result that in the late nineteenth century a gifted amateur such as Pao Chhi-Shou could publish a book, *Pi Nai Shan Fang Chi* (Pi-Nai Mountain Hut Records), which contained three-dimensional magic squares (see Fig. 41).

This account of magic squares will at least serve to show that there is no basis in the belief of some Western historians of science that magic squares were invented by the legendary Yellow Emperor, Huang Ti, who was supposed to have lived in the twenty-seventh century B.C. But the account is important, too, because the dates we now have render it possible to make a comparison with later aspects of the story. In the ninth century A.D., the magic square of three, the Lo Shu, became enormously important in Arabic alchemy, as will become evident in a later volume; indeed Thabit ibn Qurrah (died A.D. 901) was the first to discuss it. Then it came to Europe through the Byzantines due to the work of Manuel Moschopoulos (active 1295 to 1316). In Europe there seems to have been no early work of this kind, Theon of Smyrna, about A.D. 130, having only concerned himself with the first nine digits written in the form of a square, but not a magic square as was once believed.

CALCULATION WITH NATURAL NUMBERS

The number of fundamental operations which can be performed in arithmetic has not always been recognised as four (addition, subtraction,

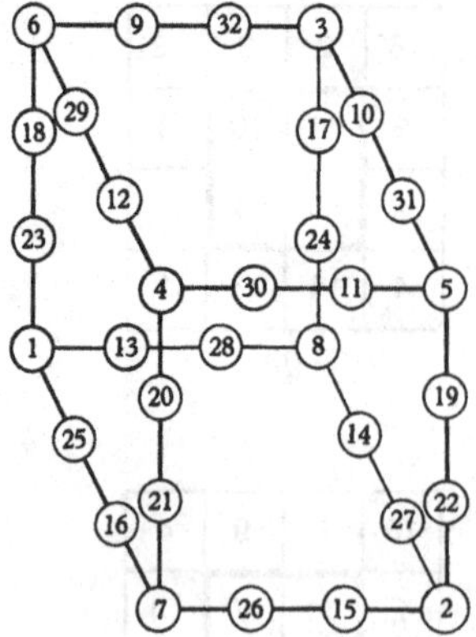

Fig. 41. A three-dimensional magic square from Pao Chhi-Shou's *Pi Nai Shan Fang Chi*.

multiplication and division). At various times in history others were included – for example, duplication and mediation (doubling and halving), and the extraction of roots. It would seem, however, that the Chinese always recognised only the modern four.

The four basic processes

About addition (*ping*, or for fractions *ho*; giving the sum, sometimes called *tu shu*) there is little to say. Texts earlier than the third century B.C. always write out the numbers in full, but it is clear that from the Warring States time addition must have been carried out with counting-rod numerals on a counting-board, using a place-value system in which blanks were left where we would put zeros. Though Chinese writing has always been run downwards from top to bottom of columns, numerals seem to have been written horizontally from left to right, and when adding the sum was placed separately below. The carrying-over process must have been very ancient, and would have been a natural outcome of counting-board operations.

Subtraction (*chien*) was carried out with rod-numerals in the same way. The remainder was placed, like the sum, separately, but above.

Just as addition was a device for obtaining results that could have been more laboriously obtained by counting, so multiplication was an 'abridgement of addition', a 'folding together' of the addends (the items to be added). The word *chhêng* seems to have been applied to a similar process of thought; its common meaning is to ride in or upon something, and here the addends were thought of as a team of horses controlled by a chariot driver. The multiplier was indeed placed in the uppermost position (*shang wei*). This was the converse of division, and the schemes were as follows:

position	*multiplication*	*division*
shang wei	multiplier	quotient
chung wei	product	dividend
hsia wei	multiplicand	divisor

The process can be followed in the accompanying diagram, set up to show the multiplication of 81 by 32.

Hundred thousands	Ten thousands	Thousands	Hundreds	Tens	Units	
				𝍰	𝍠	上位
		𝍪	𝍣			
			𝍠	𝍮		中位
				𝍫		
					𝍡	
		𝍪	𝍤	𝍱	𝍡	
				𝍫	𝍡	下位

Hundred thousands	Ten thousands	Thousands	Hundreds	Tens	Units
				8	1
		2	4		
			1	6	
				3	
					2
		2	5	9	2
				3	2

First the 8 'calls' the 3, and 24 is written two spaces to the left, representing 2,400. The 8 then calls the lower 2, and 16 is written in the middle position, representing 160. The 1 is called by the 3 and written in the tens column, then the 1 calling the 2 is set down as 2 in the units column. On addition the final product 2,592 is obtained. This process resembles one which is unusual in Europe but found in an Italian arithmetic of 1478. Probably our common modern system, known in medieval times as the 'chessboard method', originated from such calculations.

Because of the counting-board and rod-numerals, the multiplication table is naturally ancient in China. Unlike the Babylonian practice of arranging numbers in columns, the table appears in old Chinese texts written in words. Among the earliest texts in which it appears may be fragments from the fourth century B.C. which are embodied in the *Kuan Tzu* (The Book of Master Kuan). Certainly multiplication tables on bamboo tablets from about 100 B.C. have been recovered from the sand-buried city of Chü-yen in Mongolia, and portions of such tables continue to occur in texts down to the Sung. The so-called Pythagorean form of the table, with the numbers arranged in two coordinates (like a table of distances between towns given in a motoring handbook) seems to appear in China, in a triangular pattern, in the eighth century.

The square of a number has had various names. It was called *fang* (square) in the Han, *chhêng fang* (square number) in the Sung, and in modern times *tzu chhêng* (the number multiplied by itself). Another modern term is *phing fang* (the flat square), which contrasts with the name of the cube, both ancient and modern, *li fang* (the solid upstanding square thing).

Division was carried out using the rod-numerals in columns in a similar way to multiplication, and the word *chhu* was invariably used for this operation. The galley method – the technique of long division which will be familiar to most readers – was not used in China until after the time of the Jesuits. The divisor was called *fa* and the dividend *sha*. Division tables (using words) were common from the Sung onwards.

Roots

The ancient processes for the extraction of square roots (*khai fang* and *khai fang chhu chih* were two of the terms used) were essentially the same as those used today. Arabic mathematicians conceived a square number as growing out of its root, while Latin authors thought of it as the side of a geometric square; our term *radix* or root is thus derived from the Arab approach. In ancient China the procedure also arose from geometric considerations, the Han mathematicians refining the techniques so far as to lay a firm foundation for the solution of numerical equations by the great algebraists of the Sung (see page 51).

The geometrical basis of the square-root method was almost certainly pictured by Liu Hui in his edition of *Chiu Chang Suan Shu* (Nine Chapters on the Mathematical Art), although the oldest surviving diagram is in Yang Hui's *Hsiang Chieh Chiu Chang Suan Fa Tsuan Lei* (Detailed Analysis of the Mathematical Rules in the *Nine Chapters* and their Reclassification) of 1261 (Fig. 42). The figure used in the Han was similar to that given a little earlier by Euclid in about 300 B.C. The modern processes of extracting square and cube roots arose in Europe between about 1340 and 1494, and are similar to those in the *Chiu Chang Suan Chu* where the digits of the root were worked out one by one, and place-value was isolated. They seem to have arrived by way of Samarkand, but how they got there is uncertain; the influence of Chinese algebraists is a probability.

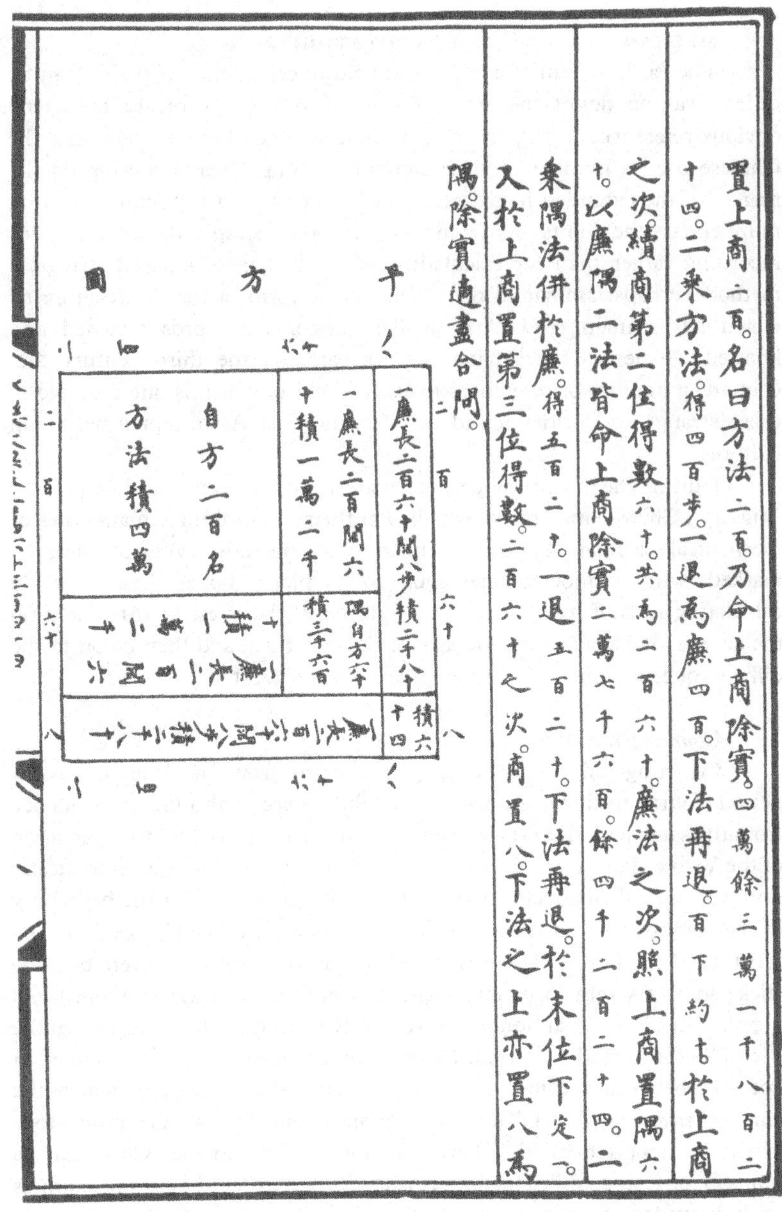

置上商二百。名曰方法二百。乃命上商除實四萬餘三萬一千八百二
十四。二乘方法。得四百步。一退爲廉四百。下法再退。百下約古於上商
之次。續商第二位得數六十。共爲二百六十。廉法之次。照上商置隅六
十。以廉隅二法皆命上商除實二萬七千六百。餘四千二百二十四。二
乘隅法併於廉。得五百二十。一退五百二十。下法再退。於末位下定一。
又於上商置第三位得數。二百六十之次。商置八。下法之上亦置八爲
隅。除實適盡合問。

Fig. 42. Diagram to illustrate the extraction of a square root, from Yang Hui's *Hsiang Chieh Chiu Chang Suan Fa Tsuan Lei* (A.D. 1261), as preserved in the manuscript *Yung-Lo Ta Tien* (A.D. 1407), ch. 16,344 (Cambridge University Library).

MECHANICAL AIDS TO COMPUTATION

The earliest form of a mechanical aid to calculation, if such it can be called, was no doubt the art of finger reckoning. There are not many obvious references to this in ancient Chinese texts, but we know that the Chinese used a number of such methods, among others counting on the fingers of the extended hand and, a more complicated technique, allotting numbers to the joints of the fingers. Another simple device, used for recording rather than for calculating, was a system of knotted strings, a method perhaps familiar to the reader in the form of the Peruvian *quipu* which has a main cord with smaller varicoloured cords attached and knotted. Ancient Chinese literature as early as the third century B.C. contains a number of references to such a device, which is one example of those strange similarities found between the East Asian and Amerindian cultures.

During the historical period, apart from the counting-board itself (Fig. 43), Chinese mathematicians had at their disposal three main types of mechanical calculating aids: (*a*) plain counting-rods; (*b*) counting-rods marked with numbers, analogous to Napier's bones (rearrangeable calculating-rods of ivory or bone introduced in the West in 1617), and (*c*) the abacus. We shall deal with each of these in turn, and then go on to the difficult question of parallel developments in the West.

Counting-rods

Counting-rod numerals appear on coins from the Warring States period (fourth to third centuries B.C.), while there is also literary evidence from the same period. Perhaps the most famous is the *Tao Tê Ching* (Canon of the Virtue (Power) of the Tao), where the sage Lao Tzu characteristically says 'Good mathematicians do not use counting-rods'. After the beginning of the Han mention of the rods is far more frequent. The *Chhien Han Shu* (History of the Former Han Dynasty), for instance, says they were bamboo sticks some 2·5 mm in diameter and 15 cm long, and 271 of them fitted together in a hexagonal bundle conveniently held in the hand. Again, in the *Shih Chi* (Historical Records), Ssuma Chhien describes a conversation in 202 B.C. between the first Han emperor and Wang Ling, in which the emperor says one of his talents was 'planning campaigns with counting-rods in the headquarters tent'. There was also a tradition that Chao Tho, a minister of Chhin Shih Huang Ti, who afterwards ruled over the south as an independent king, had several different kinds of counting-rods made before he led his army there. These were afterwards preserved in the museum of the emperor An (A.D. 397 to 419) of the Chin; they were each about 30 cm long, some white and made of bone, others black, made of horn.

Fig. 43. 'Discussions on difficult problems between Master and Pupil', a view of a counting-board (frontispiece of the *Suan Fa Thung Tsung*, A.D. 1593).

Further references would become tedious, but we may just note that Wang Jung (A.D. 235 to 306), minister of Chin and patron of water-mill engineers, 'taking his ivory counting-rods in his hands, used to calculate all through the night, as if he could hardly stop'. In the ninth century A.D. the rods were made of cast iron, and Thang administrators and engineers used to carry a bag of counting-rods at their girdle, while in the eleventh century Shen Kua, describing one of his contemporaries, the astronomer Wei Pho, says that 'he could move his counting-rods as if they were flying, so quickly that the eye could not follow their movements before the result was obtained'. This description makes one visualise rather the speed with which the abacus can be manipulated. After the late Ming less is heard of the counting-rods, no doubt because they were ousted by the abacus.

In all the above cases it is supposed the calculations were performed by the actual formation of rod-numerals by placing the rods on a counting-board; this had the advantage over writing that it was easier to cancel numbers which were no longer wanted. The rods themselves have left their traces in the writing of Chinese characters, since most of the terms for calculation have the 'bamboo' radical ⺮, and many expressions such as *thui suan*, 'to push the rods about', *chhih chhou*, 'to grasp the rods', etc. have survived as expressions for computing.

Graduated counting-rods

Counting-rods with numbers marked on them were probably a late development in Chinese mathematics. They seem to be practically identical with Napier's bones, with which multiplication is carried out by sliding one against another (Fig. 44). These sliding rods continued in use throughout the seventeenth century in the West and were quickly introduced into China and Japan, where they attracted considerable interest. The best known book on them is the *Tshê Suan* (On the Use of the Calculating-rods) of 1744 by the famous scholar and mathematician Tai Chen (volume 1 of this abridgement, pages 255 ff), and Fig. 43 shows them in their East Asian form. The set as used in China also comprised a zero rod, and rods for square and cube roots, but they retained the same name as the ancient plain computing rods, a fact which has sometimes led to confusion. This system would perhaps have been more useful if the two inventions of the slide-rule and the adding machine – both of which found their way to China (Fig. 45) – had not rapidly followed Napier's invention.

The abacus

As to the Chinese abacus, which has given rise to a large literature, it will be best if we begin with a brief description. It is called *suan phan* (calculating plate) or *chu suan phan* (ball-plate). It consists today of a rectangular wooden frame, the longer sides of which are connected by wires

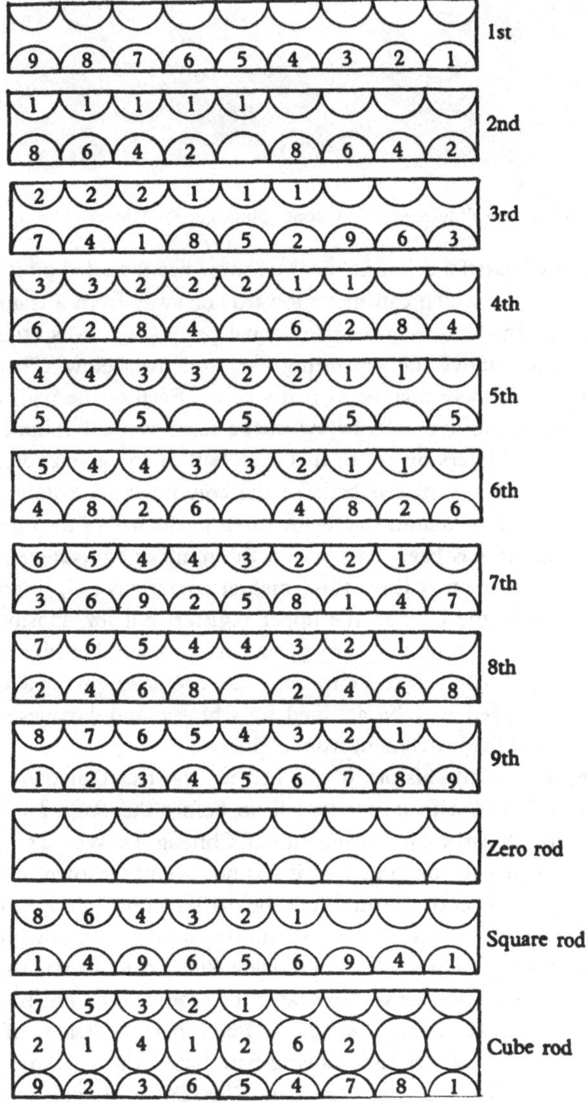

Fig. 44. The Chinese form of Napier's rods (after Chêng Chin-Tê).

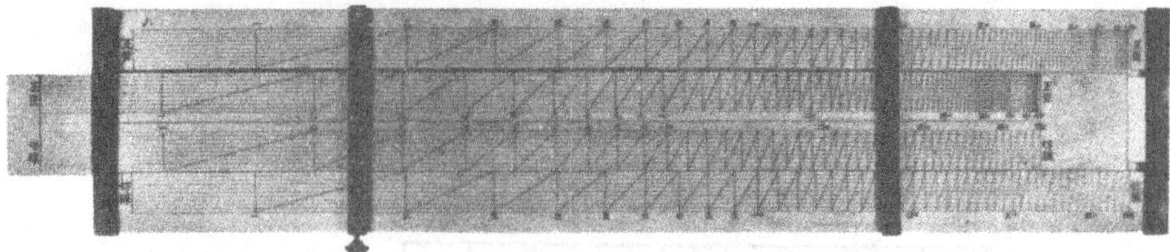

Fig. 45. A Chinese slide-rule of A.D. 1660 (photograph, Michel).

forming a series of parallel columns. Each of these has threaded on it seven slightly flattened balls which can move towards or away from a transverse bar which divides the abacus into two unequal parts (Fig. 46) so that two balls always remain above the separating bar, and five below. There are usually 12 wires, but there may be as many as 30. Each of the balls above the bar is equivalent to five balls below in the same vertical column, and each of the columns differs by 10, so that each ball in any given column is equivalent to 10 similarly placed balls in the column immediately on its right. Exactly what each column stands for varies at the will of the operator; the abacus in Fig. 46 has been set up with the number 123,346·789. It is possible to carry out the first three fundamental operations (+, − and ×) by using only one of the balls in the upper register, but for division it is convenient to be able to indicate on any given column a number higher than 10, so that two upper balls and five lower ones are provided, giving a total of 15. Extraordinary speed may be attained by Chinese and Japanese computers trained to use the instrument from childhood.

Turning now to the history of the abacus, the fact that there is no complete description of it in its modern form before the *Suan Fa Thung Tsung* (Systematic Treatise on Arithmetic) of Chhêng Ta-Wei (1593) (see Fig. 47), has led many to conclude that it did not become known in China until the end of the fifteenth century. Yet the earliest illustration of it is to be found in the oldest printed illustrated children's primer in any civilisation, the *Hsin Pien Tui Hsiang Ssu Yen* (New Revised Reader with Four Characters to the Line and Pictures to Match) of 1436, and no one seems to have noticed that the *Lu Thang Shih Hua* (Foothill Hall Essays) of 1513 describes the abacus clearly as the 'moving ball plate' which was operated according to fixed rules. Moreover, the *Shu Shu Chi I* (Memoir on some Traditions of Mathematical Art) which is attributed to Hsü Yo at the end of the Later Han (about A.D. 190), although it may have been written by its commentator Chen Luan (about 570), is in either case much the earliest work which speaks of 'ball arithmetic'. Part of the text runs as follows:

> (Text). The ball-arithmetic (method) holds and threads together (*khung tai*) the Four Seasons, and fixes the Three Powers (heaven, earth and man) like the warp and weft of a fabric.

Fig. 46. Chinese abacus (original photograph).

(Commentary). A board is carved with three horizontal divisions, the upper one and the lower one for suspending the travelling balls, and the middle one for fixing the digit. Each digit (column) has five balls. The colour of the ball in the upper division is different from the colour of the four in the lower ones. The upper one corresponds to five units, and each of the four lower balls corresponds to one unit. Because of the way in which the four balls are led (to and fro) it is called 'holding and threading together the Four Seasons'. Because there are three divisions among which the balls travel, so it is called 'fixing the Three Powers like the warp and weft of a fabric'.

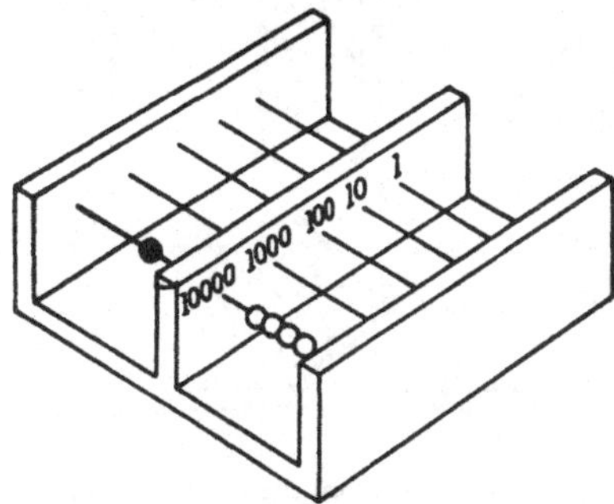

It must be admitted that this is a remarkably clear description of some kind ot abacus, obviously one in which the total number of units available in each column was 9. It could be pictured as a trough-and-ball instrument were it not for the word *tai* (belt, ribbon) which has unmistakably the sense of a wire.

The commentary on three of the other methods mentions the use of balls: one seems to have had one ball per column moving up and down a

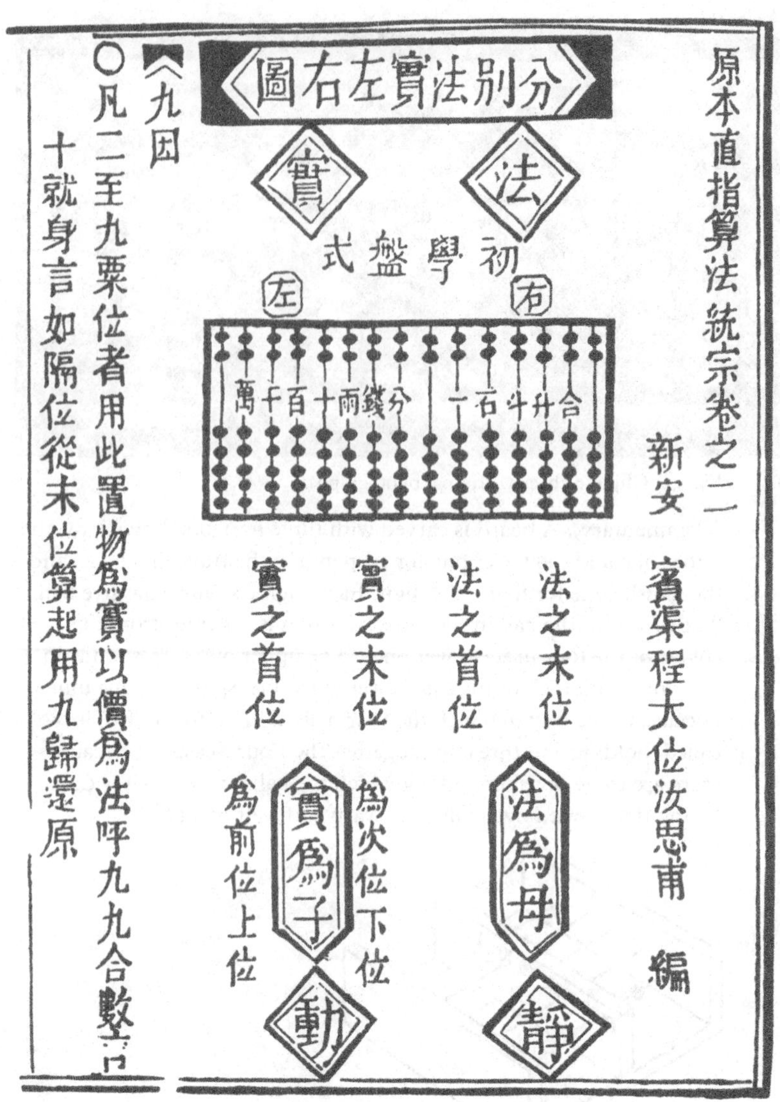
原本直指算法統宗卷之二

新安 賓渠程大位汝思甫 編

分別法實左右圖

法 實

初學盤式

右 左

合 升 斗 石 一 分 錢 兩 十 百 千 萬

法之末位 法之首位 實之末位 實之首位

法爲母 爲次位下位 實爲子 爲前位上位

靜 動

九因

○凡二至九乘位者用此置物爲實以價爲法呼九九合數言

十就身言如隔位從末位算起用九歸還原

Fig. 47. An early printed picture of the abacus, from the *Suan Fa Thung Tsung*, A.D. 1593.

line or a trough, another used balls of two different colours working in a somewhat similar way, both utilising a map-like grid of lines, or, as a mathematician would now call them, Cartesian coordinates. Yet another used balls of three different colours, but with a series of horizontal lines only. There is no doubt, however, but that all three certainly display an awareness of coordinate relationships.

Another sixth-century reference may refer to counting-rods or to the abacus, and between 1078 and 1162 there were four books which, to judge from their titles, dealt with the abacus, but none has come down to us. However, as we have already seen, there is other firm evidence from the lost book of Hsieh Chha-Wei for the eleventh-century use, although if the *Shu Shu Chi I* is indeed by Hsü Yo, then we can go back as early as the second century A.D.

What of the abacus in other civilisations? The Latin word *abacus* may well have been derived from the Semitic *abq* (dust), recalling the fact that the earliest ancestor of the abacus was probably a dust or sand tray. The next step was to have a surface ruled in lines on which pebbles (calculi) or counters could be placed, and there is some evidence that this was first used in ancient India. Some examples have been found in the West of metal plates with knobs running in slots, and there is a marble 'abacus' at Salamis, but these cannot be precisely dated. If, however, we place these specimens as late as the third or fourth century A.D., and if we accept the suggested date of Hsü Yo's text as the end of the second century A.D., then Chinese practice would slightly antedate European. But there is much uncertainty, and it is not possible to consider the question as in any way settled. Perhaps the best provisional conclusion is that independent inventions took place.

ARTIFICIAL NUMBERS

Fractions

In earliest times in all parts of the world there was a tendency to avoid fractions by creating a great number of ever smaller units of weight and measure. The relations between these were more judicious in some cases than in others; the Romans choosing multiples of 12 and 16, the Babylonians multiples of 60; while the Chinese generally chose powers of 10.

Chinese mathematics, from the time when we can first observe it, was accustomed to dealing with fractions. The early *Chou Pei Suan Ching* (The Arithmetical Classic of the Gnomon and the Circular Paths of Heaven) has problems which involve such numbers as $247\frac{933}{1,460}$ though they are given in words, not symbols. In a division such as that of 119,000 by $182\frac{5}{8}$ both numbers were first multiplied by 8, and indeed all the modern rules were used. The adoption of the bar for fractions, which seems to have been an Arabic development, was not current in China until the seventeenth century. On the other hand the facility of the Han mathematicians in using the lowest common denominators and highest common factors seems remarkably advanced when we remember that these were not employed in Europe until the fifteenth and sixteenth centuries.

Sumerian and Babylonian arithmetic was based on 60, and there can be little doubt that the sexagesimal fractions of the Greeks and Alexandrians, together with the division of the circle into 360 degrees, were derived from them. It has been speculated that the Chinese recurrent 60-day cycle, which is certainly very ancient, had the same origin, but there is no positive evidence for this. The number of degrees in the old Chinese circle was $365\frac{1}{4}$, corresponding to the days in the year, not 360, and sexagesimal fractions played no part in Chinese calculations; indeed the Chinese never had a word to express $\frac{2}{3}$, that fraction so important in Mesopotamia.

Decimals, measurement, and the handling of large numbers

When we come to discuss the Chinese development of decimal fractions we find ourselves involved in the development of the Chinese art of measurement, for their system of measures of length had steps going up by tens from a remarkably early period. Perhaps the earliest text which can be cited as showing an understanding of decimal place-value is one of the propositions of the *Mo Ching* (Mohist Canon) (see volume 1 of this abridgement, Chapter 9) dating from about 330 B.C. We have translated it as follows (C = Canon, CS = Exposition):

C 'One' is less than two and (yet) more than five. The explanation is given under establishing a position.

CS Within 'five' there is 'one' (i.e. there are 'ones', because the rod-numeral for five is |||||). (Yet) within the 'one' there is 'five' (because in the rod-numeral for six ⊤ represents five). (And one horizontal stroke in the tens position means) 'ten', (that is to say, the equivalent of) 'two' (of these symbols for five).

From this and evidence previously given (page 3), it is clear that although the idea may have been temporarily lost or not always widespread, there was an appreciation of place-value in China some fifteen hundred years before it is found in the *Sun Tzu Suan Ching* (Master Sun's Mathematical Manual).

As previously mentioned, too, on the Shang oracle-bones and in the Chou bronze inscriptions numbers from 50 upwards were expressed as in modern times by combining a figure with a word for defining its place-value. Thus there were the following special symbols:

100 or 10^2	*pai*	百
1,000 or 10^3	*chhien*	千
10,000 or 10^4	*wan*	萬

These were not a hindrance to computation, for they were not themselves numerals, as were the symbols for 18, 19, 30, 40, 500 and so on, used in other early civilisations, such as Egypt, Greece and India. The Chinese symbols were simply terms denoting particular place-values.

During the Chou period (first millennium down to 221 B.C.) measures of length were variable and not always advanced by tens. These earliest measures were based on parts of the human body – the finger, the woman's hand, the man's hand, the forearm, etc. When Chhin Shih Huang Ti first unified the empire (221 B.C.) he chose the number 6 as his emblem, and with this he carried out his standardisation of weights and measures. Yet although the double-pace was fixed at 6 *chhih*, the emperor's Legalist advisors (see volume 1 of this abridgement, Chapter 15) took advantage of the Mohist decimal notation for the principal measures of length below the *chhih*, which were henceforward arranged in tens, thus:

1 *chhih*	尺 =	10 *tshun*	寸
1 *tshun*	寸 =	10 *fên*	分
1 *fên*	分 =	10 *li*	釐
1 *li*	釐 =	10 *fa*	髮
1 *fa*	髮 =	10 *hao*	毫

As Chia I put it in his *Hsin Shu* (New Book) of about 170 B.C., these made six measures, in accordance with the system of the 'former Ruler', but in fact there was still the *chang* of 10 *chhih* and the *yin* of 10 *chang*. This system was current throughout the Han and, with slight modifications, in later ages.

The expression of decimal fractions by the use of measurement units runs through the whole of Chinese mathematics. Liu Hui in the third century A.D., in his commentary on the *Chiu Chang*, expresses a diameter of 1·355 feet as 1 *chhih*, 3 *tshun*, 5 *fên*, 5 *li*. And while, in extracting square roots, the *Chiu Chang* itself had spoken of cases where the root would not be a whole number, in which case the remainder would be left as such, Liu was concerned with these 'little nameless numbers'. He said that first 10 should be taken as denominator, then 100, and so on, thus giving rise to a series of decimal places.

After this time there was little change in the methods used, but it is worth noting that, implicitly in the *Chiu Chang* and then explicitly in the third-century A.D. *Shih Wu Lun* (Discourse on [calendrical] Time), and in the writings of the fifth century mathematician Hsiahou Yang, familiarity is shown with what we should now express as 10^{-1} and 10^{-2} for the decimals 0·1 and 0·01. Moreover, in the Thang, Han Yen, who worked between 780 and 804, seems to have made an innovation by writing down numbers as in modern decimal notation, using words of measurement to denote the last whole number, although the introduction of a unified system and terminology, together with its general application, did not occur until the thirteenth century.

Another aspect of the Chinese concern with decimals is their interest in expressing very large numbers. Some terms like 'myriads' had been used in texts of Chou date, but it is likely these did not have fixed meanings. However, in the *Shu Shu Chi I* (about A.D. 190) Hsü Yo has a very interesting passage on the matter. His statements may be represented as follows:

		Upper (*shang* 上)	Middle (*chung* 中)	Lower (*hsia* 下)
wan	萬	10^4	10^4	10^4
i	億	10^8	10^8	10^5
chao	兆	10^{16}	10^{12}	10^6
ching	京	10^{32}	10^{16}	10^7
kai	垓	—	10^{20}	10^8
tzu	秭	—	10^{24}	10^9
jang	壤	—	10^{28}	—
kou	溝	—	10^{32}	—
chien	澗	—	10^{36}	—
chêng	正	—	10^{40}	—
tsai	載	—	10^{44}	—

The interpretations of all the commentators on ancient texts show that they were following one or other of these methods for the powers of ten.

The particular interest which the Indians, especially the Buddhists, took in expressing large numbers is well known, though its significance for the progress of mathematics is somewhat doubtful. The ancient Chinese system just described cannot, however, be of Indian origin, for even if the *Shu Shu Chi I* itself is of the sixth rather than second century, the classical commentators such as Chêng Hsüan or Mao Hêng are much too early to have been affected by Buddhist influence. Indeed, the positive Chinese spirit reacted strongly against the Indian association of mathematics and mysticism. In an interesting but hitherto unnoticed passage of about the twelfth century, Shen Tso-Chê writes in the *Yü Chien* (Allegorical Essays):

> Nowadays even children learn mathematics from Buddhist textbooks which deal with the counting of infinite numbers of sand-grains, so

> that their numbers can be known . . . But how can the Buddhas know the answers when there are no definite numbers and no precise principles? That which is vague and obscure can have no place in matters connected with number and measure. Whether the numbers or dimensions be large or small, problems can all be solved and the answers distinctly stated. It is only things which are beyond shape and number which cannot be investigated. How can there be mathematics beyond the reach of shape and number?

To sum up, the use of decimal notation was extremely ancient among the Chinese, going back to the fourteenth century B.C. In this they were unique among early civilisations. In their application of the decimal system to measurement they were particularly advanced; Europe had to wait until the French Revolution. And when they applied the system to grids on maps (see pages 261–70) they antedated the Arabs and Europeans by nearly a thousand years. But the little symbol – the decimal point – which alone permitted decimalisation to revolutionise all mathematical computations had to await the Renaissance in the West.

Surds and negative numbers

In the West, surds (derived from the Latin *surdus*, 'stupid') or 'irrational' numbers were incommensurable. They were numbers like $\sqrt{2}$ or $\sqrt{3}$, which are not numbers that can be expressed by the ratio of two whole numbers: thus $\sqrt{2}$ is 1·4142136 . . . , and although approximately $\frac{7}{5}$, it is not exactly so. To the Greeks it was not so much 'unreasonable' as 'un-ratio-able'. Chinese mathematicians, with their early use of decimal fractions, seem to have been neither attracted nor perplexed by irrationals, if indeed they appreciated their separate existence.

The Chinese also found no difficulty in the idea of negative numbers. As previously mentioned, positive coefficients were represented by red counting-rods and negative ones by black counting-rods, probably as early as the second century B.C. Another variation was to use counting-rods of triangular or square section. The law of signs (– multiplied by – equals +, etc.) was known to the Sung algebraists and stated, for instance, in the *Suan Hsüeh Chhi Mêng* (Introduction to Mathematical Studies) of 1299. In India, negative numbers were not mentioned until about A.D. 630, and in Europe this did not happen until 1545.

GEOMETRY

The Mohist definitions

It is often said that all early geometry, except that of the Greeks, was intuitive: it sought facts relating to measurement without attempting to prove any theorems by deductive reasoning. Certainly deductive geometry was the central feature of Greek mathematics, and it is equally certain that the genius of Chinese mathematics lay rather in the direction of algebra. But just as Greek mathematics was not completely devoid of algebra, neither was Chinese mathematics without some theoretical geometry. The propositions in which this theoretical geometry is contained occur in the *Mo Ching* (Mohist Canon) (see volume 1 of this abridgement, pp. 116 ff), yet until recently they were almost unknown in the West. This ancient work, dating from around 330 B.C. and so contemporaneous with Euclid, has some interesting definitions; we give a selection (C = Canon, CS = Exposition):

(a) *'Atomic' definitions of a geometrical point*

C The definition of 'point' (*tuan*) is as follows: The line is separated into parts, and that part which has no remaining part (i.e. cannot be divided into still smaller parts) and (thus) forms the extreme end (of a line) is a point.

CS A point may stand at the end (of a line) or at its beginning like a head presentation in childbirth. (As to its indivisibility) there is nothing similar to it.

C That which is not (able to be separated into) halves cannot be cut further and cannot be separated.

CS If you cut a length continually in half, you go on forward until you reach the position that the middle (of the fragment) is not big enough to be separated any more into halves; and then it is a point . . . Or if you keep on cutting into half, you will come to a stage in which there is an 'almost nothing', and since nothing cannot be halved, this can no more be cut.

(b) *Lines of equal length*

C (Two things having the) same length, means that two straight lines finish at the same place.

CS It is like a straight door-bolt which can be placed flush with the edge of the door.

(c) *Space*

C Space includes all the different places.

CS East, west, south and north, are all enclosed in space.

(d) *Rectangles*

C Rectangular shapes have their four sides all straight, and their four angles all right angles.

CS Rectangular means using the carpenter's square so that the four lines all just meet (each other).

(e) *Piling up*

C As for piling up, where there is no space between (i.e. as in planes which have no thickness), they cannot mutually touch (and therefore cannot be piled up).

CS Things which have no thickness (exemplify this principle).

(f) *Centre and circumference*

C A circle is a figure such that all lines drawn through the centre (and reaching the circumference) have the same length.

CS A circle is that line described by a carpenter's compass which ends at the same point at which it started.

All these quotations have their parallel in Euclid's *Elements* (of Geometry). Nevertheless, it is evident that the Mohists were thinking along lines which, if continued, could have developed into a geometrical system of Euclidean type. We cannot indeed be certain they did not go beyond the point represented in these propositions, for the *Mo Ching* has come down to us in a very corrupt and fragmentary state, but if they did, their geometry remained the technique of a particular school, and had little or no influence on the main current of Chinese mathematics.

The Theorem of Pythagoras

There is no telling how far back the appreciation of the value of right-angled triangles in surveying and measurement may have gone in ancient China. As we have seen, such triangles appear in the *Chou Pei Suan Ching* (see page 7) and this part of the text was almost certainly written before the Mohist school came into existence, but Chinese knowledge of them may well go back to the sixth century B.C. The fact is that we are not in a position to observe much Chinese mathematics until the Warring States period.

In the third century A.D. Liu Hui called the right-angle triangle drawing (Fig. 30, page 8) 'the diagram giving the relations between the hypotenuse and the sum and difference of the other two sides whereby one can find the unknown from the known'. In the time of Liu it was coloured, the small central square being yellow, and the surrounding rectangles red, and there was an algebraic formulation (in words) of the general relationship between the hypotenuse and the other two sides. The proof is quite different from Euclid's. It is known, however, that an algebraical formulation of the Pythagoras Theorem was familiar to the Old Babylonian mathematicians who were contemporary with the Shang period (fourteenth to eleventh centuries B.C.). In the course of time the Chinese developed algebraic methods for finding any unknown side or angle, given appropriate data. However, throughout Chinese history the interest in right-angled triangles was mainly practical for use in surveying.

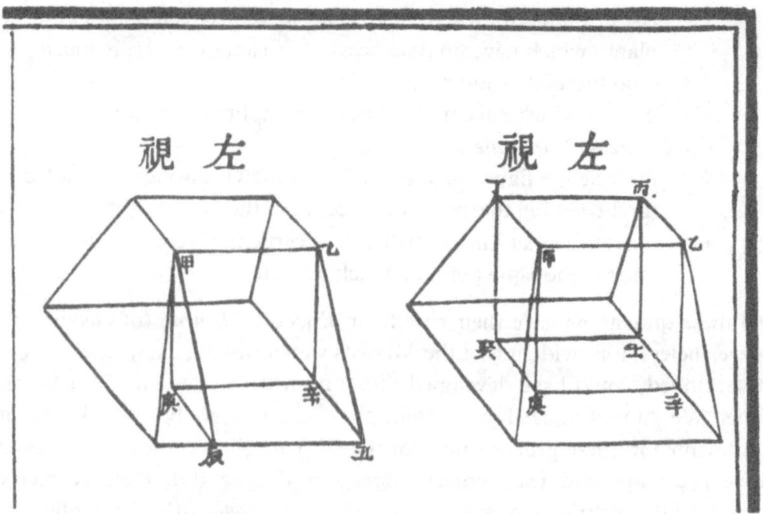

Fig. 48. Practically derived solid geometry; sections (frusta) of pyramids from the *Chiu Chang Suan Shu*.

Treatment of plane areas and solid figures

By the end of the Former Han (i.e. about A.D. 220) correct or approximately correct formulae had been worked out for determining the area of many plane shapes and the volume of many solid figures, though in all cases without any demonstrations by deductive geometry of how the formulae were obtained. Use must have been made of models, and the more complex reduced experimentally to the simpler. This knowledge was embodied in the *Chiu Chang Suan Shu* (Fig. 48), and Liu Hui, its third-century commentator, was one of the greatest expounders of this 'empirical' solid geometry. He was adept at reducing complex shapes to simple ones for which he could readily determine the volume.

Evaluation of pi (π)

Although there is evidence that the ancient Egyptians and Old Babylonians had values of 3·1604 and 3·125, the commonest practice in ancient civilisations was to take this ratio of the diameter of a circle to its circumference as 3. Our modern value is 3·1415926536. In China it was taken as 3 in the two great Han arithmetics and in two other famous books from this dynasty, and as an approximation it persisted for centuries.

The first indication that a more exact value was sought comes in the first decade of the first century A.D., when Liu Hsin prepared standard measures for the emperor Wang Mang. He used a value of 3·154, but there is no record of how he arrived at it. About A.D. 130 Chang Hêng gave a value of 3·1622 (i.e. $\sqrt{10}$), and in the third century Wang Fan recalculated it as $\frac{142}{45}$ or 3·1555, but Liu Hui, working in the Wei State further to the north, obtained a better value by inscribing a polygon with 192 sides within a circle and calculating the polygon's perimeter; he obtained $\frac{157}{50}$ or 3·14. Liu Hui also gave two other extreme values, and used a polygon of 3,072 sides for his best one, 3·14159. By about the middle of the third century, then, the Chinese had more than caught up with the Greeks, who had never achieved a value as accurate as this.

Then a great leap forward came in the fifth century with the calculations of Tsu Chhung-Chih and his son Tsu Kêng-Chih which set the Chinese ahead for a thousand years. Their final figures came out at an 'excess value' of 3·1415927 and a 'deficit value' of 3·1415926. The original work of these two mathematicians is lost, but in about 1300 Chao Yu-Chhin returned to the question and by the continued use of polygons up to 16,384 sides confirmed that Tsu's value was very accurate. It was not until about 1600 that Adriaan Anthoniszoon in Europe obtained an answer even equal to Tsu's earliest value of 3·1415929203.

Conic sections, Yang Hui and the coming of Euclid

With its general lack of theoretical demonstrations, Chinese solid geometry never stimulated any counterpart of the great Greek work on conic sections by Apollonius of Pergamon (about 220 B.C.). In China the study of the ellipse, the parabola and the hyperbola had to wait until the seventeenth century. Yet although Chinese geometry was essentially tied to the practical needs of measurement, by the thirteenth century some Chinese mathematicians were becoming dissatisfied with the mainly empirical methods on which surveying was based. Yang Hui strongly criticised those who had '... changed the name of their methods from problem to problem, so that as no specific explanation was given, there is no way of telling their theoretical origin or basis'. This was an extremely modern attitude. Yang Hui then proceeded to give a proof about parallelograms which is similar to one in Euclid. If such proofs had been extended, the Chinese might have developed an independent deductive geometry, and clearly some minds like Yang Hui were prepared to appreciate the Euclidean system. This is of great interest because there may have been at this time a translation into Chinese of Euclid's *Elements*, due to Chinese–Arabic contacts.

Coordinate geometry

Three principal steps were involved in the development of coordinate or analytic geometry (the geometry arising from expressing the positions of points, lines and curves by numbers (coordinates)). These were (*a*) the invention of a system of coordinates; (*b*) the recognition of a one-to-one correspondence between geometry and algebra (i.e. the pairing of each element in geometry with one in algebra); and (*c*) the ability to represent graphically an algebraic relationship (an equation like $y = x + 4$, or $x^2 + y^2 = 6$, for instance). Both the first two steps were taken in China

The idea of coordinates in land utilisation must be very ancient. The Egyptian hieroglyph for a district was a grid, and in Chinese there was the character *ching* (井), fields round a well, later simply a well. In Greece in the third century B.C. coordinates were extended to map-making, but were lost after about A.D. 130, when they began to flourish in China, where the grid system had always been kept alive. But there is another aspect of the use of coordinate systems besides map-making, although historians of science have paid scant attention to it, and this is the growth of systems of tabulation. About A.D. 120 the historian Pan Ku and his sister Pan Chao supplied the *Chhien Han Shu* (The History of the Former Han Dynasty) with eight chronological tables, and in one of the more remarkable of these, the names of 2,000 legendary and historical characters are arranged in a grid to give a scale of nine ratings as to virtue. Here then is a table which long antedates the chessboard which has often been called the earliest set of coordinates. Sound-tables of Chinese rhymes were another example of a coordinate system (see volume 1 of this abridgement, Fig. 2), while there were also tables of numerical quantities for summarising data. Moreover, the abacus itself is essentially a coordinate system. From the beginning of mathematics in China, geometrical propositions were expressed in an algebraic form, and when geometrical figures were used, the treatment was uniquely algebraic. In Europe this realisation came relatively late, and only around 1630 were the basic conceptions of Western analytic geometry stated by Pierre de Fermat and René Descartes.

Trigonometry

There is not a great deal to say about trigonometry in old Chinese mathematics, since the modern theory of trigonometric quantities is a Western post-Renaissance development. However, all ancient civilisations studied the properties of right-angled triangles. Having names for the sides of a right-angled triangle, the Chinese do not seem to have felt the need for naming or computing ratios of the sides, such as we use to give our trigonometric ratios, sine, cosine, etc. But by the thirteenth century the Chinese were anxious to improve their astronomical and calendar calculations, and this led Kuo Shou-Ching to his spherical 'trigonometry'. Although none of his original work survives, the methods can be reconstructed from other books, and Fig. 49 shows the most important result with which he dealt, a spherical pyramid, and Fig. 50 the kind of figure he may well have used in his calculations.

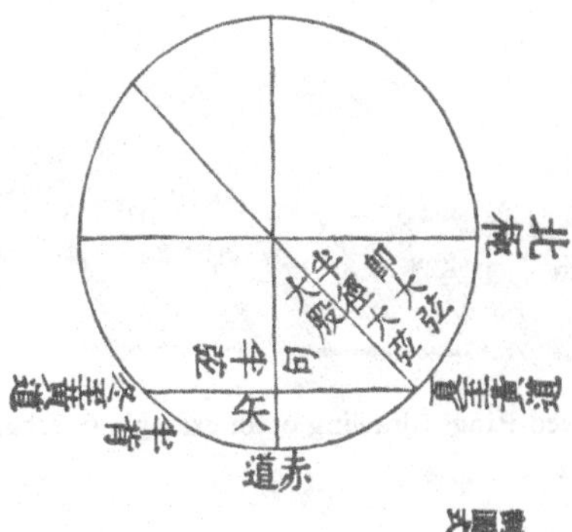

Fig. 49. Diagram illustrating problems in spherical trigonometry, from Hsing Yün-Lu's *Ku Chin Lü Li Khao* (A.D. 1600).

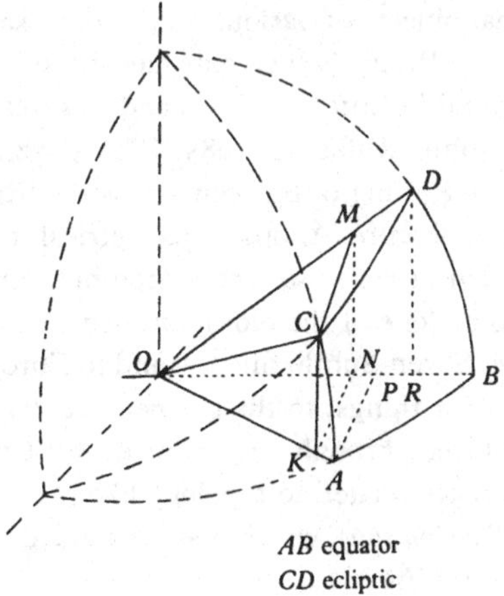

Fig. 50. Diagram to illustrate the spherical trigonometry of Kuo Shou-Ching (A.D. 1276).

To what extent, if any, Kuo was influenced by Persian astronomers, who already had a fully developed plane trigonometry and whom he probably knew at the Chinese court, is very difficult to say; it is possible that the eleventh-century work of Shen Kua on arcs and chords had given him all he needed. Certainly after Kuo's time nothing of importance was done until the Jesuits produced the first modern trigonometry in Chinese in 1607.

Problems and puzzles

For want of a better place, a word may be said here about mathematical problems and puzzles. In Europe there has long been a tendency to dub many of these 'Chinese', but it is not clear how many of them really did

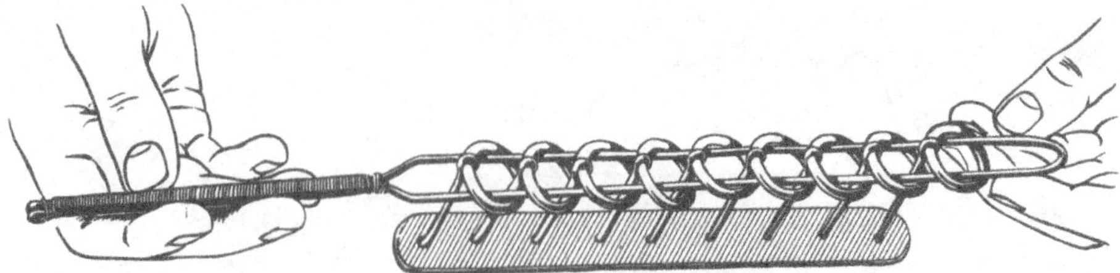

Fig. 51. The puzzle of the Linked Rings (drawing of an example purchased by Mr Brian Harland at Lanchow).

have an East Asian origin. Perhaps Europeans were inclined to ascribe to puzzles the name of what was, to them, a puzzling civilisation.

At all events, many of these puzzles involve several branches of mathematics, and relate to material objects of various kinds. For example, the topological 'Puzzle of the Chinese Rings' (which may have been derived from the abacus; Fig. 51) is first found in Europe in 1550 and was treated at some length mathematically by John Wallis in 1685. The puzzle was commonly known in China at the beginning of this century as the 'Ring of Linked Rings' but its origin is quite obscure. Another, geometrical, puzzle, the rearrangement of a set of wooden pieces (a square, a rhombus, and five triangles of different sizes), said to be 'one of the oldest amusements of the East' (known to the Chinese as the 'Seven Subtle Shapes' and to Europeans as 'tangrams'), is related, among other things, to the geometrical shapes in the lattice work of windows employed through the centuries by Chinese builders. Another widespread art, more related to topology like the Linked Rings, was that of folding paper (*chê chih shu*), which may be alluded to in a poem of the eighth-century scholar Tu Fu.

ALGEBRA

As has been well said, the early history of algebra cannot be discussed without deciding what is meant by the term. If we take it to mean the art which allows us to solve such an equation as $ax^2 + bx + c = 0$ expressed in these symbols (where a, b and c represent constant numbers and x is the unknown) it is a sixteenth-century development. If we allow other less conventional symbols, it goes back to the third century A.D.; if purely geometrical solutions are allowed, it begins in the third century B.C.; and if we are to class as algebra any problem which would now be solved by algebraic methods, then the second millennium B.C. was acquainted with it. Now algebra was dominant in Chinese mathematics as far back as we can trace it (to about the second century B.C.), yet it does not fall into any of these categories. It was in fact 'rhetorical' (i.e. fully written out in words) and positional, using symbols only rarely and late. In other words, it brought into play an abundance of abstract single-syllable technical ideograms for indicating generalised quantities (rather than specific numbers) and mathematical operations. If these were not symbols in the mathematical sense, they were more than merely words in the ordinary sense. And then, in the course of this work, the counting-board with its numbers was laid out in such a way that certain positions were occupied by specific kinds of quantities (unknowns, powers, etc.). Thus a permanent filing system of mathematical patterns was established. But since the types of equation always retained their connection with concrete problems, no general theory of equations developed. However, the tendency to think in patterns finally evolved from the counting-board a positional notation so complete that it rendered unnecessary most of our fundamental symbols. Unfortunately, though the achievement was magnificent, it led to a position from which no further advance was possible.

The Greeks solved many comparatively difficult problems by geometry. Thus Euclid solved the equivalent of $x^2 + ax + b^2 = 0$, substantially by completing the square geometrically, and neglecting negative roots. Not until five and a half centuries later did Western algebra acquire some sort of notation. Recent investigations, however, have shown that the algebra of the Babylonians was more advanced than previously realised, and included equations involving the third and fourth powers of x. In view of its great age one cannot help wondering whether this Babylonian algebra could have been transmitted in seminal forms to lay the foundations of Indian and Chinese algebra on the one hand, and of Hellenistic developments on the other.

During the decay of Western science in the early Middle Ages, Greek algebra was forgotten, and when the great days of Arabic science came its algebra was indebted rather to that of India, perhaps to a lesser extent to China. Its best known name is Muḥammad ibn Mūsa al-Khwārizmī (about 813 to 850) from whose book *Hisāb al-Jabr w'al-Muqābalah* our word algebra is derived. The two words of the title mean 'restoration' (hence the strange fact that the verb *jabara* in Moorish Spain gave rise to the Spanish word *algebrista*, meaning bone-setter), and 'confrontation'. Thus, given

$$bx + 2q = x^2 + bx - q$$

then by *al-jabr*

$$bx + 2q + q = x^2 + bx$$

and by *al-muqābalah*

$$3q = x^2$$

Thus *al-jabr* has the idea of a transposition of a negative quantity ($-q$), and *al-muqābalah* the cancellation of positive quantities (bx) to simplify both sides of the equation. In Chinese mathematics there were no technical words exactly corresponding to these, because algebraic operations were carried on entirely without the use of the equals sign (=), and the terms were arranged in tabulated columns. Nevertheless, the casting out of bx on both sides of the equation might be said to correspond to 'the subtraction of terms with the same sign' mentioned in the *Chiu Chang*. And similarly, the transposition of $-q$ to $+q$ would be a case of 'addition of terms of different signs'. Li Yeh (twelfth century) had phrases for simplification of terms by addition and by cancellation. It must be remembered, too, that the symbolism of algebra was very slow to develop in Europe and did not reach a modern level until François Viète (1580). Not until the end of the Ming dynasty was it properly established in China.

Simultaneous linear equations

These were much in evidence in the Han *Chiu Chang Suan Shu*, the counting-rods being placed in different squares in a table so as to represent the coefficients of the different unknowns. It was a question of arranging the coefficients, multiplying crosswise, and adding or subtracting. For example:

$x + 2y + 3z = 26$	1	2	3	*shang ho ping shu*
$2x + 3y + z = 34$	2	3	2	*chung ho ping shu*
$3x + 2y + z = 39$	3	1	1	*hsia ho ping shu*
	26	34	39	

the top line representing the x terms (number of bushels of first-quality cereal), the second line the y terms (bushels of second-quality cereal), the third line the z terms (bushels of third-quality cereal), and the fourth line the constant terms.

In the Han and San Kuo periods the rules for solving these equations were never divorced from specific problems; it remained for Yang Hui in the thirteenth century to state them in a generalised way.

Matrices, determinants and the Rule of False Position

The Chinese method of representing the coefficients of unknowns in simultaneous equations by means of rods on a counting-board led naturally to the discovery of simple methods of elimination. The arrangement of rods was precisely that of numbers in what today we call a matrix – a rectangular array of coefficients which is subject to special rules. Chinese mathematics therefore developed at an early date the idea of subtracting columns and rows as in the simplified matrix known today as a determinant. Yet it was not until it had been taken over by Japanese scholars in the seventeenth century that the idea of determinants assumed an independent form: surprisingly it was not evolved earlier by the Sung algebraists.

However, it should be appreciated that, before the days of mathematical symbolism, even simple equations were troublesome. The old mathematicians had to use the somewhat cumbrous method known later in Europe as the Rule of False Position. The chief form of this was the 'Double False'. For example, in an equation like $ax + b = 0$, two guesses were made of what x might be. These would give two false results (i.e. $ag_1 + b = f_1$ and $ag_2 + b = f_2$, where g is a guess and f a false result). The pair of equations obtained by these guesses and false results would now be treated as two simultaneous equations which could give the value of b/a and thus x (because in the original equation $x = -b/a$).

This rule came to Europe by way of the Arabs, but it may have had a Chinese origin, for the rule is nothing else than the Chinese method *Ying pu tsu* (Too Much and Not Enough), which is actually the title of the seventh chapter of the first-century B.C. *Chiu Chang Suan Shu*.

Indeterminate analysis and indeterminate equations

In our survey of Chinese mathematical literature mention was made several times of indeterminate analysis. When a number of equations is given involving more variables than equations, there may be an infinite number of sets of solutions. Of course, in some cases, the nature of the problem may be such that the only solutions wanted are those which have positive whole numbers. However, indeterminate analysis was always a marked mathematical interest of the Chinese, at least from the fourth century A.D., when the *Sun Tzu Suan Ching* (Master Sun's Mathematical Manual) gave the following problem:

> We have a number of things, but we do not know exactly how many. If we count them by threes we have two left over. If we count them by fives we have three left over. If we count them by sevens we have two left over. How many things are there?

Sun Tzu determined the 'use numbers' 70, 21 and 15; these are multiples of 5×7, 3×7 and 3×5, and have the remainder 1 when divided by 3, 5 and 7, respectively. The sum $2 \times 70 + 3 \times 21 + 2 \times 15 = 233$ is one answer, and by casting out a multiple of $3 \times 5 \times 7$ (= 105) as many times as possible (in this case, twice) the least answer, 23, is obtained.

In the eighth century A.D. I-Hsing used the method in his work on the calendar, and five centuries later Chhin Chiu-Shao gave a full explanation. But the commonest form taken by indeterminate problems in Chinese mathematics was that of the 'Hundred Fowls', which first appeared in A.D. 475. Hsieh Chha-Wei of the eleventh century has it thus: 'If a cock is worth 5 cash, a hen 3 cash, and 3 chickens together only 1 cash, how many cocks, hens and chickens, in all 100, may be bought for 100 cash?' In the fifth century, Chang Chhiu-Chien solved his problem by incompletely expressed indeterminate equations, but others found they could get an answer by simpler methods and did so. Analysis was not applied to this problem until Chhing times.

Quadratic equations and the Method of Finite Differences

Quadratic equations (equations with x^2) were used early in Chinese mathematics. The *Chiu Chang* has a problem which was solved by taking the positive root of such an equation, although quadratics were sometimes solved by reducing them to linear form (i.e. with terms of x not x^2) and then solving for the square root involved, a process already employed by the Old Babylonian mathematicians.

One of the most interesting problems in which quadratic expressions were involved was the process of finding arbitrary constants in formulae for celestial motions. It was the same as that now called the Method of Finite Differences. How far it goes back is not clear, but it was certainly used in A.D. 665; it may, however, have been late fifth century. Here, in the work of Li Shun-Fêng, who was concerned with finding a formula to express irregularities in the sun's motion across the sky, it is based on a quadratic equation. Writing it in modern terms as $Ax + Bx^2 = C$, the problem was to find the constants A and B, since x was known (it was the time interval between successive observations of the sun) and also C (the number of degrees the sun had moved during each interval). From the successive data gathered he had:

$$Ax_1 + Bx_1^2 = C_1$$
$$Ax_2 + Bx_2^2 = C_2, \text{ and so on.}$$

Then, subtracting to obtain differences,

$$A(x_2 - x_1) + B(x_2^2 - x_1^2) = C_2 - C_1$$

or $A + B(x_2 + x_1) = \dfrac{C_2 - C_1}{x_2 - x_1}$

and similarly

$$A + B(x_3 + x_2) = \frac{C_3 - C_2}{x_3 - x_2}$$

and subtracting

$$B(x_3 - x_1) = \frac{C_3 - C_2}{x_3 - x_2} - \frac{C_2 - C_1}{x_2 - x_1}$$

This gives a numerical answer for B, and by a similar process, a numerical answer for A. Greater accuracy could be achieved by an equation with higher powers of x and a third arbitrary constant, and this was recognised by Kuo Shou-Ching in 1281. The method he used was related to a procedure used by Chu Shih-Chien in 1303 for finding the sum of certain series, and this seems to have been a remarkable anticipation on the part of the Chinese for it was not taken up and thoroughly worked out in Europe until the seventeenth and eighteenth centuries.

Cubic and numerical higher equations

Though Kuo Shou-Ching used cubic equations (i.e. equations containing x^3) at the end of the third century A.D., they had first been considered by the Chinese more than six hundred years earlier, although mostly restricted to numerical values with positive terms only. Equations of the fourth to the ninth degree (i.e. with $x^4, x^5, \ldots, x^9$) were handled in Sung China, provided they were numerical. Indeed, the solution of numerical higher equations for approximate values of roots begins, as far as we know, in China. It has been called the most characteristic Chinese mathematical contribution.

It has long been known that the technique was well developed in the Sung, but it is possible to show that if the text of the Han *Chiu Chang Suan Shu* is very carefully followed, the essentials of the method are already there at a time which may be dated as the first century B.C. Even in its earliest form, it was similar to a procedure evolved in nineteenth-century Europe. Early Greek and Indian mathematics seem to have contributed little or nothing to the solution of higher numerical equations. The first noteworthy work on them in Europe was done by Leonardo Fibonacci at the beginning of the thirteenth century. Yet in the solution he gave of an equation in 1225, the equation itself is exactly characteristic of those which Wang Hsiao-Thung used to solve in Thang China (seventh century A.D.), and since Fibonacci gave no details of how he obtained his result, it may well be that, as communication with the East was possible at this time, he had learned of the solution during his extensive travels in Europe, Algeria, Egypt and Syria.

The Thien Yuan *notation*

We come now to the general system of notation used by the Sung algebraists for the expression of numerical equations. It was of 'square' or matrix character. The centre compartment was occupied by the absolute term, if any; it was called *thai* (太).

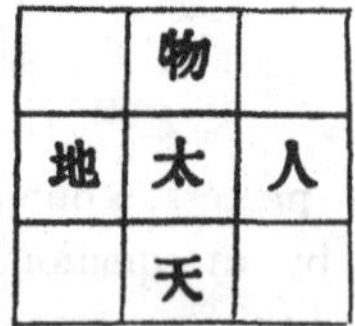

Jen (z and its powers) were written to the right of *thai*; *ti* (y and its powers) to the left of it; *thien* (x and its powers) were written below it, and *wu* (u and its powers) written above it. Proceeding outwards from *thai* in any straight direction, the first compartment was for the insertion of the simple term (e.g. $10x$); the second compartment outwards was for the square, the third for the cube, the fourth for the fourth power, and so on. Proceeding outwards from *thai* diagonally, the first compartments were for multiplied terms, such as xy or xz. But since, with four unknowns, there could be six of these, it was necessary to insert a small number in the centre compartment alongside *thai*. The diagram on the left below illustrates the method of writing $x + y + u + z$. The middle diagram illustrates writing the more complex expression $x^2 + y^2 + z^2 + u^2 + 2xy + 2xz + 2xu + 2yu + 2yz + 2zu$; the right-hand one the expression $2y^3 - 8y^2 - xy^2 + 28y + 6yx - 2x - x^2$. It will be remembered that the negative signs were represented by a diagonal line crossing a rod-numeral.

	1	
1	太	1
	1	

		1		
	2	0	2	
1	0	2 太 2	0	1
	2	0	2	
		1		

2	−8	28	太
0	−1	6	−2
0	0	0	−1

Zeros, as always, indicate that the terms for the compartments they occupy do not occur in the expression.

In the books of the Sung algebraists, as they have come down to us, there are never any of these diagrams in the text, except in so far as they have been inserted by modern editors (Fig. 52 shows some), yet their nature is certain from the description given. We do find, however, calculations in

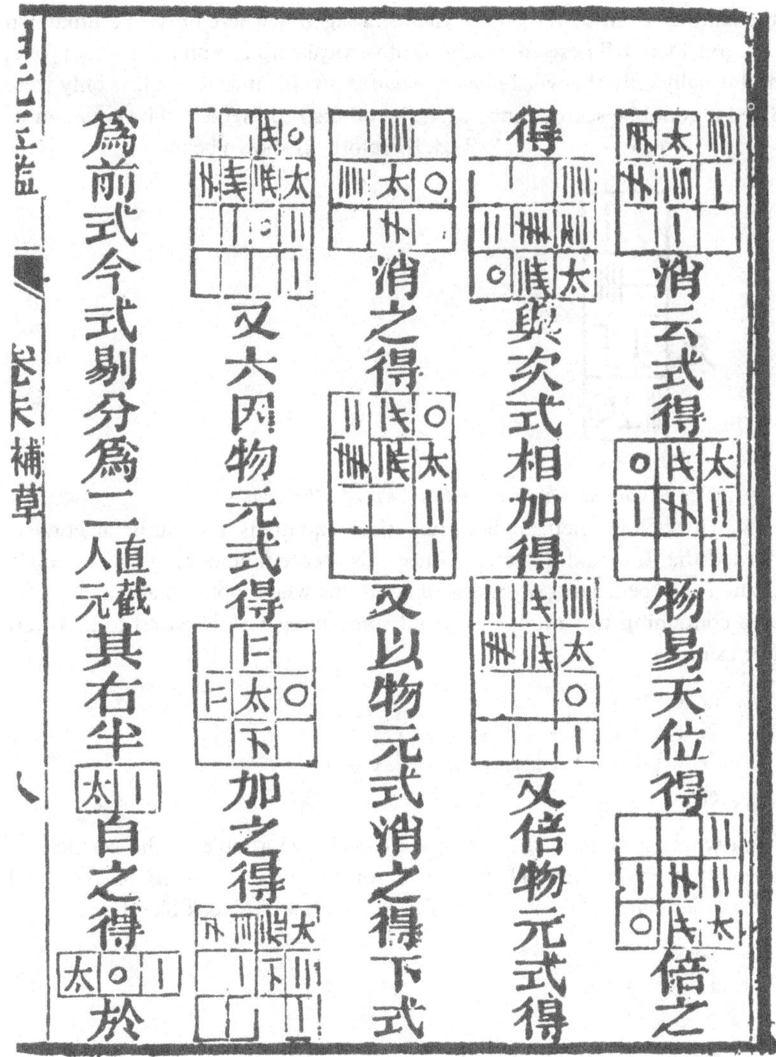

Fig. 52. A page from Ting Chhü-Chung's edition of the *Ssu Yuan Yü Chien* of Chu Shih-Chieh (A.D. 1303) showing the 'matrices' of the *Thien Yuan* algebraic notation. The middle frame in the far right-hand column is similar to the example given in the right-hand diagram on page 52; it shows $xy^2 - 120y - 2xy + 2x + 2x^2$.

the simpler form of the *Thien yuan shu* algebra where only one unknown was considered. These are arranged in vertical columns on the printed page, and usually only the word *thien* or *yuan* is given, since if one line only were fixed it could be seen at once what the others represented. The equation $x^3 + 15x^2 + 66x - 360 = 0$ looked, therefore, as shown below.

The Binomial Theorem and the 'Pascal' Triangle

In solving their higher numerical equations the Sung algebraists needed the Binomial Theorem, since this theorem amounts to a device for finding the coefficients of intermediate terms when a binomial – an expression containing two numbers (x and some integer) – is raised to a power. For example

$$(x + 1)^2 = x^2 + 2x + 1$$
$$(x + 1)^3 = x^3 + 3x^2 + 3x + 1$$
$$(x + 1)^4 = x^4 + 4x^3 + 6x^2 + 4x + 1$$

etc.

from which it is easy to see that we can derive a table of the coefficients formed when a binomial expression of the type $(x + a)$ is expanded (multiplied out). The beginning of such a table would look like this:

power						
2	1	2	1			
3	1	3	3	1		
4	1	4	6	4	1	
5	1	5	10	10	5	1
etc.						

This array has been known since the seventeenth century in Europe as Pascal's Triangle, since it was in A.D. 1665 that Blaise Pascal's thorough investigation of it in his *Traité du Triangle Arithmetique* was posthumously published. Actually it had already appeared in print for the first time more than a century earlier, on the title page of Apianus' *Arithmetic* (1527), and in the sixteenth century had been fairly widely known. But Apianus, Pascal

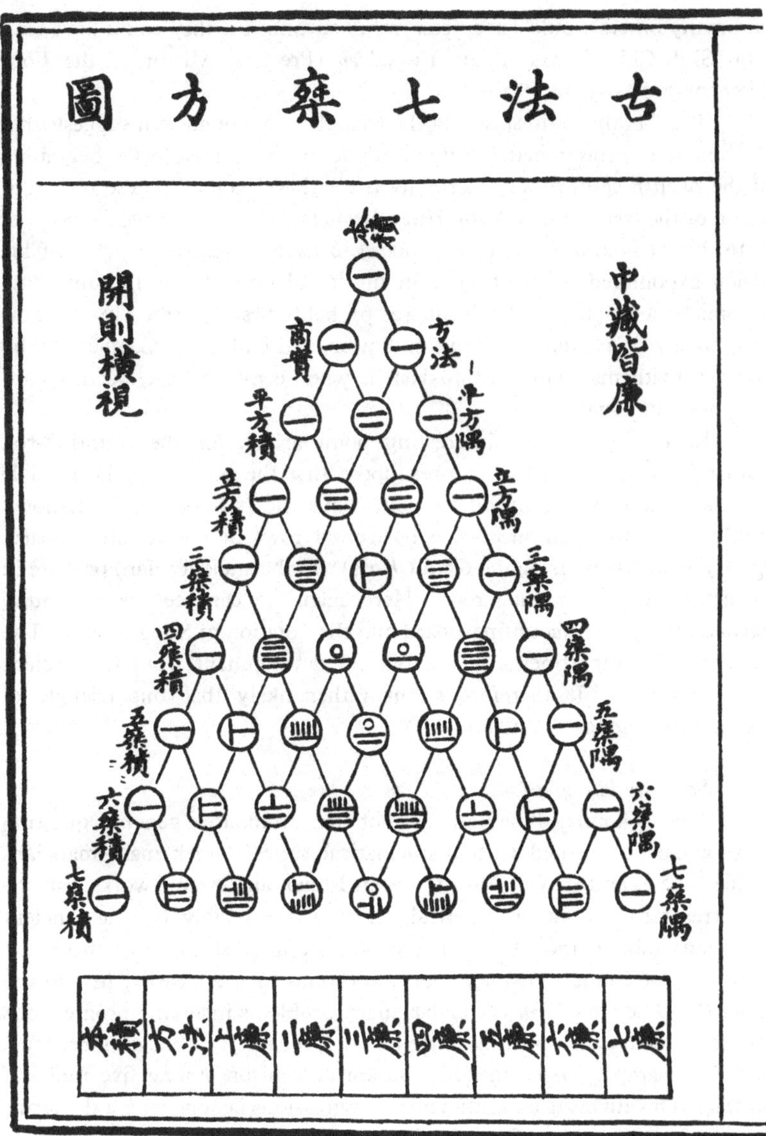

Fig. 53. The 'Pascal' Triangle as depicted in A.D. 1303 at the front of Chu Shih-Chieh's *Ssu Yuan Yü Chien*. It is entitled 'The Old Method Chart of the Seven Multiplying Squares' and tabulates the binomial coefficients up to the sixth power.

and many others would have been rather surprised if they could have seen Chu Shih-Chieh's *Ssu Yuan Yü Chien* (Precious Mirror of the Four Elements) of 1303 (Fig. 53).

The fact that Chu speaks of the triangle as old or ancient suggests that the binomial theorem had already been understood at least by the beginning of the twelfth century A.D. Certainly the earliest extant Chinese representation of the triangle is in Yang Hui's *Hsiang Chieh Chiu Chang Suan Fa* of 1216, but it is known (from this book) to have existed long before. Chia Hsien expounded it about 1100 in his 'tabulation system for unlocking binomial coefficients', although it was probably first described in the (now lost) book *Ju Chi Shih So* (Piling-up Powers and Unlocking Coefficients) by another mathematician, Liu Ju-Hsieh, who seems to have been a contemporary of Chia's.

In Fig. 52 a very interesting point arises, for the counting-rod numerals are so turned as to presuppose that the bottom of the triangle originally stood vertically on the left. Thus the powers of the unknown would have stood on successive horizontal rows of the counting-board, exactly as we know (from the *Chiu Chang*) was the ancient (Han) practice in extracting square or cube roots. Here again we can see the continuity between the ancient counting-board and the notation of Sung algebra. The network of compartments was a natural development from the ancient horizontal rows. It therefore seems rather likely that this triangle of coefficients originated in China.

Series and progressions

The Binomial Theorem is not far from the general question of problems concerned with mathematical series. Greek mathematicians studied these, and they are also found in Indian and Arabic works, but the first treatment of an arithmetical series was probably by the Ancient Egyptians (about 1700 B.C.). There was an interest in series from the beginning of Chinese mathematics, indications of it appearing first in the *Chou Pei*. The Han *Chiu Chang* has many problems involving progressions (series in which each term is related to its predecessor by a uniform law), one, for example, about the distribution of venison among five ranks of officials. This involves an arithmetical progression (i.e. one where the terms increase by a constant difference) for its solution. In the Han and San Kuo periods there was evidently considerable interest in textile production, for several problems concerned the output of women weavers. Both the *Chiu Chang* and the *Sun Tzu* have a question running thus: 'A girl skilful at weaving doubles each day the output of the previous day. She produces 5 metres of cloth in 5 days. What is the result of the first day, and of the successive days respectively?' Here a geometrical progression is needed to obtain a solution.

Chang Chhiu-Chien in the fifth century generalised to some extent the procedures for solving problems of this kind, working out a formula for the total or sum of each type of series. However, way back in the first century A.D., it was known that 271 counting-rods fitted into a hexagonal bundle – an example not only of a figurate number but also of an arithmetical progression, and towards the end of the thirteenth century, Chu Shih-Chieh, in his *Ssu Yuan Yü Chien* (Precious Mirror of the Four Elements), discussed series at an advanced level, dealing with bundles of arrows made into various cross-sections, such as circles and squares, and with balls gathered into triangles, pyramids, cones, etc. Yet this seems to mark the end of the Chinese development of series, which progressed no further until after the coming of the Jesuits.

Permutations and combinations

In view of the pattern or matrix character of the way the Chinese wrote down their mathematics, one would expect that chessboard problems would have reached Europe from China, and also, of course, from India, where chess in its modern military form was developed. Some of these problems involved series, others permutations and combinations (i.e. questions of the ordering and selection from a set of objects of items taken so many at a time), and even questions of probability. One famous problem is associated with the eighth-century Thang monk I-Hsing, for in the eleventh-century *Mêng Chhi Pi Than* (Dream Pool Essays) Shen Kua tells us:

> The story-tellers say that I-Hsing once calculated the total number of possible situations in chess, and found he could exhaust them all. I thought about this a good deal, and came to the conclusion it was quite easy. But the numbers involved cannot be expressed in the commonly used terms for numbers. I will only briefly mention the large numbers which have to be used. With two rows and four pieces the number of probable situations will be of 81 different kinds. With three rows and nine pieces the number will be 19,683... For five rows and twenty-five pieces, the number will be 847,288,609,443... Above seven rows we do not have any names for the large numbers involved. When the whole 361 places are used, the number will come to some figure (of the order of) $10{,}000^{52}$...

Shen Kua goes on to enumerate all I-Hsing's methods, saying they are capable of specifying all possible changes and transformations occurring on the chessboard.

In any discussion of permutations and combinations in China one immediately thinks of the third-century B.C. *I Ching* (Book of Changes) with its eight trigrams and 64 hexagrams (see volume 1 of this abridgement, pages 182 ff). Some mathematical study of all its possible arrangements is to be expected, and there are reasons for thinking that this was carried out in some (perhaps Taoist) circles as an esoteric doctrine. Here the *Shu Shu Chi I* of A.D. 190, with its Taoist background, is interesting; in connection with divination it mentions arrangements of the trigrams and what may also have been the throws of a dice. It is also concerned, as we have seen (page 33), with balls of different colours being placed on numbered or graduated coordinates, and in view of the skill of mathematicians in the use of counting-rods, their value is not obvious unless what they were really concerned with was the study of permutations and combinations. A more detailed study of ball arithmetic shows that it could readily be used to solve questions like 'how many numbers can be made out of 9,183?', or 'In how many ways can eight people seat themselves at a round table?' The hexagrams, too, clearly contain patterns of repetition: this was what the seventeenth-century European mathematician Gottfried Leibniz recognized when he said they contained the numbers 1 to 64 written in binary arithmetic (volume 1 of this abridgement, page 189). Since some of the thirteenth-century Chinese mathematical schools were, we know, concerned with divination and fate-calculation, so, perhaps, the Chinese use of permutations and combinations was primarily developed for these very purposes.

The calculus

There seem to have been four stages in the development of what is generally called the calculus. The first, found among the Greeks of the fifth century B.C. and the Chinese of the third century A.D., consisted in passing from commensurable to incommensurable magnitudes by the method of exhaustion. For example, when the early seekers after the true value of π inscribed polygons inside circles, it was an attempt to 'exhaust' the residual area between the outside of the polygons and the inside of the circumference. The second step was the method of infinitesimals, which began to attract attention in the seventeenth century, and was used by Newton and Leibniz. The third was the fluxions of Newton, and the fourth, that of limits, is also due to him.

Among the Greeks the nearest approach to that part of calculus known as integration was the work of Archimedes, about 225 B.C., in determining the area of a segment of a parabolic curve. Knowing the area of a triangle, he found the area of the segment by inscribing inside the parabola the largest triangle he could; he then inscribed a new triangle with the same base and the same height as the space that was left, and continued to do this so that he ended up with triangles that were extremely – one might say infinitesimally – small and thin. In China, too, there were certain roots of the idea of infinitesimals, exhaustion and integration which are worth looking at. Shen Kua, in his 'Dream Pool Essays', spoke of the art of piling up very small things, by which he certainly had in mind something almost equivalent to the summation of an infinite number of indivisibles or infinitesimals which Bonaventura Cavalieri developed in Europe in 1653, six centuries later. For areas Shen Kua spoke of an art of cutting and making to meet, while for volumes he spoke of 'interstice-volumes' or 'crack-volumes', i.e. exactly the residual space which it was the object of exhaustion to assess, and it seems clear he must have known that the smaller the units were, the better it would be for exhausting any given area or volume. This too was the method Liu Hui had used for finding π (page 43). Moreover, the ideas of forming a line by piling up 'atomic' points, and a plane by piling up lines, had been present from the dawn of Chinese philosophy in the Mohist definitions (already quoted, page 40). The idea of continuity, of taking infinitely small pieces, had also been expressed by the logical paradoxers, the friends of Hui Shih (early fourth century B.C.) (see volume 1 of this abridgement, Chapter 9). But of course these arguments were covered by the dust of centuries before being appreciated in our own time.

Questions concerning the piling up of small units within a given volume continued to attract the attention of Chinese mathematicians down to the sixteenth century. Chou Shu-Hsüeh in his *Shen Tao Ta Pien Li Tsung Suan Hui* (Assembly of Computing Methods connected with the Calendar) of 1558 gave illustrations, showing, for example, the piling up of spheres in ten layers within a pyramid; and in the seventeenth century the Japanese mathematicians produced a good deal of work similar to Cavalieri's (Fig. 54).

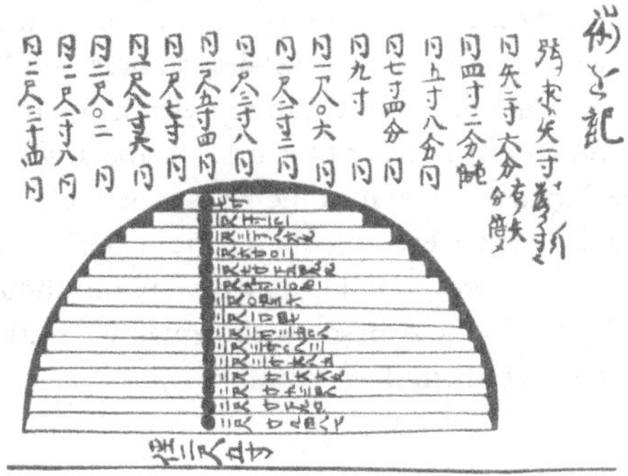

Fig. 54. Integration of thin rectangles in circle area measurement, from the *Kaisan-ki-Kōmoku* (A.D. 1687) of Michinaga Toyotsugu and Ōhashi Takusei, derived from Sawaguchi Kazuyuk's *Kokon Sampō-ki* (c. A.D. 1670).

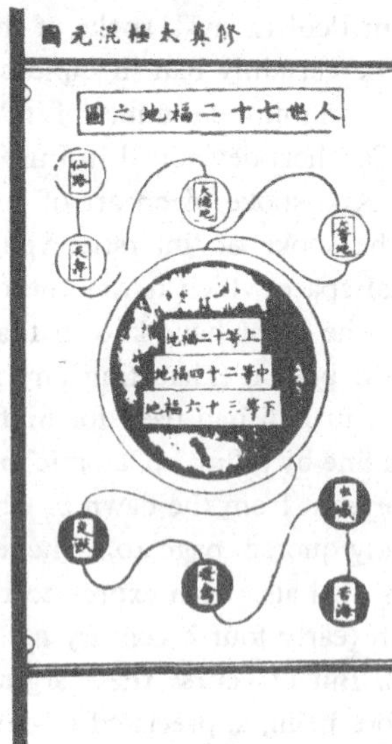

Fig. 55. Rectangles inscribed within a circle (from Hsiao Tao-Tshun's *Hsiu Chen Thai Chi Hun Yuan Thu*, *c.* eleventh century A.D.).

While it has been suspected that this exhaustion method began in China, it has not so far been possible to point to any instance of it in a Chinese mathematical work. However, an illustration (Fig. 55) from a Taoist work of the Sung (perhaps eleventh century?) attributed to Hsiao Tao-Tshun, the *Hsiu Chen Thai Chi Hun Yuan Thu* (Restored True Chart of the Great-Ultimate Chaos-Origin), shows rectangles inscribed in a circle. It may be that here, under Taoist and Neo-Confucian auspices, we are in the presence of one of the germs of the idea of the exhaustion of a circular area by the inscribing of rectangles. Shen Kua's problems were essentially concerned with what modern physicists and crystallographers would call 'packing', and this problem became one of the ruling passions of Japanese mathematicians from the seventeenth century onwards. In the seventeenth and eighteenth centuries the Japanese were in some ways more cut off from European influence than the Chinese were, but the precise extent to which their work was original is a difficult question. The essential point, though, is that all the indigenous Chinese and Japanese work remained static: the dynamic approach was due to Newton and Leibniz alone.

INFLUENCE AND TRANSMISSION

We may pause here to glance for a moment at what little has been gathered about contacts between Chinese mathematics and that of other great culture-areas of the Old World. In the first place, we have already seen (page 36) that little detailed influence seems to have come from Mesopotamia or, for that matter, from Egypt, where the treatment of fractions was radically different. When, however, we ask what mathematical ideas seem to have radiated from China southwards and westwards, we find a considerable list, which contains the following:

(*a*) the extraction of square and cube roots;
(*b*) expressing fractions in a vertical column;
(*c*) the use of negative numbers;
(*d*) an independent proof of Pythagoras' Theorem;
(*e*) geometrical questions like the areas of circles and some solid figures;
(*f*) the Rule of Three for determining proportions;
(*g*) the Rule of False Position for solving equations, and the solution of cubic and higher equations;
(*h*) indeterminate analysis;
(*i*) the Pascal triangle.

There are also the questions of the development of place-value notation and the written symbol for zero. As far as the first is concerned, this seems to have grown quite independently in China, deriving nothing from that form of it used by the Old Babylonian mathematicians except, just possibly, the bare idea. And although evidence is uncertain, it seems likely that India not only received place-value notation later, but obtained it from China rather than Mesopotamia. A written symbol for zero seems to have been the one mathematical invention we have mentioned that occurred outside China, although even in this case it was a development which took place at a mutual cultural area on the Chinese–Indian border.

It seems probable, then, that in spite of the 'isolation' of China and various social factors which made transmission difficult, between 250 B.C. and A.D. 1250 a good deal more came out than went in. Only at about the latter date did some influence from the south and west begin to be noticeable, and even then little took root. There was a little trigonometry, some small change in writing numbers and in the way multiplication was displayed, but that is all.

MATHEMATICS AND SCIENCE IN CHINA AND THE WEST

The conclusion of the account of Chinese mathematics brings us to what might be described as the focal point in the plan of this present work. What exactly were the relations of mathematics to science in ancient and medieval China? What was it that happened in Renaissance Europe when mathematics and science joined in a new combination that was destined to transform the world? And why did this not happen elsewhere?

First we must get the perspective right. Few mathematical works before the Renaissance were at all comparable in achievement with the wealth and power of the developments which took place afterwards. It is pointless, therefore, to subject the old Chinese contributions to the yard-stick of modern mathematics. We have to put ourselves in the position of those who had to take the earliest steps and try to realise how difficult it was *for them*. Measured in terms of human labour and intellectual attack, one can hardly say that the achievements of the writers of the *Chiu Chang Suan Shu* or the Sung algebraists with their *Thien Yuan* notation for equations were less arduous than those of the men who opened new mathematical fields in the nineteenth century. The only comparison that can be made is between old Chinese mathematics and the mathematics of other ancient peoples, Babylonian and Egyptian, Indian and Arabic.

From the description given it is clear that Chinese mathematics was quite comparable with the pre-Renaissance achievements of the other medieval peoples in the Old World. Greek mathematics was on a higher level, if only on account of its more abstract and systematic character – as is shown in Euclid – although it was weak just where the mathematics of India and China were strong, namely, in algebra. But historians of science are beginning to question whether the predilection of Greek science and mathematics for the abstract, the deductive and the pure over the concrete, the empirical and the 'applied' was wholly a gain. Certainly, in the flight from practice into the realms of the pure intellect, Chinese mathematics did not participate.

Chinese mathematics, for all its originality, displayed certain weaknesses. There was an absence of the idea of rigorous proof, possibly as a result of the mental outlook which avoided the development of formal logic in China and which allowed associative or organic thinking to dominate (volume 1 of this abridgement, pages 161 ff). The *Thien Yuan* notation had beautiful symmetry but extreme limitations; after an initial burst of advance Sung algebra experienced no rapid and extended growth.

Social factors were also involved, and it is striking that throughout Chinese history the main importance of mathematics was in relation to the calendar. For reasons connected with the ancient corpus of beliefs about the cosmos, the establishment of the calendar was the jealously guarded prerogative of the emperor, and its acceptance by tributary States signified loyalty to him. When rebellions or famines occurred, it was often concluded that something was wrong with the calendar, and mathematicians were asked to reconstruct it. Possibly this preoccupation fixed them irretrievably to concrete number, and prevented consideration of abstract ideas; in any case the practical genius of the Chinese tended in that direction. But there are reasons for thinking the mathematics of the calendar also had unorthodox Taoist connections. Hsü Yo, in the second century, was under Taoist influence, in the fifth this influence affected the strange figure Hsiao

Tao-Tshun; in the eleventh it inspired Li Yeh. Lastly, a factor of great importance lies in the Chinese attitude to the 'laws of Nature' (volume 1 of this abridgement, Chapter 16). It will be recalled that there was no belief in the idea of a creator deity, and hence of a supreme law-giver; this, combined with the conviction that the whole universe was an organic, self-sufficient system, led to the concept of an all-embracing Order in which there was no room for Law, and hence few regularities to which it would be profitable to apply theoretical mathematics in the mundane sphere.

Mathematics in China was therefore utilitarian, its social origins bound up with the bureaucratic government system, and devoted to the problems the ruling officials had to solve. Of mathematics for the sake of mathematics there was very little. This does not mean that Chinese calculators were not interested in truth, but it was not that abstract systematised academic truth which the Greeks sought. And during this time the masses remained illiterate, having no access to the manuscript books which the government commissioned, copied and distributed. Artisans, no matter how greatly gifted, flourished on the other side of an invisible wall which separated them from the scholars of literary training. When Taoists and Buddhists inspired the invention of printing, this without doubt fostered the second flowering of mathematics in the Sung, but the upsurge did not last. As soon as the Confucians swept back in to power with the nationalist reaction of the Ming, mathematics was again confined to the back rooms of government offices and ministries.

What was it then that happened at the Renaissance in Europe whereby mathematised natural science came into being? Why did this not occur in China? If it is difficult enough to find out why modern science developed in one civilisation, it may be even more difficult to find out why it did not develop in another. Yet the study of an absence can throw strong light upon a presence. If we contrast Galileo (1564 to 1642), who must be considered the central figure in the mathematisation of natural science, with Leonardo da Vinci (1452 to 1519), we see that in spite of all the latter's deep insight into Nature and brilliance in experimentation, no theoretical development followed because of his lack of mathematics. Leonardo was not an isolated genius; he was the most outstanding of a long line of practical men in the fifteenth and sixteenth centuries, who busied themselves with the investigation of the natural world, yet the mathematics they all used was purely utilitarian: they were men of measure and rule. There was no sense of a mathematical approach to the problems they studied.

No one has yet understood the inner mechanism of the mathematising of sixteenth- and seventeenth-century science, although it has been said that while algebra and geometry had developed separately in the past, what Galileo and his successors did was to apply algebraic methods to the field of geometry. This was the great step that made all the difference. Certainly the Chinese had always considered geometrical problems algebraically, but that was a different thing, as we can see if we dissect the Galilean method. We find that the new experimental investigation of the natural world involved the following steps:

(*a*) Selecting those specific aspects which could be measured.
(*b*) Forming an hypothesis which involved a mathematical relationship between the quantities observed.
(*c*) Deducing certain consequences from this hypothesis which are in the range of practical verification.
(*d*) Observing, followed by change of conditions, followed by further observing, i.e. experimentation; embodying as far as possible measurement expressed in numbers.
(*e*) Accepting or rejecting the hypothesis.
(*f*) Once an accepted hypothesis has been found, using it as a starting-point for fresh hypotheses and submitting them to test.

This was the 'new, or experimental philosophy'. A world of quantity was substituted for the world of quality.

The advance into abstraction went further than this, however. Motion, for instance, was considered apart from any particular moving bodies. It was also recognized as being the same everywhere in the universe. Space was geometrised; it was abstract and expressible in numbers. The cosmos was no longer to be considered a hierarchy of differentiated parts in an organic whole: now it was a universe held together only by the universal applicability of simple fundamental laws. There was nowhere in the universe, for example, where the writ of the law of gravitation would not run – once the idea of gravitation had been expressed. This was indeed a fundamental change.

This break up of the organic unity of the cosmos removed the idea that everything had its natural place and function. The quality of objects – their compact properties of shape, weight, colour and movement – which had seemed so obvious a unity – was now no longer valid. But it needed a mind of the highest originality, goaded by centuries of frustration, to reject this unity, and maintain that a wooden ball and a planet of unknown substance had more in common than all the different qualities of the ball (colour, texture, smell, etc.) had with each other. That the hypothesis then formed must be a mathematical one was of enormous importance. As Galileo put it:

> Philosophy is written in that great book which ever lies before our gaze – I mean the universe – but we cannot understand it if we do not first learn the language and grasp the symbols in which it is written. The book is written in the mathematical language, and the symbols

> are triangles, circles, and other geometrical figures, without the help of which it is impossible to conceive a single word of it, and without which one wanders in vain through a dark labyrinth.

Since practical mathematics had long been known to the technician, perhaps the Galilean innovation may best be described as the marriage of craft practice with scholarly theory. This had happened before but only on a lower plane: for example in China with the alliance of the hermit philosophies of the Tao of Nature and the shamanist magician-technologists (volume 1 of this abridgement, page 107).

If we pay due regard to times and places, there was not much to choose between China and Europe regarding the heights of mastery achieved; no Westerners surpassed the bronze-founders of the Shang and Chou, or equalled the ceramicists of the Thang and Sung. The inhibition to the development of modern science lay in the realm of hypothesis-making. For instance, while Leonardo da Vinci could sketch advanced machines such as helicopters or centrifugal pumps, he still talked in terms of things having their natural place and qualities. With Leonardo, as well as with others, remarkable technical achievements could be achieved but no adequate scientific theory. This throws light on the Chinese situation, and defines the point reached by Chinese science and technology as Vincian, not Galilean.

If we look more broadly, we see that the development of modern science in sixteenth- and seventeenth-century Europe was no isolated happening. It occurred step by step along with the Renaissance, the Reformation, and the rise of mercantile capitalism followed by industrial manufacture. It may well be that these concurrent social and economic changes in Europe formed the milieu in which natural science could rise at last above the level achieved previously by the highest artisans, the semi-mathematical technicians. The reduction of all quality to quantities, the affirmation of mathematical reality behind all appearances, the proclaiming of a space and time uniform throughout all the universe: was it not analogous to the merchant's standard of value? There were no goods or commodities, no jewels or moneys, but such as could be computed and exchanged in number, quantity or measure.

Of this there are abundant traces among European mathematicians. The first literary exposition of double-entry book-keeping is contained in the sixteenth century's best mathematical textbook; the first application of double-entry book-keeping to public finance and administration was in an early-seventeenth-century book by an engineer-mathematician, while even Copernicus wrote on monetary reform. And these are not isolated examples: commerce and industry were in the air as never before.

Put in another way, we can say there came no vivifying demand from the side of natural science. Interest in Nature was not enough, controlled experimentation was not enough, empirical induction was not enough, eclipse-prediction and calendar-calculation were not enough. All of these the Chinese had. Apparently mercantile culture alone was able to do what agrarian bureaucratic civilisation could not – bring to fusion point the formerly separated disciplines of mathematics and Nature-knowledge.

2

The sciences of the heavens: (i) Astronomy

Astronomy was a science of cardinal importance to the Chinese, since it arose naturally out of that 'cosmic religion', that sense of the unity and even 'ethical solidarity' of the universe, which led philosophers of the Sung to their great organic conceptions (volume I of this abridgement, pages 227 ff and 291 ff). Moreover, the establishment of a calendar by the imperial ruler of an agricultural people, and its acceptance by all those who paid allegiance to him, are threads which run continuously through Chinese history from the earliest times. Correspondingly, astronomy and calendrical science were always orthodox, 'Confucian', sciences, unlike alchemy, for example, which was typically Taoist and 'heterodox'. It has been well said that while among the Greeks the astronomer was a private person, a philosopher, a lover of the truth, as often as not on uncertain terms with the priests of his city; in China, on the contrary, he was intimately connected with the sovereign pontificate of the Son of Heaven, part and parcel of an official government service, and sometimes ritually accommodated within the very walls of the imperial palace.

This is not to say that the Chinese astronomers of ancient and medieval times were not also lovers of truth, but it did not appear necessary to them that it should be expressed in the highly theoretical and geometrical form which was characteristic of the Greeks. Apart from the Babylonian records, those of the Chinese show that they were the most persistent and accurate observers of celestial phenomena anywhere in the world before the Arabs. Even today (as we shall see later) those who seek information about celestial events in the past are obliged to have recourse to Chinese records, since over long periods they are the only ones we have, or if not so, then much the best. This was not only because of the meticulous records of the Chinese observers, but also because they regularly noted phenomena such as sun-spots which Europeans not only ignored, but would have found inadmissible because of their preconceptions about the cosmos (in this case that all celestial bodies were perfect and changeless). And the importance of Chinese observations is not lessened one whit by the fact that the observations of earlier centuries were often due to the belief in the importance of the heavens as foretelling State affairs. After all, astrology in Europe was long lasting and in a narrower, individual form it persisted until the time of Johannes Kepler in the seventeenth century; indeed it is still a popular superstition today.

If, then, we may adopt the general conclusion that Chinese astronomy participated in the fundamental idea of observation characteristic of all Chinese science, this had, at any rate, the effect of allowing a suspension of judgement when it came, for instance, to the movements of the planets. When Matthew Ricci discussed astronomy with Chinese scholars, after the Jesuits arrived in the late sixteenth century, their ideas, preserved in his own account of the conversations, sound in many ways more modern today than his own views of a stationary earth surrounded by circling planets pinned to solid celestial spheres.

The Western literature on Chinese astronomy is much more voluminous than that on Chinese mathematics. Unfortunately it is confused, controversial, and repetitive. From the first, European understanding of Chinese astronomy was affected by the advantages that the Jesuit missionaries saw they could gain by acquainting the Chinese with the advances of the European Renaissance, and introducing themselves into official circles by their superior calendrical calculations and eclipse predictions. They then strove on the one hand to impress the Chinese and win them over to Christianity by belittling indigenous Chinese astronomical knowledge, while on the other hand in many European publications they praised it, the better to fortify their own position in the mission field. Moreover, from the first, the Jesuit understanding of Chinese astronomy, very honestly undertaken in itself, was vitiated by fundamental misconceptions. These mostly stemmed from the fact that Chinese astronomy depended so much, as we shall see, on observation of the stars near the pole and the equator, while Greek and European astronomy relied on motions measured with reference to the sun's path across the constellations of the zodiac. The Jesuits were naturally quite unprepared for the possibility that another entire system of astronomy might have existed that was equal in scope and value to that of the Greeks, but different in method. This led to a series of misapprehensions that was not cleared up until the end of the nineteenth century.

The enormous difficulties which confronted the Jesuit pioneers must not be underestimated. They came to China at a standstill period of indigenous science during the Ming dynasty and the early Chhing. It was a time when exceedingly few Chinese scholars were available with enough learning to expound with clarity the traditional Chinese system, and to find and translate the essential passages in old books, some of which, as we have already seen for mathematics, had been lost and were not recovered until the eighteenth century. There was also a language difficulty, for sinology hardly existed and no good dictionaries had been compiled. The surprising thing is that the Jesuits got as far as they did.

When we come to later times, we find a better understanding from the middle of the last century, and an improvement at the beginning of the present one. Studies of the zodiac, however, led scholars to take firm positions and make categorical statements about the passage of ideas between China, India and the Arab world, even though they generally knew only one of the languages concerned. In brief we may distinguish two types of interest taken by Westerners in Chinese astronomy – first, its history as an aspect of the history of all science, and concern with Chinese observations for the help they provide in finding how different astronomical quantities have changed over the centuries. When the historical records were reliable, as from the Former Han period onwards, the results were valuable; but when excessive weight was placed on texts of the semi-legendary period, where dating is difficult, or on alleged observations by Chou Kung (the Duke of Chou) about 1000 B.C., the results only discredited Chinese material. The basic facts of Chinese astronomy are not so difficult to disentangle and are not now in dispute, though there are, of course, still many unsolved problems, particularly those connected with the historical vicissitudes of the Chinese calendar.

DEFINITIONS

Before proceeding further, we must be clear on how to describe the positions in the sky of all celestial bodies, and it will therefore be as well if we define some basic terms. Since the sky appears as a dome centred on the observer, it is convenient, from a descriptive point of view, to consider the entire heavens set out on the inside of a giant sphere, centred on the earth. We can then draw out coordinates on the sphere analogous to those we use to define the positions of places on the surface of the earth, and can describe the apparent motions of bodies like the sun and moon by drawing circles on the sphere (technically 'great circles', that is to say circles whose diameters pass through the centre of the sphere, and thus the centre of the earth). Since our imaginary sphere of the heavens is so large, we can for our purposes consider the observer fixed at the centre of the sphere itself, and take little account of the size of the earth.

Fig. 56 illustrates the celestial sphere. At the top is the *north celestial pole* P: this is directly above the earth's north pole – in other words it is the projection of the earth's north pole on the celestial sphere. At the bottom,

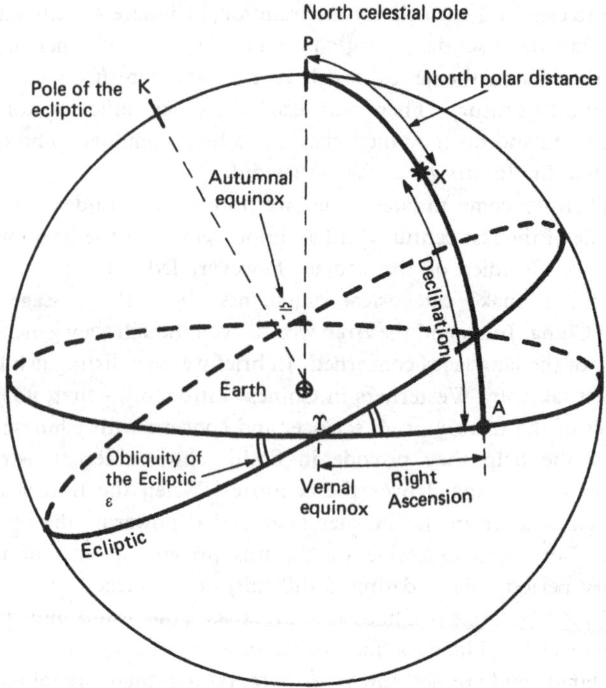

Fig. 56. Diagram of the circles of the celestial sphere.

directly under P, is P′, the south celestial pole; it is the projection on the celestial sphere of the earth's south pole. The daily rotation of the earth causes the whole celestial sphere to appear, to an observer on the earth, to rotate about the points P and P′. The great circle half way between the poles is the *celestial equator*; it is the projection on the sphere of the terrestrial equator. The second great circle is the *ecliptic* and to an observer on earth it is along this that the sun appears to move during the day and the year. The two places where the ecliptic and celestial equator cross are known as the *equinoxes*, since when the sun reaches these points day and night are equal in length everywhere on the earth's surface. The equinox at which the sun crosses the celestial equator in a south to north direction is known as the *First Point of Aries* and often denoted by the sign ♈; this point is also called the *vernal* or *spring* equinox, since the sun reaches it on or about 21 March. The opposite or *autumnal* equinox is also known as the *First Point of Libra*, . The point S marks the sun's highest point north of the celestial equator; it is known as the *summer solstice* or summer standstill point. The opposite point W marks the sun's southernmost point below the celestial equator, and marks the *winter solstice*. (The seasonal adjectives spring, summer, etc.

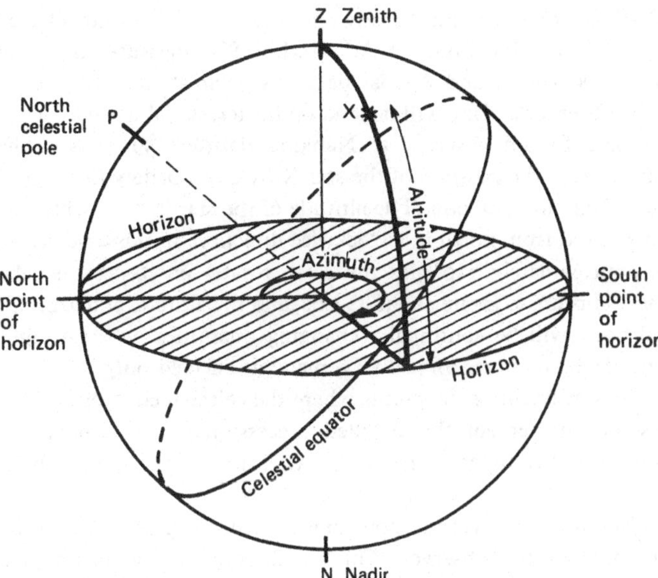

Fig. 57. Diagram of the Arabic system of coordinates.

are, of course, really only valid for observers living in the northern hemisphere, since those on the earth's southern hemisphere will experience the seasons reversed.)

The positions of stars in the sky may be referred to by coordinates analogous to those used on earth, but the celestial equivalents have special names: they are known as *right ascension* and *declination*. Right ascension (the equivalent of terrestrial longitude) is measured eastwards from ♈; the measurement is usually made in hours, minutes and seconds of time, one hour being equivalent to 15°, since 24 hours = 360°. Declination (the equivalent of terrestrial latitude) is the angular distance north or south of the celestial equator. Thus the star X in Fig. 56 has a right ascension measured by ♈A (about 6 hours or 90°) and a declination of AX (about 45°). Declinations are counted positive when running north of the celestial equator, and negative when south.

There are other coordinates which may be used to define the position of a celestial body. For instance, the ecliptic may be used as a reference instead of the celestial equator. Then, beginning from ♈ we have *celestial longitude*, and angles north or south of the ecliptic measure *celestial latitude*. This was the Greek system. There is also a set of coordinates which depends on the observer's position on earth. This was the Arabic system. Fig. 57 illustrates these. The hatched area represents the plane of an observer, its edge representing his horizon. The point Z is the observer's *zenith* and N the *nadir*, the point directly underneath. X is the star referred to in Fig. 56, ♈ the First Point of Aries, while ♈A measures the right ascension of X as before, and AX is the star's declination. The precise position of the observer's plane will depend on his terrestrial latitude: it has been drawn here for an observer at Nanking (latitude 39° 55′N). The observer can measure the position of the star X by a completely local set of coordinates, *altitude* and *azimuth*. The altitude of the star is $A_z X$, while the azimuth is measured from north in a clockwise direction, in this case giving an angle of approximately 230°. As the earth rotates on its axis, so the celestial sphere appears to rotate about its axis, and the altitude and azimuth of a celestial body will be continually changing quantities, whereas right ascension and declination are virtually permanent, affected only by a very small westward movement of the points where the celestial equator and the ecliptic cross. This movement, the so-called 'precession of the equinoxes', is due to a long-term movement of the earth's axis: it amounts only to about 1·4° per century.

The elevation of the celestial pole above the horizon depends on the terrestrial latitude of the observer. For an observer at the north pole (latitude 90°) the celestial pole will have an elevation of 90°: it will coincide with the zenith. For an observer on the equator (latitude 0°) the celestial pole will have an elevation of 0°: it will lie on the observer's horizon. At intermediate latitudes the elevation of the celestial pole above the horizon will be equal to the latitude of the observer. And since the whole celestial sphere appears to rotate about the celestial poles once every 24 hours, the elevation of the celestial pole above the horizon will result in some stars never rising or setting. The higher the latitude of the observer, the higher the elevation of the celestial pole, and the more of these circumpolar stars there will be. To be precise, at any given latitude, the circumpolar stars will be those whose declination is greater than the complement of that latitude or co-latitude (i.e. whose declination is (90° – latitude) or more; Fig. 58).

ASTRONOMICAL LITERATURE

European

The Interpreter-General and Father Superior of the history of Chinese astronomy was Antoine Gaubil. Born in 1689, he was a Jesuit missionary from 1723 until his death in 1759, having received a considerable astronomical training at the Paris Observatory. His Chinese was excellent and he was often called upon by the emperor to act as verbatim interpreter at State interviews. Yet in spite of his evident erudition, his writings are somewhat confused in presentation and constitute a mine in which one must know how to dig. The first to realise something of the true

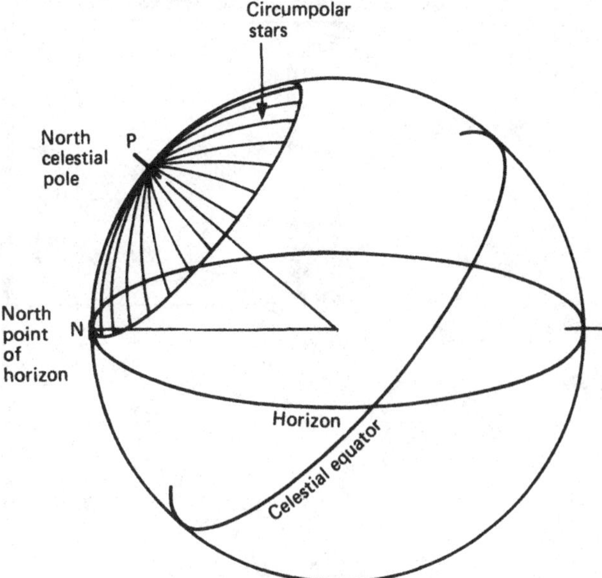

Fig. 58. Diagram showing circumpolar stars for the latitude of Nanking.

nature of Chinese astronomy was the many-sided French astronomer and chemist Jean-Baptiste Biot (1774 to 1862), and a few scholars have followed him during this century, but descriptions of Chinese astronomy in works on the history of astronomy or history of science are nearly all unsatisfactory.

The principal Chinese sources

In considering the principal original Chinese literature available, it must be recalled that in ancient and medieval China astronomy had an intimate connection with the government and the bureaucracy. The *locus classicus* of its official character is the famous commission of the legendary emperor Yao to the 'astronomers' Hsi and Ho (Fig. 59). It is found in the first chapter of the *Shu Ching* (Historical Classic), and concerns two (or rather six) astronomers. Only the vaguest guess can be made about the date, but it would seem likely to be about seventh or eighth century B.C. The text runs:

> (Yao) commanded the (brothers) Hsi and Ho, in reverent accordance with the august heavens, to compute and delineate the sun, moon and stars, and the celestial markers, and so to deliver respectfully the seasons to be observed by the people.
>
> He particularly ordered the younger brother Hsi to reside among the Yü barbarians (at the place called) Yang-Ku and to receive

Fig. 59. A late Chhing representation of the legend of the Hsi and Ho brothers receiving the government commission from the emperor Yao to organise the calendar and pay respect to the celestial bodies. From *Shu Ching Thu Shuo* (imperial illustrated edition of *Shu Ching* (Historical Classic)).

> as a guest the rising sun, in order to regulate the labours of the east (the spring).
>
> He further ordered the youngest brother Hsi to go and live at Nan-Chiao in order to regulate the works of the south and pay respectful attention to the (summer) solstice.
>
> He particularly ordered the younger brother Ho to reside in the west (at the place called) Mei-Ku, and to bid farewell respectfully to the setting sun, in order to regulate the western (autumnal) accomplishment.
>
> He further ordered the youngest brother Ho to go and live in the region of the north (at the place called) Yu-Tu, in order to supervise the works of the north.

For nearly three thousand years this remained, as it were, the foundation charter of Chinese official astronomy, but modern research has thrown quite new light on its legendary basis. It is now known that everywhere else in pre-Han literature Hsi-Ho is not the name of two or six persons, but a binome, the name, in fact, of the mythological being who is sometimes the mother and sometimes the chariot-driver of the sun. The name then in some way became split up and applied to four magicians or cult-masters who were charged by the mythological emperor to proceed to the four 'ends' of the world in order to stop the sun and turn it back to its course at each solstice, and to keep it going on its way at each equinox. This legend must have derived quite naturally from fears which primitive minds might well entertain that winter might go on getting continually colder, or summer intensify to lethal heat. Similar fears are found in other ancient civilisations, and in some cultures it took a terrible form, as, for instance, the mass human sacrifices which the Aztecs thought were necessary to prevent the standstill or death of the sun.

In China, in addition to their concern with the course of the sun, the mythological magicians were charged with the prevention of eclipses. A later chapter in the *Shu Ching* deals precisely with this, and purports to record a punitive expedition sent out against the magicians by the emperor owing to their failure to prevent an eclipse. What lies behind this is probably some ancient ritual directed against the semi-divine magicians at the four ends of the world who were thought to have neglected their duties.

In any case, the astronomical character of ancient Chinese State-religion comes out clearly in the above. An astronomical observatory was from the beginning an integral part of the Ming Thang, that cosmological temple which was also the emperor's ritual home. For an agricultural economy, astronomical knowledge as a regulator of the calendar was of prime importance. He who could give a calendar to the people would become their leader. More especially was this true for an agricultural economy which depended so largely upon artificial irrigation; it was necessary to be forewarned of the melting of the snows and the consequent rise and fall of the rivers and their derivative canals, as well as of the beginning of the rainy monsoon season.

All this can be seen in the importance given to astronomy in the *Chou Li* (Record of the Rites of the Chou), put together, no doubt, in the Former Han dynasty, about the second century B.C. In the opening sentence of the whole book the emperor is named as he who fixes the four cardinal points (by observations of the Pole Star and of the sun). Of the imperial astronomer (Fêng Hsiang Shih) we read:

> He concerns himself with the twelve years (the time Jupiter takes to complete a circuit of the sun), the twelve months, the twelve (double-) hours, the ten days, and the positions of the twenty-eight stars (the stars which determine the lunar mansions). He distinguishes them and orders them so that he can make a general plan of the state of the heavens. He takes observations of the sun at the winter and summer solstices, and of the moon at the spring and autumn equinoxes, in order to determine the succession of the four seasons.

The office was important and hereditary; the title implied that the holder had to keep watch at night from the astronomical tower or platform. Of the imperial astrologer (Pao Chang Shih) we likewise read:

> He concerns himself with the stars in the heavens, keeping a record of the changes and movements of the planets, the sun and the moon, in order to examine the movements of the terrestrial world, with the object of distinguishing (foretelling) good and bad fortune ... All the fiefs and principalities are connected with distinct stars, and from this their prosperity or misfortune can be ascertained. He makes prognostications, according to the twelve years (of the Jupiter cycle) of good and evil in the terrestrial world. From the colours of the five kinds of clouds, he determines the coming of floods or drought, abundance or famine. From the twelve winds he draws conclusions about the state of harmony of heaven and earth, ...

The office was again important and hereditary; its title refers specifically to the keeping of records. Both these officials had (or were supposed to have) a considerable staff. A third official, the Shih Chin, was charged with observations of a more meteorological character, which may, however, have included eclipses. Lastly, there was the official in charge of the clepsydras (water-clocks), the Chhieh Hu Shih, whom we shall meet again when Chinese astronomical instruments are described (page 151).

For more than two millennia these officials were organised in a special government department, the Astronomical Bureau or Directorate. To the end it retained high prestige, even when this was hardly justified by the scientific competence of the scholars who held office, as in the time of the Jesuits. The official astronomers enjoyed special privileges equivalent in a way to our 'benefit of clergy', since, as late as the last century, exceptionally light punishments in case of offences by members of the Astronomical Bureau were provided for in the Chhing penal code. Probably the most ancient title of their Director or President was Thai Shih Ling. Though quite conscious of his astrological functions, we feel that the true astronomical and calendrical element in the work of his department was amply sufficient to warrant the translation 'Astronomer-Royal'.

Under his control, of course, in all ages, was the imperial observatory and its equipment. More remarkable (and less well known) is the fact that in some periods it was customary to have two observatories at the capital, both furnished with clepsydras and other apparatus. Phêng Chheng, one of Shen Kua's contemporaries, tells us about these in the Northern Sung (eleventh century A.D.). One, the Astronomical Department of the Hanlin Academy, was situated within the imperial palace itself, while the Directorate of Astronomy and Calendar, presided over by the Astronomer-Royal, was outside. The data of the two observatories, especially concerning unusual phenomena, were supposed to be compared each night and presented jointly so as to avoid false or mistaken reports. This is a remarkable example of the sceptical and indeed scientific temper of the Confucian bureaucracy in medieval China. Phêng Chhêng also tells us that astronomy was in a bad way by the middle of the century. When he himself became Astronomer-Royal in 1070, he found the observers of the two institutions had been simply copying each others' reports for years past and seeing nothing strange in it. Nor did they make any use of the observatory equipment, but contented themselves with presenting ephemerides (tables of positions of celestial bodies) very roughly computed. Phêng Chhêng had six officials punished, but could not achieve full reform. Shen Kua, who succeeded him as Astronomer-Royal, was more successful, and no less critical of the former neglect of astronomy. He wrote:

> In the Huang-Yu reign period (between A.D. 1049 and 1053) the Ministry of Rites arranged for the examination-candidates to be asked to write essays on the instruments used for gaining knowledge of the heavens. But the scholars could only write confusedly about the celestial globe. However, as the examiners themselves knew nothing about the subject either, they passed them all with a high class.

But Shen Kua himself was the sign of a new period of brilliance.

The observatories in the capital were not the only ones; astronomical observations in many ages were certainly made in outlying parts of the country and the results forwarded to the Bureau. Moreover, astronomy had its place with mathematics in the Imperial University. To the two degrees in classics which the Sui dynasty (A.D. 581 to 618) had established, the Thang (A.D. 618 to 906) added four others, including one in mathematics, but few wanted to take it since it was not likely to lead to high advancement in the bureaucracy.

With this background in mind, one begins to see why it is rather hard to talk about chief landmarks in Chinese astronomical literature, as we did at the beginning of the chapter on mathematics. A great portion of the surviving Chinese astronomical literature is to be found in the dynastic histories in their chapters dealing with astronomy, the science of the calendar, and unusual natural events, and the identity of the writers is not always certain. Apart from these the majority of the ancient books on astronomy have perished, except for a few brief fragments. The reason for this is probably that whereas mathematical texts were of general use, those on astronomy were regarded as more technical and of interest only to the restricted circles connected with the Astronomical Directorate. They were probably copied in far fewer numbers, and were concentrated in palace and governmental libraries where they disappeared in one or other of the successive holocausts which accompanied the disturbances at changes of dynasty. After the invention of printing the situation changed somewhat, but not entirely.

Furthermore, owing to the close association between the calendar and State power, any imperial bureaucracy was likely to view with alarm any independent investigation of the stars, since this might lead to calendrical calculations which could be of use to rebels interested in setting up a new dynasty. Indeed, there were exhortations to security-mindedness, as for instance in the ninth century A.D. when:

> ... an imperial edict was issued ordering that the observers in the imperial observatory should keep their business secret. 'If we hear', it is said, 'of any intercourse between the astronomical officials or their subordinates and officials of other government departments or miscellaneous common people, it will be regarded as a violation of the security regulations which should be strictly adhered to. From now onwards, therefore, the astronomical officials are on no account to mix with civil servants and common people in general. Let the Censorate look to it.

As a sociological phenomenon, there was nothing new about Los Alamos or Harwell. But whether or not the best scientific achievements happen under such conditions is another question. From the earliest times Chinese astronomy had State support, but its semi-secrecy was a disadvantage, as some Chinese historians themselves felt. In the seventh-century *Chin Shu*

(History of the Chin Dynasty) we read:

> Thus astronomical instruments have been in use from very ancient days,... closely guarded by official astronomers. Scholars have therefore had little opportunity to examine them, and this is the reason why unorthodox cosmological theories were able to spread and flourish.

Nevertheless, it would be a mistake to push this too far. In the Sung dynasty, at any rate, there are clear indications that the study of astronomy was quite possible in scholarly families connected with the bureaucracy. Still, the general tradition explains well enough why Matteo Ricci's mathematical books were confiscated when he was on his way to the capital in 1600. Yet although there is not so clear a succession of treatises in astronomy as in mathematics, this does not mean that the Chinese astronomical literature is not very large.

Ancient calendars

Astronomical data are contained in the two oldest calendars that have come down to us. One of these, the *Hsia Hsiao Chêng* (The Lesser Annuary of the Hsia Dynasty) has nothing to do with the Hsia dynasty itself; it is substantially a farmer's calendar, but includes comments on the weather, the stars, and animal life, arranged in a twelve-month lunar calendar. Its date is uncertain, but the fifth century B.C. may be a fair estimate. The other is the *Yüeh Ling*, which later came to be incorporated in the *Hsiao Tai Li Chi* (Record of the Rites of Tai the Younger) of the first century B.C., and is much longer. It gives the astronomical characteristic of the (lunar) months with details of the appropriate musical notes, sacrifices, etc., the bulk of each chapter describing the imperial ceremonies to be performed, and ending with the prohibitions of various activities, and warnings of what will happen if the proper rites are not followed. These monthly observances may date from the third century B.C., or even be as early as the fifth.

Later on we shall have more to say about calendrical science. But this may be the place to mention that between 370 B.C. and 1851 no less than 102 'calendars' were calculated and promulgated in China, generally at the beginnings of particular reigns. It has not been sufficiently realised that these were in effect ephemerides, something like our 'nautical almanacs', giving expected positions of many celestial bodies; and they therefore constituted an unequalled register of astronomical data. The gradual and progressive advances in accuracy of the various astronomical constants can be clearly seen in them, and indeed they have been drawn upon to demonstrate how well ancient and medieval Chinese astronomers appreciated the concept of progressive approximations to a true description of reality.

Astronomical writings from the Chou to the Liang (sixth century A.D.)

At the time of Mencius (Mêng Kho), Confucius' disciple, there were then living two of China's greatest and earliest astronomers, Shih Shen of the State of Chhi and Kan Tê of the State of Wei. It was they, together with a third astronomer whose name is unknown but whose work was attributed to Wu Hsien, a legendary minister of the Shang, who drew up the first star-catalogue. This is the *Hsing Ching* (Star Manual). There has been much debate about the actual time of the quantitative data found in it now, and the latest view is that the measurements belong to some time around 70 B.C. The original work, which was quite comparable to that of the famous Greek astronomer Hipparchus (about 134 B.C.), was thus carried out two centuries earlier.

After this period of preliminary observational work in the late Chou time, the Han period was notable particularly for the interest taken in theories about the cosmos. However, questions about the cosmos were not then new; they had been raised in the fourth century B.C. by the School of Naturalists in the north and by the famous poet Chhü Yuan in the south. Since cosmological theories flourished in the Han, one would expect to find something of them in the apocryphal writings, already mentioned in connection with magic squares (page 20). This is indeed the case, for a considerable amount of two relevant texts has come down to us in the *Ku Wei Shu* (Old Mysterious Books) collection made in the Ming. There is also a very important chapter in Ssuma Chhien's *Shih Chi* (Historical Record), finished in 90 B.C. The author, who had himself filled the highest astronomical and astrological offices of State, first reviews the stars and constellations of the five 'Palaces' or sections of the sky, then discusses planetary motions, and follows this with remarks about the astrological associations of the lunar mansions with specific terrestrial regions, the interpretation of unusual appearances of the sun and moon, comets and meteors, clouds and vapours (including the aurora or 'Northern Lights'), earthquakes and harvest signs. He ends with his reflections as an historian, and the whole chapter is of the highest importance for ancient Chinese astronomy.

The first of the official dynastic histories, the *Chhien Han Shu* (History of the Former Han Dynasty) also contains astronomical chapters, the author of which was probably Ma Hsü, about A.D. 100. Here we find details of how to calculate synodic periods (i.e. the times planets take to reach the same apparent place in the sky with respect to the sun when viewed from the earth) and how to predict the appearance of eclipses. The ecliptic had played very little part in Chinese astronomy before the first century, but about A.D. 85, when Chia Khuei initiated a calendar reform, instruments were made with which to measure its position. This new knowledge together with a measure of the obliquity of the ecliptic, that is, the angle between the ecliptic and the celestial equator (Fig. 56), appeared in 178 in the *Lü Li Chih* (Memoir on the Calendar) by Liu Hung and Tshai Yung. A new step came in the fourth century A.D. with the discovery of the precession of the equinoxes by Yü Hsi, who worked mainly between 307 and 338, and announced the result in his *An Thien Lun* (Discussion on the Conformation of the Heavens).

Astronomical writings from the Liang to the beginning of the Sung (tenth century A.D.)

In the Sui dynasty, at the end of the sixth century, came the work of compilation of Wu Mi, and an important astronomical poem by Wang Hsi-Ming. Its title *Pu Thien Ko* (The Song of the March of the Heavens) well justifies its good description of the great constellations recognised at this time. In the Thang, in the seventh century, histories were prepared which had astronomical chapters containing a mine of astronomical information, and in the eighth century appeared the *Khai-Yuan Chan Ching* (The Khai-Yuan Reign Period Treatise on Astrology (and Astronomy)) which preserves many ancient astronomical writings in spite of its general concern with astrology. This was also the period of activity of the Tantric Buddhist monk I-Hsing and certain Indian astronomers resident in China. Though most of their writings failed to survive, the *Ta Yen Li* (Great Extension Calendar), an important work by I-Hsing, has recently been translated in full. Their work on planetary motions seems to have been valuable, though it had little effect on the course of Chinese astronomy. But there were other influences at work. Persian astronomers also came to China and astrological -astronomical literature was enriched by their presence during the eighth and ninth centuries, for they were one of the channels through which Babylonian mathematics and astronomy, especially the computation of eclipses and tables of future planetary positions, passed to China for further development.

Sung, Yuan and Ming

In view of the great flourishing of natural sciences during the Sung dynasty, we should expect astronomy to have flourished then also, and that this would be reflected in a proliferation of astronomical literature. There does, indeed, seem to have been a rich harvest at this time, and the second Sung emperor (976 to 997) had a large astronomical library. Yet little of it all has survived, although we are lucky in having one Sung astronomical work of capital importance, the *Hsin I Hsiang Fa Yao* (New Description of an Armillary Clock) by Su Sung, begun in 1088 and finished seven years later. The clock described is of prime importance in the history of time-keeping, but a description of it must wait until the discussion of Chinese mechanical engineering (the next volume of this abridgement). There is also a substantial work on astronomy and calendrical science, the *Liu Ching Thien Wen Pien* (Treatise on Astronomy in Six Classics), written by Wang Ying-Lin, which seems not to have been noticed by modern scholars.

The Yuan (Mongol) period, naturally enough, was one of close collaboration between Chinese and Muslim (Persian and Arab) astronomers, and brought new instruments to Chinese astronomy. After this, literature was less voluminous until the arrival of the Jesuits. Unfortunately, none of the writings of Kuo Shou-Ching, the greatest astronomer of the Yuan dynasty, has survived except his *Shou Shih Li* (Calendar of the Works and Days) of A.D. 1281 which is now being translated. It may be noted, though, that three years after his death, in 1319, there appeared the *Wên Hsien Thung Khao* (Historical Investigation of Public Affairs) of Ma Tuan-Lin, which collected, among other things, elaborate lists of the appearances of comets, meteors, novae (new stars), etc.

Astronomy seems to have suffered in the general standstill of science during the Ming. There was only the work of Wang Wei (about 1445), and Wang Kho-Ta somewhat later. After the coming of the Jesuits, astronomy re-awakened and there was a burst of publications, all of which echo the stimulation of the new Western ideas.

ANCIENT AND MEDIEVAL IDEAS ABOUT THE UNIVERSE

As we have already seen, the period of the late Warring States and the Earlier and Later Han was one of intense speculation in astronomy and cosmology – the study of the universe as a whole. The chief schools of thought which formed were thus described about A.D. 180 by Tshai Yung, himself a skilled astronomer, in a memorial to the emperor:

> Those who discuss the heavens form three schools. The first is that known as the Chou Pei school, the second is the Hsüan Yeh school, and the third is the Hung Thien school. The teaching of the Hsüan Yeh school has been interrupted ... As for the Chou Pei theory, though its methods and computations still remain, it proved incorrect ... Only the Hun Thien theory approximates to the truth.

The fifth-century astronomer and mathematician, Tsu Kêng-Chih, says the same thing, but gives to the Chou Pei school its alternative name, Kai Thien. The words Chou Pei are the same as those of the title of the oldest Chinese astronomical-mathematical work and may be taken to mean 'the gnomon and the circular paths of heaven'; the alternative Kai Thien means 'Heavenly Cover'. The words Hsüan Yeh were later explained as meaning 'brightness and darkness' though modern scholars suggest 'all-pervading night'. As for Hun Thien, there is no doubt that the term means 'celestial sphere'.

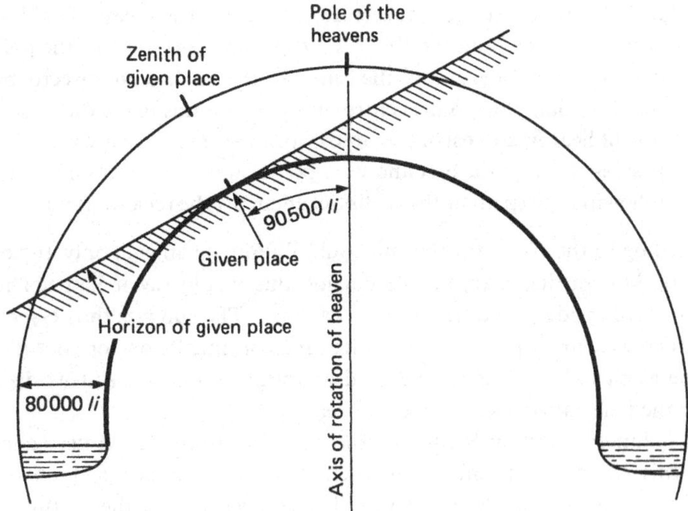

Fig. 60. Reconstruction of the Kai Thien cosmology (after Herbert Chatley).

The Kai Thien (hemispherical dome)

Internal evidence suggests that this is the oldest theory. The heavens were imagined as an inverted bowl covering the earth, which was itself thought of as another inverted bowl (Fig. 60). The distance between the two was 80,000 *li* or about 46,000 km (1 Chinese *li* is approximately equal to 0·576 km). The constellation of the Great Bear was in the middle of the heavens (it appears overhead at times in the latitude of Hangchow), and man's kingdom was in the middle of the earth. Rain falling on the earth flowed down to form the rim-ocean, the Great Trench; the earth itself was square. The heavens were round, and rotated like a mill from right to left, carrying with them the sun and moon, which nevertheless had a motion of their own from left to right, but much slower than the vast dome to which they were attached. Risings and settings of the heavenly bodies were only illusions; in fact they never passed below the base of the earth. In about 265, Yü Sung, a forebear and probably the grandfather of Yü Hsi who discovered the precession of the equinoxes, wrote in his *Chhiung Thien Lun* (Discourse on the Vastness of Heaven):

> The shape of the heavens is lofty, and concave like the membrane of a hen's egg. Their edges meet the surface of the four seas (the rim-ocean). It is like a bowl upside down which swims on water without sinking because it is filled with air ... Heaven has a pole just as a cover has a domed top. The northern heaven is lower than the earth by 30°. (The axis of the celestial) pole inclines towards the north, also

> making an angle of 30° (as seen from) the due east–west line. Now men live over a hundred thousand *li* (57,600 km) south of the pole's due east–west line. Hence the centre of the earth is not directly below the (celestial) pole. (This centre) just corresponds to the due east–west line of heaven and earth (i.e. to the 'prime vertical' or great circle passing through the east and west points and the zenith). The sun, following the (path of the) ecliptic, encircles the (celestial) pole...

According to the *Chou Pei* the sun could illuminate an area only 167,000 *li* (96,000 km) in diameter; people outside this would say it had not risen, while those inside would be enjoying daylight. The sun was thus essentially regarded as a circumpolar star, illuminating continually one or another part of the earth's surface, as if by a kind of searchlight-beam. But its distance from the pole varied according to the season.

Chinese tradition holds that the Kai Thien universe is very ancient, and this claim is certainly strengthened by the fact that a double-vault theory existed in Babylonia. It would have been one of the culture-traits which passed both westward to the Greeks and eastwards to the Chinese, to be developed later in both civilisations into the theory of the celestial sphere. Rather characteristically Chinese, however, was the insistence that the heavens were circular and the earth square, an idea which would arise naturally enough from the circles of the celestial sphere on the one hand and the four cardinal points of earthly space on the other. And the inclination of the celestial pole must have been among the earliest of astronomical observations. It is not surprising to find, therefore, among the Chinese myths:

> In ancient times Kung Kung (one of the legendary rebels) strove with Chuan Hsü (one of the legendary emperors) for the Empire.
> Angered, he smote the Unrotating Mountain,
> Heaven's pillars broke, the bonds with earth were ruptured,
> Heaven leaned over to the north-west;
> Hence the sun, moon, stars and planets were shifted,
> And earth became empty in the south-east.

There were other, later, ideas about the pole and rotation, but one thing is clear from all of them; the heavens rotated about the polar axis. However, it is evident that the question of the nature of the bearings of the polar axis must have caused much difficulty. This can be seen from what remains of Yao Hsin's *Hsin Thien Lun* (Discourse on the Diurnal Revolution), which dates from about A.D. 250.

> At the winter solstice the pole is low (on the polar axis), and the heavens in their rotation move near the south, so the sun is far from

> man and the Great Bear near him. Thus the *chhi* of the northern sky arrives and (the weather) becomes icy cold. At the summer solstice the pole rises (on the polar axis), and the heavens in their rotation move near the north, so that the Great Bear is far from man and the sun near him. The *chhi* of the southern sky arrives, so (the weather) becomes steamy and hot.

He explains a theory of 'two earths', the lower one of which furnished bearings on which the celestial axis could revolve, and he accounts for its inclination on a universe-and-man or macrocosm–microcosm analogy, the chine of the human head pointing forwards and downwards but not backwards. The most curious part of his scheme was that the celestial vault not only revolved on the polar axis, but slid up and down along it, the pole being much further away from the earth in summer than in winter. This theory of the lifting and sinking pole had, however, been current in the first century A.D.

If some interpreters are right, a theory very similar to this was expounded in the *Timaeus* of Plato (fourth century B.C.). The earth was supposed to slide up and down the polar axis, oscillating about a central point. But the passage is notoriously obscure.

The Hun Thien school (the celestial sphere)

The school of the celestial sphere corresponded to the idea of spherical motions centred on the earth, an idea which had also been developing slowly among the Greeks and came to be associated particularly with Eudoxus of Cnidus (about 409 to 356 B.C.). While it must have been known in China at least as early as the fourth century B.C., when Shih Shen was making his star lists, the earliest exponent of it whose name has come down to us was Lohsia Hung (*c.* 140 B.C. to 104 B.C.). The oldest full description of it comes from the pen of the astronomer Chang Hêng of the late first century A.D. in his *Ling Hsien* (Spiritual Constitution of the Universe) and his *Hun I Chu* (Commentary on the Armillary Sphere). In the latter he is at his most explicit:

> The heavens are like a hen's egg and as round as a crossbow bullet; the earth is like the yolk of the egg, and lies alone in the centre. Heaven is large and earth small. Inside the lower part of the heavens there is water. The heavens are supported by *chhi* (vapour), the earth floats on the waters.
>
> The circumference of the heavens is divided into $365\frac{1}{4}°$; hence half of it, $182\frac{5}{8}°$, is above the earth, and the other half is below. This is why of the 28 *hsiu* (equatorial star groups) only half are visible at one time. The two extremities of the heavens are the north and south

> poles, the former, in the middle of the sky, is exactly 36° above the earth, and consequently a circle with a diameter of 72° encloses all the stars which are permanently visible. A similar circle round the south pole encloses stars which we never see. The two poles are distant from one another 182° and a little more than half a degree. The rotation goes on like that around the axle of a chariot.

These words of Chang Hêng are precious for several reasons. His natural philosophy is of interest in connection with the beginnings of the idea of laws of Nature, and in astronomy he attributes the visualisation of the celestial sphere to a much earlier time than his own, clearly showing how the conception of a spherical earth, with antipodes, would arise naturally out of it. So also would the first astronomical instruments, armillary rings and armillary spheres (page 164). Moreover Chang Hêng realises that space must be infinite, and he is able, so to say, to look through the immediate mechanism of sun and stars to the unknown lying beyond.

There are numerous later expositions of the Hun Thien theory and some criticisms of it. At the end of the first century A.D., Wang Chhung, though advancing some arguments against the Kai Thien hypothesis, found himself still more unable to accept the Hun Thien, as he felt it would imply that the sun, the fiery essence of Yang, would have to move through water. The alchemist Ko Hung exerted himself to prove, in favour of the Hun Thien hypothesis, that this was not impossible, for dragons, which are very Yang, can live in water. But soon the archaic conceptions of 'waters under the earth' fell into abeyance.

The Hsüan Yeh teaching (infinite empty space)

The earliest name associated with this view is a relatively late one, Chhi Mêng, who lived during the Later Han and may have been an older contemporary of Chang Hêng, though we do not know his exact dates or much else about him. Somewhat over a century later, Ko Hung wrote:

> The books of the Hsüan Yeh school were all lost, but Chhi Mêng, one of the librarians, remembered what its masters before his time had taught concerning it. They said that the heavens were empty and void of substance. When we look up at it we can see that it is immensely high and far away, having no bounds. The (human) eye is (as it were) colour-blind, and the pupil short-sighted; this is why the heavens appear deeply blue. It is like seeing yellow mountains sideways at a great distance, for then they all appear blue. Or when we gaze down into a valley a thousand fathoms deep, it seems sombre and black. But the blue (of the mountains) is not a true colour, nor is the dark colour (of the valley) really its own.

> 'The sun, moon, and the company of stars float (freely) in the empty space, moving or standing still. All are condensed vapour. Thus the seven luminaries sometimes appear and sometimes disappear, sometimes move forward and sometimes retrograde, seeming to follow each a different series of regularities; their advances and recessions are not the same. It is because they are not rooted (to any basis) or tied together that their movements can vary so much. Among the heavenly bodies the pole star always keeps its place, and the Great Bear never sinks below the horizon as do other stars.

This cosmological view is surely as enlightened as anything that ever came out of Greece. The vision of infinite space, with celestial bodies at rare intervals floating in it, is far more advanced (and the point is worth emphasising) than the rigid conception of Aristotle and Ptolemy, with its concentric crystalline spheres, which fettered European thought for more than a thousand years. And the Hsüan Yeh school pervaded Chinese thought to a greater extent than would at first sight appear. For instance, Chang Hêng, explaining the Hun Thien view, could nevertheless write 'These things can all be calculated, but what is beyond (the celestial sphere) no one knows, and it is called the 'cosmos' (*yü chou*)', thus paying tribute to Hsüan Yeh ideas. So even beyond the great circles of the celestial sphere they conceived an infinity of space.

Chinese astronomy is often reproached for its overwhelmingly observational bias, but the lack of theory was an inevitable result of the Chinese lack of a deductive geometry. It may be argued, though, that the Greeks had too much. As will become evident later (page 213), Hsüan Yeh ideas persisted until the coming of the Jesuits, and it has been said that Chinese scholars saw a close resemblance between this system and the one introduced by Europeans after Jesuit times. They were not far wrong.

The Hsüan Yeh system has a particularly Taoist flavour, which may account for the disappearance of the oldest writings concerning it. One senses a connection with the 'great emptiness' of the Taoist Lao Tzu, and with the idea of heaven as 'piled-up *chhi*' in the *Lieh Tzu*. Buddhism also contributed. It had conceptions of infinite space and time, and of a plurality of worlds; from the Chin (third century A.D.) onwards, the Hsüan Yeh must have received much support from this Indian source. But it was the twelfth-century Neo-Confucian philosopher, Chu Hsi, who gave these views great philosophical authority, claiming that the heavens are bodiless and empty.

To support the important conclusion that the Hsüan Yeh world-picture and the Hun Thien spherical motions came to form the background of Chinese astronomical thinking, it is necessary to look very briefly at the subsequent history of theoretical cosmology in China. The Kai Thien system lasted on into the sixth century A.D. There was an attempt to reconcile it with the concept of a spherical universe, but the attempt was unsuccessful, and from that time onwards the official histories consider the Hun Thien sphere as the only correct view. But running through the centuries was the additional conception of a 'hard wind'; this helped the Chinese (who could not know of the absence of atmosphere in outer space) to imagine that stars and planets could be borne along without being attached to anything – the hard wind was responsible. This idea is generally thought to be of Taoist origin, and it may well go back to the earliest use of bellows in metallurgy, when technicians noticed the thrust possible with a powerful jet of air. This is something which the Taoist (but not the Confucian) would have known.

Among the Neo-Confucian philosophers, Chang Tsai gave particular attention to cosmological theory. When he spoke of 'the great emptiness without substance' of the heavens, he was following the Hsüan Yeh tradition. Earth, he said, consisted of pure Yin, solidly condensed at the centre of the universe, the heavens of buoyant Yang, revolving anti-clockwise (looking up towards the pole). The fixed stars are carried round endlessly with this floating, rushing *chhi*. In order to explain the opposite direction of the annual movements of the sun, moon, and five planets, Chang Tsi used the interesting conception of viscous drag or resistance to motion; these bodies, he thought, were so much nearer to us that the earth's *chhi* impeded their forward motions. Chang Tsai's ideas, and others like them, retained their vitality until the time of fusion with modern science after the coming of the Jesuits.

In a word, the Chinese astronomers were practically free from the cramping orthodoxy of the Greek and medieval European idea that the heavenly bodies were fixed to a series of spheres nesting one within another: spheres which were essentially nothing but a materialisation of spherical geometry. Indeed, it is rather paradoxical that the Chinese, whose views are so often accused of being excessively materialistic and concrete, should have remained comparatively free of explanation of this kind. They had no deductive geometry, but they had no crystalline spheres either.

If the celestial bodies moved on the hard wind, perhaps the earth moved also? Not a few ancient Chinese thought it did, although at first they envisaged an oscillation rather than a rotation. Chang Tsai and other Neo-Confucians combined the oscillation of the earth up and down the polar axis of the universe with tides of Yang and Yin force within it to explain the seasonal heat and cold. They also linked it with the phenomenon of sea tides. But the main point of interest for us is that the deeply held conviction of man and an immovable earth in the centre of the universe, which so dominated European thinking, was not marked in Chinese thought.

Other systems

There were other Chinese views about the universe besides those just described. About 120 B.C. a theory was described which is obscure, but seems to have been based on observations made with a series of gnomons, and held that the sun at the meridian is five times further away from the earth than at its rising and setting; an idea which would involve a very elliptical cover or shell.

We also find attempts made in some of the Han apocryphal books (first century B.C.) to measure the distances of celestial bodies and the size of the earth, and these were further elaborated in the third century A.D. by Wang Fan. They were paralleled by attempts which the Greeks were making at about the same time. But to all this one must add a story which has come down to us, intriguing because it illustrates the chilling lack of interest of Confucianism in scientific problems, a social factor against which the Greeks did not have to contend. The passage is in *Lieh Tzu* (The Book of Master Lieh) and may date from any time between the fourth and first centuries B.C.:

> When Confucius was travelling in the east, he came upon two boys who were disputing, and he asked them why. One said 'I believe that the rising sun is nearer to us and that the midday sun is further away'. The other said, 'On the contrary, I believe that the rising and setting sun is further away from us, and that at midday it is nearest'. The first replied, 'The rising sun is as big as a chariot-roof, while at midday the sun is no bigger than a plate. That which is large must be near us, while that which is small must be further away'. But the second said, 'At dawn the sun is cool but at midday it burns, and the hotter it gets the nearer it must be to us'. Confucius was unable to solve their problem. So the two boys laughed him to scorn saying, 'Why do people pretend that you are so learned?'

This story belongs to a whole corpus of legend and folk-tale centring on a quasi-Taoist theme of 'an old head on young shoulders', in which Hsiang Tho, together with other small boys, always defeats Confucius in argument or riddle. But though the Taoists or Naturalists might laugh at the Confucians, it was the latter who became more and more the dominant social group, and indifference to natural philosophy grew with their dominance.

There are records of real discussions on these matters continuing into the first century A.D., as well they might, for the changes in apparent size of the sun and moon, when close to the horizon and when high in the sky, are not easy to explain. From the modern standpoint, they are found to involve psychological as well as physical factors. There were also discussions about the nature of celestial bodies themselves. That the sun was of a fiery Yang (male) nature and that the moon was Yin (female) and watery was a commonplace from the earliest phases of Chinese science. And it was from an early time, too, that the sun was termed Thai Yang (the Greater Yang) and the fixed stars Hsiao Yang (the Lesser Yang), while the moon (Thai Yin) corresponded to the planets (Hsiao Yin); distinction thus being made quite correctly between those bodies which shine by their own light and those which shine by reflected light. But perhaps this was only a coincidence; we do not know of any early or medieval text which distinctly states that the fixed stars were of the nature of distant suns. In Greece, Parmenides of Elea who flourished in the fifth century B.C. and was therefore a young contemporary of Confucius, was apparently the first Greek to state definitely that the moon shines only by reflected sunlight, and by the time of Aristotle (fourth century B.C.) this was accepted as a matter of course. Presumably the oldest mention of it in Chinese literature is that of the *Chou Pei*: 'The sun gives the moon her appearance, so the moonlight shines brightly forth'. This cannot be later than early Han, and may well be of the fourth century B.C., if not as old as the sixth. In the latter half of the first century B.C., Ching Fang extended this to the planets. Erroneous theories arrived in the sixth century A.D., with the translation of an Indian work, but they did not affect the general acceptance of the correct view, which is stated over and over again.

Another erroneous Indian theory which travelled to China with Buddhism was that of the existence of two imaginary invisible planets, which 'personified' the crossing-points of the moon's orbit with the ecliptic and were doubtless devised to account for lunar eclipses. However, the Chinese themselves had imagined, from ancient times, the existence of a 'counter-Jupiter' which moved round diametrically opposite the real planet; and there was a Greek parallel to this in the theory of a 'counter-earth' (fifth century B.C.) which was devised to bring the number of the heavenly bodies up to the perfect number 10, or to explain eclipses. Perhaps both originated from a more ancient Babylonian theory.

THE POLAR AND EQUATORIAL CHARACTER OF CHINESE ASTRONOMY

It is now established beyond question that ancient and medieval Chinese astronomy was based upon a system quite different from that of the Egyptians, Greeks and later Europeans, though in no way less logical or useful. This was not understood by the first Jesuits, and it requires some explanation.

Early astronomers faced the great difficulty that the star which determines the seasons (the sun) dims the other stars to invisibility by its brilliance, so that its position among them cannot easily be found. There are, in consequence, only two ways to determine this. One is to find what stars the sun is near – by observing the stars visible just before sunrise or just after sunset – and the other to find what stars lie opposite to the sun in the sky. The Egyptians and the Greeks adopted the first method, observing the 'heliacal risings and settings' of stars at dawn and dusk: indeed one of the most famous of all ancient scientific observations was that of the heliacal rising of Sirius, which warned the ancient Egyptians of imminent flooding by the Nile. Such observations needed no knowledge of the pole, celestial equator or even the meridian: they required only the recognition of the constellations lying along the ecliptic – the constellations of the zodiac. Attention in Egyptian and Greek astronomy was concentrated, therefore, on the horizon and the ecliptic.

The Chinese, on the contrary, adopted the method of opposability, finding what stars lay opposite to the sun. This they achieved by concentrating their attention on the celestial pole and the constellations around it – the circumpolar constellations, which never rise and set, but are always above the horizon. Their system was therefore intimately associated with the meridian (that great circle which passes through the celestial pole and the zenith and thus through the north and south points of the horizon). They observed systematically, noting the times when the circumpolar constellations culminated (i.e. reached their highest points in the sky), and when they achieved lower transit (which happened when they lay 'underneath' the celestial pole).

The Greeks, of course, knew of the circumpolar constellations – Homer refers to them – and there is even a Greek story that the sentinels at the siege of Troy changed their guard according to the vertical or horizontal positions of the tail of the Great Bear. There are, conversely, indications that some heliacal risings and settings were noted by the ancient Chinese. But the emphasis in China was quite different from that in Greece. The pole was the fundamental basis of Chinese astronomy, and was connected with a background of thinking about the microcosm and the macrocosm, the correspondence between man and the universe. Thus the celestial pole corresponded to the emperor on earth, around whom the vast system of a bureaucratic agrarian state naturally and spontaneously revolved.

The meridian was a very understandable derivative from what was probably the most ancient astronomical instrument of all, the gnomon, a post stuck vertically in the ground for measuring the length of the sun's shadow. Looking south by day, the observer measured the shadows at noon;

looking north at night, he measured the times at which the circumpolar stars made their upper and lower transits. Thus the 'Artificer's Record' section of the *Chou Li* (Record of the Rites of Chou):

> By day they collected observations of the length of the sun's shadow, and by night they investigated the culmination of stars, so that they might set in order mornings and evenings.

In measuring times of the transits of celestial bodies, one makes use of the concept of hour-circles, those equally spaced circles drawn through the celestial pole and through a celestial body, going on to meet the equator and then the horizon. These too had their symbolic microcosmic–macrocosmic meaning; just as the influence of the emperor, the Son of Heaven on earth, radiated in all directions, so the hour-circles radiated from the pole 'like the spars of an umbrella' as Shen Kua said in 1086. And during the first millennium B.C., the Chinese built up a complete system of equatorial divisions, defined by the points at which the hour-circles cut the equator; these were the 28 *hsiu* or 'lunar mansions'. One has to think of them as segments of the celestial sphere (like segments of an orange) bounded by hour-circles and named by the constellations which provided the stars lying on these hour-circles, and from which the unequal number of degrees in each *hsiu* could be counted.

Once having determined the boundaries of the *hsiu*, the declinations north or south of the celestial equator of the key stars matter not at all – the Chinese were in a position to know their exact locations, even when invisible below the horizon, simply by observing the meridian passages of the circumpolar stars keyed to them. And this is the way they solved the problem of finding the position of the sun among the stars, since they knew too that the full moon is always opposite in the sky to the sun. Here, indeed, was the real essence of the Hun Thien theory mentioned above (page 85); once having gained a clear understanding of the daily rotation of the heavens, then the culminations and lower transits of the circumpolar stars would fix the position of every point on the celestial equator. Hence the position of the sun among the stars could be known; solar and stellar coordinates could be married together.

Circumpolar stars and equatorial mark-points

That the Chinese did key the invisible *hsiu* to the transit of circumpolar stars may be illustrated most clearly from the 'Celestial Officials' chapter of the first-century B.C. *Shih Chi* (Historical Records). Ssuma Chhien says:

> Piao is attached to the Dragon's Horn (Chio; *hsiu* no. 1). Heng hits the Southern Dipper (Nan Tou; *hsiu* no. 8) in the middle. Khuei is

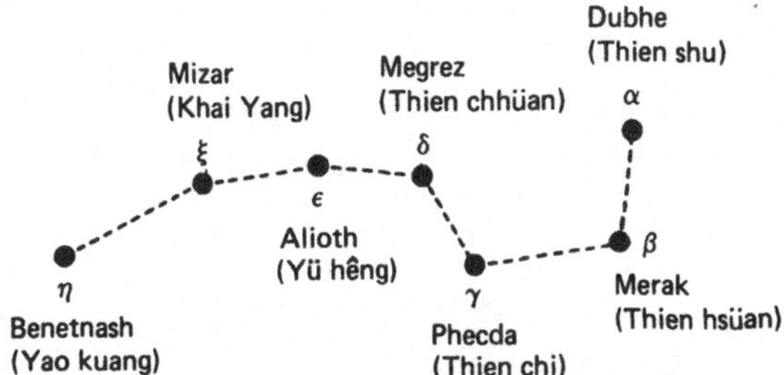

Fig. 61. The stars of the Great Bear (Ursa Major).

> pillowed on the head of Orion (Shen; *hsiu* no. 21). The dusk indicators (those of which transits are noted at dusk) are the Piao stars. The midnight indicator (the star of which the midnight transit is noted) is Heng. The dawn indicators (those whose transits are noted at dawn) are Khuei stars...

This passage is immediately clear once we know the names of the stars in the Great Bear (Ursa Major), and of some other constellations. As is well known, astronomers now refer to the stars of a constellation visible to the naked eye either by Greek letters or, in the case of dimmer ones, by a number or a Roman letter. Some of the brighter stars also have names, often of Arabic origin. By using this system and inserting Chinese names, we can draw up a table of the brighter stars of the Great Bear, i.e. the group known either as the Plough or the Dipper. This the Chinese divided into two parts, what we should call the blade (of the Plough) or the bowl or box of the Dipper, which they called Khuei (The Chiefs), and the handle of the Plough or Dipper, which was their Piao or 'Spoon'. The stars, illustrated in Fig. 61, are as follows:

(*a*) The 'bowl' or 'box', Khuei ('The Chiefs'):

α (alpha)	Dubhe	Thien shu, 'Celestial pivot'.
β (beta)	Merak	Thien hsuan, 'Celestial template'.
γ (gamma)	Phecda	Thien chi, 'Celestial armillary'.
δ (delta)	Megrez	Thien chhuan, 'Celestial balance'.

(*b*) The 'handle', Piao ('The Spoon'):

ε (epsilon)	Alioth	Yu heng, 'Jade sighting tube'.
ζ (zeta)	Mizar	Khai Yang, 'Opener or introducer of heat, or of the Yang'.
η (eta)	Benetnash	Yao kuang, 'Twinkling brilliance'.

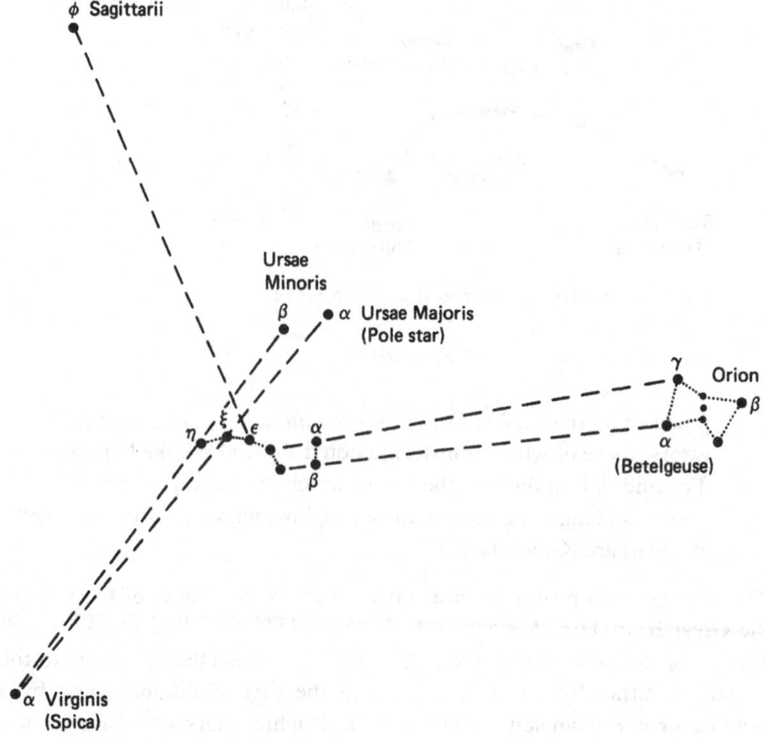

Fig. 62. Diagram to illustrate the keying of circumpolar with other stars.

With this information, and by using Fig. 62, we can interpret the passage. The first sentence explains that Chio (our *a* Virginis; Spica) can be found from the last two stars of the handle. In effect what has been done is to draw two lines, one through the Pole Star (*a* Ursae Minoris or *a* U Mi) (in Chinese, Thien huang ti or Thien chi) and Mizar (*ζ* Ursae Majoris or *ζ* U Ma) (in Chinese, Kai Yang), and another through Thien ti hsing (*β* U Mi) and Yao kuang (*η* U Ma) to meet at Chio. The next sentence states that a line from Yu heng (*ε* U Ma) parallel with that between Thien Chi (*γ* U Ma) and Thien chhuan (*δ* U Ma) will indicate the position of Nan Tou (our *ϕ* (phi) Sagittarii). And the prolongation of the 'top' and 'bottom' of the bowl (i.e. *γ* and *β* U Ma, and *δ* and *a* U Ma), will give lines meeting in Shen (Orion).

The system of the 28 *hsiu* in its full development is given in Table 22. By comparing columns 4 and 7*a* it will be seen that there is practically no parallel between the ancient names of the corresponding constellations in China and the West. As is shown by column 7*d* some of the key or determinative stars were quite far from the celestial equator, e.g. Wei as much as 37° south, and Mao as much as 23° north. In the second century B.C. there was considerable variation of accuracy in the measurement of how far the *hsiu* extended along the celestial equator. Some, such as Niu, were remarkably accurately observed, but in other cases, such as Shen and Pi, they were out by more than a degree; this would hardly be surprising since the observers were confined to measurements with the naked eye, with comparatively simple instruments. An extremely important point which appears from column 7*b* is that the determinative stars were chosen largely irrespective of magnitude (i.e. apparent brightness). Remembering that the higher the magnitude number the dimmer the star, with magnitude 6 referring to the dimmest stars visible on a clear night to the naked eye, this column shows that only one was first magnitude (Spica; α Virginis) and no less than four of fourth magnitude, while one was almost sixth magnitude. This shows that what the ancient Chinese astronomers were interested in was a geometrical division of the heavens, and they neglected bright stars if they were not useful for their purposes. The *hsiu*-determinative stars had to have the same right ascensions as the constantly visible circumpolar stars. And in this connection it might be pointed out that this delineation of groups by coordinates, rather than by visual patterns, distinctly foreshadows the way constellation boundaries are fixed by astronomers today.

Why was the number of the *hsiu* (lunar mansions) just 28? The question is not quite so simple as it looks. The most ancient forms of the character *hsiu* indicate a shed made of matting. These segments of the heavens must thus have been thought of as temporary resting places of the sun, moon, and planets, like the tea-houses scattered along roads on earth; but especially of the moon, the greatest night luminary. The line of mansions was thus a graduated scale on which the motion of the moon could be measured, and probably their number was an ancient compromise between the time spans of its two basic periods. The moon takes just over $29\frac{1}{2}$ days to complete its cycle of phases (the lunation or synodic month), but it takes only $27\frac{1}{3}$ days to return to the same place among the stars (sidereal month), and so 28 days would be a convenient average, even though both periods would always be out of step.

In Table 22 the *hsiu* are grouped, seven each, in the four equatorial 'Palaces'. The symbolic names for these palaces, which corresponded to the seasons, are given later (page 101). Here the principle of opposability gave a curious result, namely the apparently incorrect positions of the spring and autumn palaces. Hsin receives the visit of the sun in autumn, but is associated with the spring, Shen receives it in the spring, but is associated with autumn. This is because the results sought were not of conjunction – as would be the case with heliacal risings and settings – but of opposition. Spring full moons, on the other hand, do appear in the 'spring' *hsiu*, and

Table 22. *Table of the* hsiu *('lunar mansions', equatorial divisions, or segments of the celestial sphere bounded by hour-circles)*

Explanation

Col. 1. The 'Palace' (*kung*). The central, or circumpolar, palace is of course not involved, since the constellations which give their name to the *hsiu* are all on one or other side of the equator, though they may be as much as 35° away from it.
Col. 2. The number of the *hsiu* according to the order in, for example, *Huai Nan Tsu*.
Col. 3. Name of the *hsiu* (romanised and Chinese).
Col. 4. Probable ancient significance of the name.
Col. 5. Number of stars in the constellation which gives its name to the *hsiu*.
Col. 6. (*a*) Equatorial extension of the *hsiu* in Chinese degrees (365¼), as given by *Huai Nan Tzu*.
(*b*) Conversion of the *Huai Nan Tzu* figures to modern degrees (360).
(*c*) True extensions along the equator calculated for 450 B.C. by Nōda.
Col. 7. (*a*) Identification of the determinative star of the *hsiu*, i.e. that lying on the hour-circle with which it begins.
(*b*) Magnitude of the determinative star.
(*c*) Right ascension of the star for A.D. 1900.
(*d*) Declination of the star for A.D. 1900.
Col. 8. Correlated circumpolar phenomena. This list gives the culminations and lower transits of circumpolar stars (mostly in Ursa Major, Ursa Minor and Draco) which would closely correspond to the invisible meridian transits of the *hsiu* below the horizon. It should be noted that this series was calculated by Biot for a region between the 34th and 40th parallels of terrestrial latitude and for the year 2357 B.C. There would be modifications if a similar list was to be constructed for the 4th century B.C., for example, but the general principle would not be affected. For the Chinese characters of these circumpolar asterisms see Table 23 immediately following.

1	2	3 Name		4	5	6 Equatorial extension			7 Determinative star				8
				Probable	No.						*c* R.A.	*d*	Correlated
				ancient	of	*a*	*b*	*c*	*a*	*b*	(A.D. 1900)	Decl.	circumpolar
Palace	No.	rom.	Ch.	significance	stars	Ch.	mod.	calc.	ident.	mag.	h. m. s.	(A.D. 1900)	phenomena
Eastern	1	CHIO	角	Horn	2	12°	11·83°	11·70°	*α* Virginis (Spica)	1·2	13 19 55	−10° 38′ 22″	Lower transits of *α* Ursae Minoris (Thien huang ta ti); and of a 3233 Ursae Minoris (Shu tzu) Culmination of i Draconis (Thien i)
	2	KHANG	亢	Neck	4	9°	8·87°	8·81°	*κ* Virginis	4·3	14 07 34	−09° 48′ 30″	Transits of *α* and *β* Centauri (Nan mên)[a]
	3	TI	氐	Root	4	15°	14·78°	14·46°	$α^2$ Librae	2·9	14 45 21	−15° 37′ 35″	None
	4	FANG	房	Room	4	5°	4·93°	5·25°	*π* Scorpii	3·0	15 52 48	−25° 49′ 35″	Culmination of *α* Draconis (Yu shu)
	5	HSIN	心	Heart	3	5°	4·93°	4·14°	*σ* Scorpii	3·1	16 15 07	−25° 21′ 10″	None

	6	WEI	尾	Tail	9	18°	17·74°	18·95°	μ^1 Scorpii	3·1	16 45 06	−37° 52′ 33″	None
	7	CHI	箕	Winnowing-basket	4	$11\frac{1}{4}$°	11·0°	10·22°	γ Sagittarii	3·1	17 59 23	−30° 25′ 31″	Lower transit of κ Draconis, and upward perpendicular position of the tail of the Great Bear (Pei tou)
Northern	8	NAN TOU	南斗	Southern Dipper	6	26°	25·8°	26·54°	ϕ Sagittarii	3·3	18 39 25	−27° 05′ 37″	Culmination of ι Draconis (Tso shu)
	9	NIU or CHHIEN NIU[b]	牛 牽牛	Ox Herd boy	6	8°	7·89°	7·90°	β Capricorni	3·3	20 15 24	−15° 05′ 50″	Lower transits of α Ursae Majoris (Thien shu) and of β Ursae Majoris (Thien hsüan) Culminations of α Lyrae (Chih nü) (Vega) and of β Lyrae (Chien thai)
	10	NÜ or HSÜ NÜ[c]	女 須女	Girl Serving-maid	4	12°	11·83°	11·82°	ε Aquarii	3·6	20 42 16	−09° 51′ 43″	None
	11	HSÜ	虛	Emptiness	2	10°	9·86°	9·56°	β Aquarii	3·1	21 26 18	−06° 00′ 40′	Lower transits of γ Ursae Majoris (Thien chi) and of δ Ursae Majoris (Thien chhüan)
Northern	12	WEI	危	Rooftop	3	17°	16·76°	16·64°	α Aquarii	3·2	22 00 39	−00° 48′ 21″	Lower transits of δ to ε Ursae Majoris (Yü hêng) and of 42 and 184 Draconis (Thai i)[d]
	13	SHIH or YING SHIH	室 營室	House Encampment	2	16°	15·77°	16·52°	α Pegasi (Markab)	2·6	22 59 47	+14° 40′ 02″	Lower transits of ε Ursae Majoris (Yü hêng) and of 42 and 184 Draconis (Thai i)

Table 22. (*contd*)

	14	Pi or Tung Pi	壁 東壁	Wall Eastern Wall	2	9°	8·87°	8·44°	γ Pegasi	2·9	00 08 05	+14° 37′ 39″	Culmination of β Ursae Minoris (Thien ti hsing), lower transit of ζ Ursae Majoris (Khai Yang)
Western	15	Khuei	奎	Legs	16	16°	15·77°	15·66°	η Andromedae	4·2	00 42 02	+23° 43′ 23″	Culmination of a 3233 Ursae Minoris (Shu tzu)
	16	Lou	婁	Bond	3	12°	11·83°	10·83°	β Arietis	2·7	01 49 07	+20° 19′ 09″	Lower transit of η Ursae Majoris (Yao kuang) and of i Draconis (Thien i)
	17	Wei	胃	Stomach	3	14°	13·8°	15·2°	41 Arietis	3·7	02 44 06	+26° 50′54″	None
	18	Mao	昴	graph of a group of stars (Pleiades)	7	11°	10·84°	10·44°	η Tauri	3·0	03 41 32	+23° 47′ 45″	Lower transit of α Draconis (Yu shu)
	19	Pi	畢	Net (Hyades)	8	16°	15·77°	17·86°	ε Tauri	3·6	04 22 47	+18° 57′ 31″	None
	20	Tsui or Tsui Chui	觜 觜觿	Turtle	3	2°	1·97°	1·47°	λ^1 Orionis	3·4	05 29 38	+09° 52′ 02″	Culmination of κ Draconis
	21	Shen	參	graph of 3 stars	10	9°	8·87°	6·93°	ζ Orionis	1·9	05 35 43	−01° 59′ 44″	Downward perpendicular position of the tail of the Great Bear (Pei tou)
Southern	22	Ching or Tung Ching	井 東井	Well Eastern well	8	33°	32·53°	32·60°	μ Geminorum	3·2	06 16 55	+22° 33′ 54″	Culmination of α Ursae Majoris (Thien shu) and of β Ursae Majoris (Thien hsüan) these form the end of this very broad *hsiu*

23	Kuei or Yü Kuei	鬼 輿鬼	Ghosts Ghost-vehicle	4	4°	3·94°	4·46°	θ Cancri	5·8	08 25 54	+18° 25′ 57″	The same as the preceding
24	Liu	柳	Willow	8	15°	14·78°	15·16°	δ Hydrae	4·2	08 32 22	+06° 03′ 09″	Culmination of β Ursae Majoris, as for the preceding Culminations of α Puppis (Lao jen) (Canopus)[e]
25	Hsing or Chhi Hsing	星 七星	Star Seven Stars	7	7°	6·9°	6·86°	α Hydrae	2·1	09 22 40	−08° 13′ 30″	Culminations of γ Ursae Majoris (Thien chi) and of δ Ursae Majoris (Thien chhüan)
26	Chang	張	Extended net	6	18°	17·74°	17·13°	μ Hydrae	3·9	10 21 15	−16° 19′ 33″	None
27	I	翼	Wings	22	18°	17·74°	17·81°	α Crateris	4·2	10 54 54	−17° 45′ 59″	Lower transit of γ Ursae Minoris (Thai tzu) Culminations of 42 and 184 Draconis (Thai i) and of ε Ursae Majoris (Yü hêng)
28	Chen	軫	Chariot platform	4	17°	16·76°	16·64°	γ Corvi	2·4	12 10 40	−16° 59′ 12″	Lower transit of β Ursae Minoris (Thien ti hsing) Culmination of ζ Ursae Majoris (Khai Yang)

[a] Stars of the southern hemisphere, not circumpolars; mentioned in *Hsia Hsiao Chêng*.
[b] Properly speaking, Chhien Niu is α Aquilae (Altair).
[c] Not to be confused with Chih nü, the Weaving Girl (see Table 23 below).
[d] There is some doubt about the identification of the small star or stars near the pole which corresponded to Thai i.
[e] This star is supposed to have been just visible above the Chinese horizon, at any rate in the 3rd millennium B.C., but whether its transit was noted in connection with the position of Liu is very uncertain. Like Nan mên, it is, of course, a star of the southern hemisphere, and not a circumpolar.

Table 23. *Stars referred to in col. 8 of Table 22*

Chinese name			Identifications made in 1956 by Chhen Tsun-Kuei
Character	Romanisation	Translation	
漸臺	Chien thai	Clepsydra terrace	β Lyrae
織女	Chih nü	Weaving girl	α Lyrae
開陽	Khai Yang	Introducer of heat, or of the Yang	ζ Ursae Majoris
老人	Lao jen	Old man	α Carinae
南門	Nan mên	Southern gate	α^2 and ε Centauri
北斗	Pei tou	Northern dipper (Great Bear)	Ursa Major
庶子	Shu tzu	Son of (imperial) concubine	5 Ursae Minoris
太一	Thai i	Great unity, or the Great first one	—
太子	Thai tzu	Imperial prince	γ^2 Ursae Minoris
天權	Thien chhüan	Celestial balance	δ Ursae Majoris
天璣	Thien chi	Celestial armillary (see below, p. 159)	γ Ursae Majoris
天璿	Thien hsüan	Celestial template (see below, p. 163)	β Ursae Majoris
天皇大帝	Thien huang ta ti	Great emperor of august heaven	H Cephei
天乙	Thien i	Celestial unity, or the Heavenly first one	i Draconis
天樞	Thien shu	Pivot of heaven	α Ursae Majoris
天帝星	Thien ti hsing	Star of the heavenly emperor, or Sovereign star	β Ursae Minoris
左樞	Tso shu	Pivot of the left	ι Draconis
搖光	Yao kuang	Twinkling brilliance	η Ursae Majoris
玉衡	Yü hêng	Celestial sighting-tube (see below, p. 164)	ε Ursae Majoris
右樞	Yu shu	Pivot of the right	α Draconis

autumn ones in the 'autumn' *hsiu*. Such difficulties were bound to arise when the cardinal point system was extended to the celestial equator. The addition of a fifth (central or circumpolar) palace was extremely characteristic of Chinese cosmology, important for all questions of mutual influence with respect to other cultures (e.g. the Persian), and brought the fields of the heavens into line with all other five-fold divisions such as the five elements, which have already been discussed (volume 1 of this abridgement, page 153 ff). In view of the obvious analogy between the emperor and the Pole Star, and the peculiar respect enjoyed by the circumpolar stars in the Chinese system, it was entirely natural that the latter should be viewed as the residences or offices of the principal members of the (celestial) imperial bureaucracy. Fig. 63, taken from the Wu Liang tomb-shrine reliefs (Later Han) shows one of them seated in the 'bowl' of the Dipper.

The development of the system of the hsiu

The first question which arises is that of the antiquity of the *hsiu*, and this has been resolved in recent times by the discovery of the oracle-bones at Anyang. These date from the Shang period (about 1500 B.C. onwards), so we can now be sure that the system of the *hsiu* was growing up gradually from the middle of the Shang, since its nucleus can be found in the fourteenth century B.C. This was the time of the ruler Wu-Ting (1339 to 1281 B.C.), and on inscribed bones of his reign (as, indeed, on others earlier, and later) there are mentions of stars. Of greatest importance were Niao hsing, the Bird Star or Constellation, to be identified with Chu chhiao (the Red Bird), i.e. the 25th *hsiu*, Hsing (the red star, α Hydrae), central to the southern palace (the Vermilion Bird) – and the Huo hsing, the Fire Star or Constellation, to be identified with the red coloured star Antares (α Scorpii) and the fourth and fifth *hsiu*, Fang and Hsin, central to the eastern palace. These names probably indicate the beginning of the scheme of dividing the heavens along the celestial equator into four main palaces (the Blue Dragon in the east, the Vermilion Bird in the south, the White Tiger in the west, and the Black Tortoise in the north).

Besides the two star-names already mentioned, the bones refer to an important star presumably pronounced Shang, which has not yet been identified, and to another, the 'Great Star', Ta hsing. Possibly these would complete the four *hsiu* for the cardinal points. Fig. 64 shows one of the oracle-bones which mentions the Bird Star. The nucleus of the *hsiu* system can be seen further constituted by the stars mentioned in the *Shih* Ching (Book of Odes), a collection of folksongs which dates from the eighth or ninth century B.C. One speaks of the culmination of Ting (an old name for Pegasus, the 13th and 14th *hsiu*), another mentions Mao (the Pleiades, the 18th *hsiu*) under its old name of Liu, and Shen (Orion, the 21st *hsiu*):

> As often as a ewe has a ram's head,
> As often as Orion is in the Pleiades,
> Do people today, if they find food at all,
> Get a chance to eat their fill.

Fig. 63. The Great Bear carrying one of the celestial bureaucrats; a relief from the Wu Liang tomb-shrines (*c.* A.D. 147). On the original relief, the spirit with the isolated star is in line with, and beyond the last of the stars of, the 'handle', which is represented in a straighter line. The isolated star must therefore be Chao yao (γ Boötis).

In all, at least eight out of the 28 *hsiu* are found in the *Shih Ching*.

In discussing the legendary commissioning of the astronomers Hsi and Ho (page 73), reference was made to the *Shu Ching* (Historical Classic). The commissioning text is interwoven with another about 'stars and seasons', which should also be considered in connection with two other texts, the *Hsia Hsiao Chêng* (Lesser Annuary of the Hsia) and the *Yüeh Ling* (Monthly Ordinances of the Chou). The text runs as follows:

> The day of medium length and the (culmination of the) star Niao (serve to) adjust the middle of spring ... The day of greatest length and the (culmination of the) star Heo (serve to) fix the middle of the summer ... The night of medium length and the (culmination of the) star Hsu (serve to) adjust the middle of autumn ... The night of greatest length and the (culmination of the) star Mao (serve to) fix the middle of the winter ... The year has 366 days. The four seasons are regulated by means of intercalary (i.e. inserted) months.

To say the year had 366 days was not a very inspired remark, since we know that the Shang people in the thirteenth century B.C. were aware that a year was approximately $365\frac{1}{4}$ days long. But perhaps the figure of 366 days simply meant that it was over 365. More significant is that the 25th, 5th, 11th and 18th *hsiu* are mentioned, each one central to its palace or

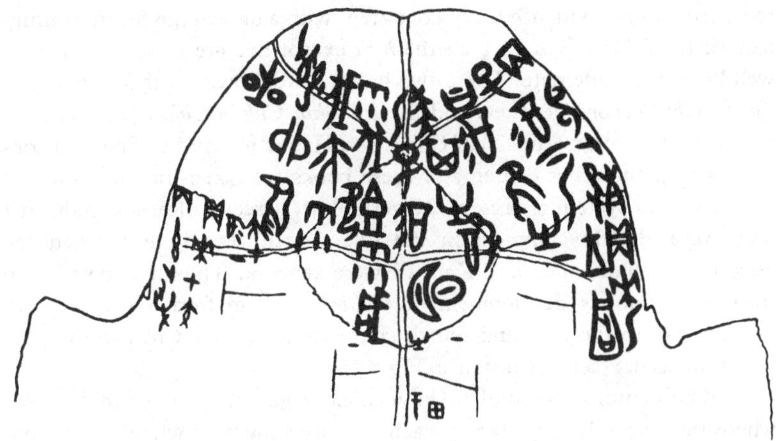

Fig. 64. Oracle-bone inscription which mentions the Bird-Star (α Hydrae). The pictograph of a bird can be seen at the bottom of the penultimate column of characters on the left, and the character for 'star' follows it alone. The date of the inscription is between 1339 and 1281 B.C.

equatorial quadrant. At first sight, they seem to be associated with the wrong seasons, but not if they refer to the situation in the second millennium B.C., when the precession of the equinoxes would have shifted the constellations to the positions indicated in the text. Nevertheless, the precise date of the quotation is not easy to determine, though in view of all we know of Chinese astronomy, it is unlikely to be earlier than 1500 B.C. on even the most generous estimate. But the possibility remains open that the text is a remnant of a very ancient observational tradition, not Chinese at all but Babylonian.

The text describes a situation which makes it clear that one of the basic observations of the old Chinese astronomers was that the quarters of the daily rotation of the skies corresponded every three months with the quadrants of the annual revolution of the celestial sphere. For example, the *hsiu* which culminates at 6.0 p.m. at the winter solstice (Mao, in the text quoted) could be identified as that in which the sun would stand at noon on the following spring equinox (three months later), and so successively all the year round. This procedure was entirely in character for ancient Chinese astronomy, which solved the problems of finding the position of the sun among the stars by deducing the positions of unobservable bodies from those of observable ones, all being firmly held in a network of coordinates based on the celestial poles and celestial equator.

Other mentions of the *hsiu* occur in the *Hsia Hsiao Chêng* (Lesser Annuary of the Hsia) and in the *Yüeh Ling* (Monthly Ordinances of the Chou). In the first, six *hsiu* are given, two of which appear for the first time, the astronomical evidence being consistent with a date in the fourth century B.C. In the fuller *Yüeh Ling* all the *hsiu* except five are given, and it may well be of the same date. These two books were, however, incorporated in the *Li Chi* (Record of Rites) and the *Lü Shih Chhun Chhiu* (Master Lu's Spring and Autumn Annals) of Chhin and Han date (third to first centuries B.C.), and if the other contents of these books are taken into account, the complete *hsiu* system is present. Nevertheless, it seems almost certain that they were already complete in the mid fourth century B.C. when the astronomers Shih Shen and Kan Tê were at work. Thus it is possible to trace a continuous development of the *hsiu* system from the fourteenth century B.C. to the fifth and fourth centuries B.C., after which no further variation occurs (see, for instance, Fig. 65).

The completed round of hour-angle segments is seen in Fig. 66, where the determinative stars of each *hsiu* are shown by white circles, and the other stars of each constellation in black. At first sight the distribution of the *hsiu*-determining stars seems strange, with wide variations in their declinations so that many lie far above or below the celestial equator. However, if we take account of the precession of the equinoxes and draw in the celestial equator as it was in 2400 B.C. (the dotted circle in Fig. 66) many more of the *hsiu* fall upon it. This date corresponds to a time well before the Shang period. Moreover the Mao and Fang constellations lie very close to what would then have been the equinoxes, and Hsing and Hsü close to where the solstices would have been. And it is precisely Hsing and Fang which are mentioned in the most ancient Chinese astronomical documents, the oracle-bones of about 1300 B.C. As the Shang people certainly had two of these points at 90° to each other, they must surely have been aware of the other two also. The *Shu Ching* text mentions all four. The significant thing is that they occupy the four 90° points of the equator, which also satisfies the majority of the *hsiu*-determinative stars.

One is tempted to go back to the third millennium B.C. for the establishment of the *hsiu* system, but the great difficulty is that all the Chinese archaeological and literary evidence is against such a date. It may of course have started in Babylonia. The question is also complicated by the fact that the declination of a star was irrelevant to its selection as one for determining the *hsiu*; indeed several *hsiu* will not fit into any likely position of the celestial equator. The right ascension, not the declination, was the important factor. It was the right ascension which decided whether the determinative star could be keyed to a circumpolar one or not, and most of these exceptions could. That some of the other *hsiu* appear to fit well on to the celestial equator of Shang times may be a coincidence; their selection may have been much later and due to other motives.

There is, however, another point. The positions of the *hsiu* Mao and Fang at opposite equinoxes would have been shared, at a different time, by

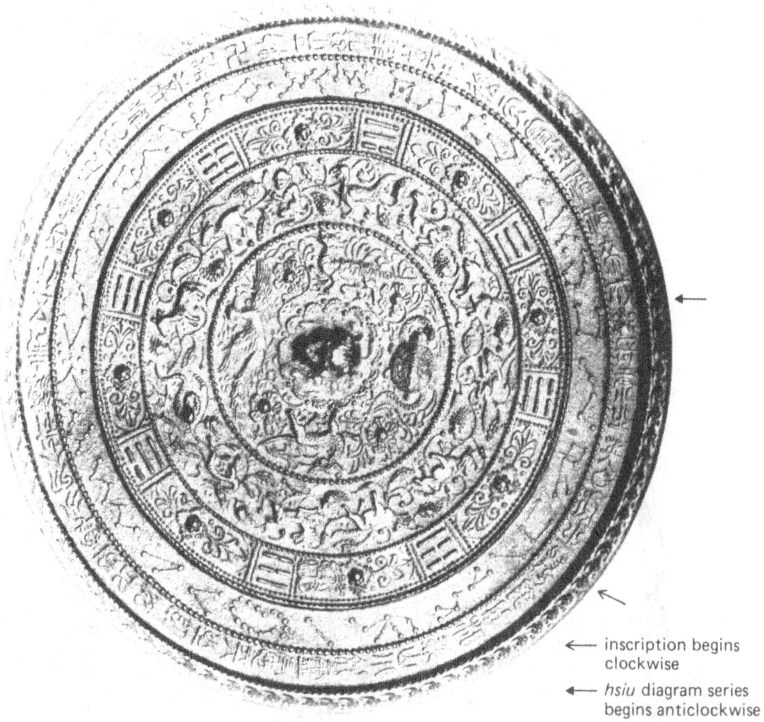

Fig. 65. Bronze mirror of the Thang period (between A.D. 620 and 900) showing constellation diagrams of the 28 *hsiu* (second circle from the outside), the eight trigrams (next circle), the twelve animals of the animal cycle (next innermost circle), and the four symbolic animals of the Celestial Palaces (innermost disc). From the collection of the American Museum of Natural History. The outer circle is inscribed with a poem beginning at the floret on the right (at 3.0 p.m.).

(This mirror) has the virtue of *Chhang-kêng* (the Evening Star, Hesperus, Venus)
And the essence of the White Tiger (symbol of the Western Palace),†
The mutual endowments of Yin and Yang (are present in it)
The mysterious spirituality of Mountains and Rivers (is fulfilled in it).
With due observance of the regularities of the Heavens,
And due regard to the tranquillity of the Earth,
The Eight Trigrams are exhibited upon it,
And the Five Elements disposed in order on it.
Let none of the hundred spiritual beings hide their face from it;
Let none of the myriad things withhold their reflection from it.
Whoever possesses this mirror and treasures it,
Will meet with good fortune and achieve exalted rank.

†Both planet and palace preside over metal in the system of symbolic correlations.

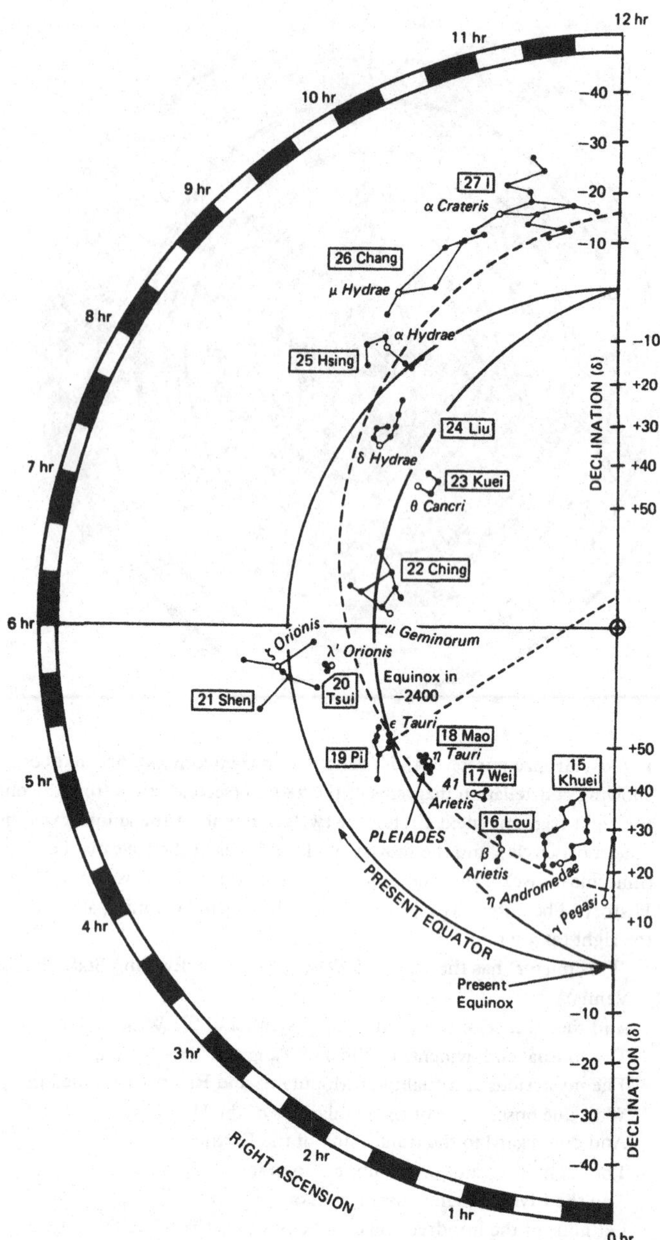

Fig. 66. Chart of the 28 *hsiu*, the lunar mansions. The Chinese names of the *hsiu* are those printed within rectangles. Western star names and designations are given to the determinative star of each *hsiu*, each determinative star being indicated by an unfilled circle.

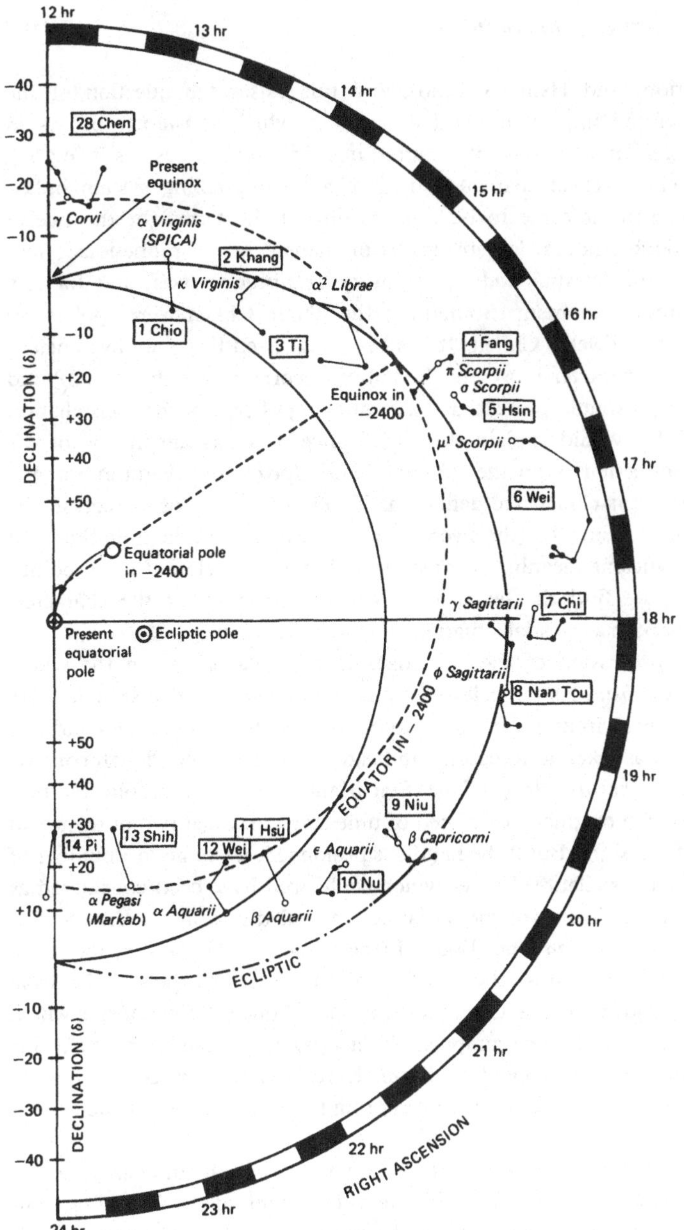

12 hr
13 hr
14 hr
15 hr
16 hr
17 hr
18 hr
19 hr
20 hr
21 hr
22 hr
23 hr
24 hr
RIGHT ASCENSION
DECLINATION (δ)
28 Chen
Present equinox
γ Corvi
α Virginis (SPICA)
κ Virginis
2 Khang
α^2 Librae
1 Chio
3 Ti
4 Fang
π Scorpii
σ Scorpii
5 Hsin
Equinox in −2400
μ^1 Scorpii
6 Wei
Equatorial pole in −2400
γ Sagittarii
7 Chi
Present equatorial pole
Ecliptic pole
φ Sagittarii
8 Nan Tou
EQUATOR IN −2400
9 Niu
β Capricorni
11 Hsü
12 Wei
13 Shih
14 Pi
ε Aquarii
10 Nu
α Pegasi (Markab)
α Aquarii
β Aquarii
ECLIPTIC

Shen (Orion) and Hsin (Scorpio), and this raises the question of the meaning of the important word *chhen* (辰), which constantly occurs in ancient texts. In later usage it acquired many meanings; besides being one of the twelve cyclical characters used in a fortune-telling system linked to the calendar, it came to be used as one of the terms for the twelve standard double-hours, for auspicious or inauspicious conjunctions (close approaches) of celestial bodies, for lucky or unlucky stars, and for any definite time or moment. Sometimes the Three Chhen were spoken of, sometimes the Twelve Chhen. It has been suggested that the most ancient forms of the character represent the tail of a scorpion or a dragon (辰) and that the graph should be regarded as a drawing of part of the constellation Scorpio. This would explain the significance of a passage in one of the ancient commentaries on the *Chhun Chhiu* (Spring and Autumn Annals) where the Great Chhen are defined as Ta Huo (Antares or α Scorpii, the central star of Hsin), Fa (the sword of Orion, Shen), and the Pole Star (Pei Chi). The ancient meaning of *chhen* would thus be 'celestial mark-point'. Here again we find the texts, and even the structure of the character, referring back to a celestial situation earlier than 2000 B.C.

The precession of the equinoxes had a great effect on the whole system of the *hsiu*. As the celestial equator and celestial pole shifted, stars which had been circumpolar ceased to be so. The direction of the handle of the Plough was taken as a seasonal indicator and the whole Plough consists of seven stars, yet the *Hsing Ching* (Star Manual) mentions an old tradition that the Plough originally consisted of nine stars, and that two of them had been lost from sight. But if the handle is prolonged, we come to a number of stars in the constellation Boötes which could well have been considered as belonging to it. Again, in the social-ceremonial directions of the second-century B.C. *Huai Nan Tzu* (Book of (the Prince of) Huai-Nan) there is a whole chapter which links these to the point indicated by the star Chao yao (Twinkling indicator). As Chao yao is probably our γ Boötis, a star which must have ceased being circumpolar about 1500 B.C., it can be seen that this text points to a very ancient tradition. Here is another piece of evidence which should make us cautious in accepting too late a dating for the origins of Chinese astronomy.

With the passage of time, precession also brings about changes in the right ascensions of stars. This will affect the keying of the *hsiu* to circumpolar stars, and there seems to be evidence that this did indeed happen. For instance, the bright Chhien niu (the Herd-boy, our Altair) was later replaced by the dimmer Nu (ε Aquarii), while Chih nu (the Weaving girl, our Vega) gave place to the weak Niu (β Capricornii). Again Ta chio

(Arcturus), which may have been one of the stars from Boötes which earlier formed part of the handle of the Plough, seems to have been replaced by Chio (Spica). And in connection with Ta chio it is worth noting that either side of it are two small groups of three stars each, the 'left' and the 'right' She-thi or 'Assistant Conductors', and that the 'rule of the She-thi' was associated with the Jupiter cycle (see below). This is one of a few cases where the Chinese denoted a heliacal rising. And in this connection, it has been pointed out that the word *Lung* (朧), which means the rising of the moon, must have originated from the fact that, at the beginning of the year in Han times, it rose somewhere between the horns of the Spring Dragon. The character combines the radicals for moon and dragon. Hence the oft-repeated art motif of the dragons and the pearl – the moon (Fig. 67).

The origin of the system of the hsiu

The replacement of Vega and Altair and other bright stars as mark-points by weaker ones in more strategic positions did not occur in other civilisations. Particularly it did not occur in India, and this is where we come to a problem which has caused much controversy, namely, the relation of the Indian and Arabic 'moon-stations' to the Chinese *hsiu*.

There are some similarities between the Indian and Chinese systems. Nine out of the 28 *hsiu*-determinative stars and the Indian *yogatara* (junction-stars) are identical, while a further 11 share the same constellation, though not the same determinative star. Only eight *hsiu*- determinative stars and *yogatara* are in quite different constellations, and of these, two are Vega and Altair, which are possible *hsiu*-determinatives of earlier times. Both in China and India the new year was reckoned from the mansion corresponding to Spica, and in both cultures the Pleiades was one of the four constellations at the 90° points (i.e. the two equinoxes and solstices).

However, the Chinese is the only civilisation in which it is possible to trace (as has been done above) the gradual development of the system. In China we find the 'coupling' of *hsiu* of greater spread along the celestial equator with those of lesser spread along the equator on the opposite side of the celestial sphere (Fig. 66), but this is not so for the Indian moon-stations. Again, in India there is no coupling of the moon-stations with the circumpolar constellations, which was the essence of the Chinese system, while the Indian positions follow even less closely the celestial equator of 2400 B.C. Finally, as far as documentary evidence is concerned, India has not much to yield, and such as there is leads one to conjecture that in ancient times the two systems developed separately, but were brought into relation

Fig. 67. The symbolism of the dragon and the moon; a portion of the Nine Dragon Screen wall in the Imperial Palace at Peking (photograph, A. Nawrath, *Indien und China; Meister werke der Baukunst und Plastik* (album of photographs), Schroll, Vienna, 1938).

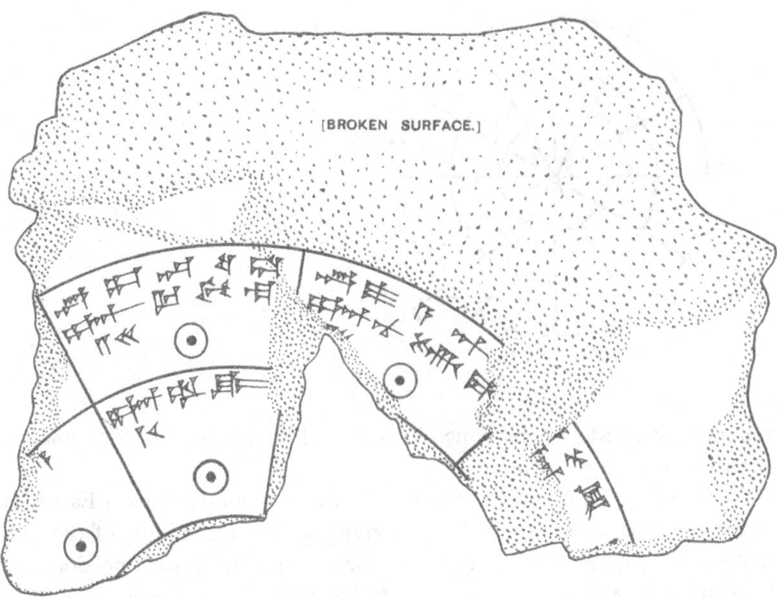

Fig. 68. Fragment of Babylonian planisphere (*c.* 1200 B.C.). From E.A. Wallis Budge, (ed.), *Cuneiform Texts from Babylonian Tablets etc. in the British Museum*, 41 vols., 1896–1931.

to some extent in later ages when Sino-Indian cultural contacts intensified.

It has been suggested that both systems were derived from a Babylonian 'lunar-zodiac' which was received by all Asian peoples. Certainly in the library of the Assyrian king Assurbanipal (668 to 612 B.C.) at Nineveh, there are some cuneiform tablets whose contents date from the second millennium B.C. These show three concentric circles, divided into 12 sectors. In each of the 36 fields thus obtained there are constellation names and certain numbers (Fig. 68). Perhaps these may be regarded as primitive planispheres (flat charts of the celestial sphere) showing both the circumpolar stars and the 'moon-stations' corresponding to them. Such planispheres are not unknown; they belong to a series of some 7,000 astrological omens, of a date contemporary with the Shang (about 1400 to 1000 B.C.), and consist of three 'roads', each marked with 12 stars, one for each of the months according to the times of their heliacal risings and settings. Those of the central road (the equatorial belt) were the Stars of Anu, those of the outer road (south of the celestial equator) the Stars of Ea, and the inner road was travelled by circumpolar stars and stars north of the equatorial belt, the stars of Enlil. About 700 B.C., the planispheres were replaced by the famous Mul Apin lists of stars. In these texts there is never any mention of any

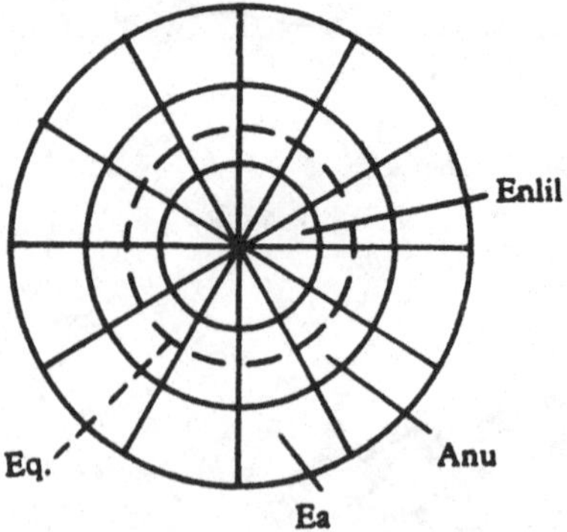

zodiac or of constellations lying along the ecliptic; those did not come for almost another two centuries.

One might well surmise that the equatorial moon-stations of East Asia originated from Old Babylonian astronomy before the middle of the first millennium B.C., and probably a long time before. In this connection it is interesting (though hitherto unnoticed) that the *Chou Pei Suan Ching* (The Arithmetical Classic of the Gnomon and the Circular Paths of Heaven) of Sung times contains a diagram closely resembling the Babylonian planispheres, and the text accompanying it closely parallels that associated with the Babylonian 'Hilprecht' cuneiform tablet of about 1400 B.C. The significance of this is evident when we recall that the *Chou Pei* represents the most archaic of the ancient cosmological theories. Again, the Altar of Heaven and the Temple of Heaven at Peking both retain to this day three circulating terraces as if to symbolise the Three Roads of the Ea–Anu–Enlil. Here the officiating emperor naturally occupied the central polar position. If Chinese equatorial astronomy derived, at least in part, from Old Babylonian equatorial astronomy, such traces are just what we might expect to find.

The pole and the pole-stars

The basic importance of the celestial pole for Chinese astronomy has by now been made clear. But the effect of precession of the equinoxes on the position of the pole is considerable. The pole moves in a circle around the pole of the ecliptic (K in Fig. 56). At present it is close to Polaris (α U Mi), the Pole Star of contemporary astronomy, but some 11,000 years hence it will be at the other extremity of its circular orbit about K and will be close to Vega (α Lyrae). In 3000 B.C. it was close to Thuban (α Draconis). It is therefore a fact of the greatest interest that we find, along the whole length of the path it has traversed since that date, stars which have preserved Chinese names indicating that they were at various times pole-stars, but later ceased to be so.

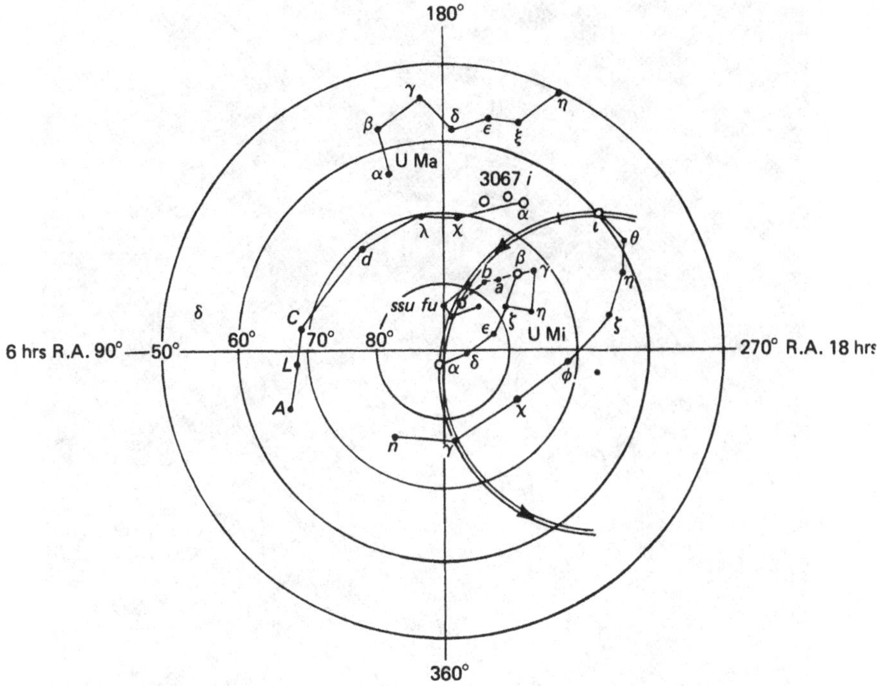

Fig. 69. Polar projection showing the trajectory of the north celestial pole, and the ancient pole-stars.

Investigation shows that the Chinese had a zone around the present Pole Star which was bounded by two 'barriers' of stars, enclosing the 'Purple Forbidden Enclosure' (an analogy with the imperial court). And the significant point is that on each side of the space, the end stars on one side were called the 'Left Pivot', and on the other the 'Right Pivot'. Precisely between them lies the point which the celestial pole occupied about 3000 B.C. Near by, furthermore, are two stars in Draco – the 'Celestial Unique' and the 'Great Unique' – both of which are given on the thirteenth-century planisphere reproduced in Fig. 77. They are dim stars, and might have been considered pole-stars in the earlier and later part of the second millennium B.C. Admittedly they are not very close to the path of the celestial pole, but they are almost as close as the 'Celestial Emperor' (Kochab or β U Mi) which it seems was used about 1000 B.C. Ursa Minor was not a constellation recognised by the Chinese. Instead they used only two brighter stars of this constellation, together with some dim ones and one from the adjoining constellation Camelopardus (4339 Camelopardi; Fig. 69). The star from Camelopardus was the Pole Star of the Han.

Fig. 70. A time-exposure photograph of circumpolar stars. (Photograph, Peter Gill)

Table 24. *Breakdown of star totals in ancient star lists*

	'Chairs'	Stars	'Chairs'	Stars
Shih Shen				
Internal (*chung*), i.e. north of the equator	64	270	—	—
External (*wai*), i.e. south of the equator	30	257	—	—
The 28 *hsiu*	28	282	—	—
Total of 'red' stars	—	—	122	809
Kan Tê				
Internal	76	281	—	—
External	42	230	—	—
Total of 'black' stars	—	—	118	511
Wu Hsien				
Internal and External; total of 'white' or 'yellow' stars	—	—	44	144
	—	—	284	1464

Since none of these stars was exactly on the polar path, we should expect to find some efforts to determine the position of the pole more accurately. The earliest suggestion of this appears in the *Chou Pei*, which speaks of the 'four excursions' of the Pole Star. The 'great star in the middle of the North Pole constellation' was to be observed through what was probably a sighting-tube (page 159) fitted with a template, and its displacement in the four directions measured. With the passage of time the Pole Star of the Han moved; by the fifth century A.D. it was recognised as more than 1° away from the true pole, and by the eleventh century Shen Kua was describing his measurement of the displacement. In the West, a century later, this displacement was also described. In due course our present Pole Star was adopted: in China this occurred at the end of the Ming. Fig. 70 shows a photograph of the present situation.

THE NAMING, CATALOGUING AND MAPPING OF STARS

Star-catalogues and star coordinates

Catalogues of stars were drawn up in the fourth century B.C. by three astronomers – Shih Shen, Kan Tê and Wu Hsien – who have already been mentioned in connection with Chinese astronomical literature (page 80). These catalogues were still in use a thousand years later, but in the fourth century A.D. they were at last combined when the astronomer Chhen Cho constructed a star-map from them. Then in the fifth century another astronomer, Chhien Lu-Chih, made an improved planisphere, marking the

Table 25. *Probable dates of observations in the star-catalogues*

	Probable epoch of observations
28 Hsiu	
6 (Chio, Hsin, Fang, Chi and Chang; perhaps also Tou)	350 B.C.
17 other *hsiu*	200 A.D.
2 data for north polar distances lacking (Khang & Shen)	—
3 aberrant (Ti, Liu and Hsing)	—
62 Northern hemisphere	
27 stars	350 B.C.
13 stars	180 A.D.
6 stars, north polar distances data lacking	—
4 stars,	150 A.D.
remainder uncertain	
30 Southern hemisphere	
10 stars	350 B.C.
16 stars	200 A.D.
4 aberrant	—

stars determined by the three early astronomers in different colours so as to differentiate between the astrological systems they had employed. At least the second of these maps existed in A.D. 715 when the *Khai-Yuan Chan Ching* (Khai-Yuan reign period Treatise on Astrology) – partly a pre-Thang compilation of Chou and Han star-catalogues – gives us the fullest information about observations from the fourth century B.C.

There is evidently a long and continuous Chinese tradition of celestial map-making (Table 25). The total number of stars in the ancient lists is 1464 grouped into 284 'chairs' or constellations. And all the lists adopt a similar layout and give same kind of information: the name of the group of stars comes first, then the number of stars it contains, followed by its position with respect to other star groups, and then a measurement in Chinese degrees for its principal or determinative star. The measurements invariably include (*a*) the hour-angle of the principal star measured from the first point of the *hsiu* in which it lies, and (*b*) the north polar distance. These are illustrated in Fig. 71. Alone of the texts, however, the *Khai-Yuan Chan Ching* gives, in addition, (*c*) the celestial latitude. This was a typically Greek measurement, which may have found its way to China through the compilation of Chhüthan Hsi-Ta who, if not himself born in India, had close Indian connections, significant because the Indians used a form of the Greek coordinates. The mainstay of Chinese astronomy was, nevertheless,

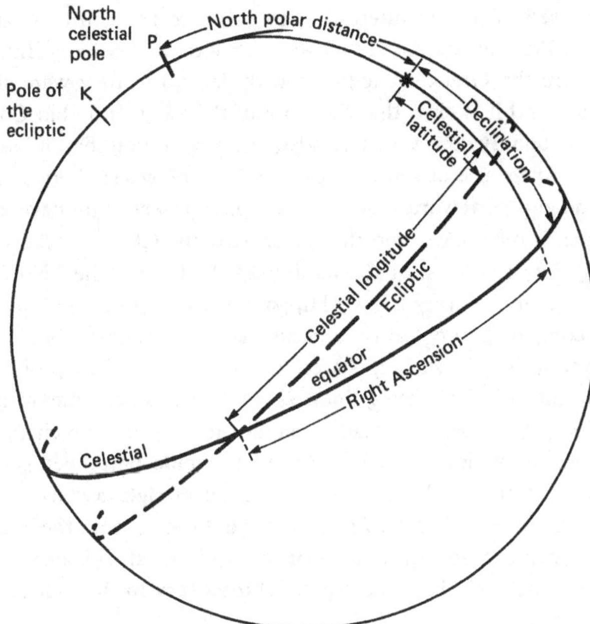

Fig. 71. Diagram of celestial coordinates.

the system of coordinates based on the celestial equator, which the West was to adopt during the Renaissance, and especially after the accurate celestial measurements in the latter part of the sixteenth century by the Danish astronomer Tycho Brahe. This is the right ascension and declination system, although the Chinese preferred the north polar distance of a star rather than its declination. There was, as we saw (page 71), a third system of coordinates based on the horizon and the zenith; this used the altitude and azimuth of a celestial body, and was rather characteristic of Arabic astronomy. It had the great disadvantage that it is applicable only to a particular point on the earth's surface; and this method is never found in China. It was, however, particularly associated with the astrolabe, an instrument of which no Chinese examples are known.

It is possible to use the north polar distances given in the Chinese catalogues to date the catalogues themselves, since this distance (as well as the declination) will depend upon the pole star in use. While formerly dates between 350 B.C. and A.D. 200 were envisaged, the most recent estimate is, that it is more likely to be about 70 B.C. This also shows how wrong is the oft-repeated statement that the star-catalogue given by the Greek astronomer Ptolemy in his *Almagest* (second century A.D.) was, until the time of Tycho Brahe, the only source of information on star positions that the world possessed. Yet it is interesting to compare the Chinese catalogues with those of Ptolemy and his second-century B.C. predecessor, Hipparchus. Not only were the Greek catalogues mostly later, but they were also one-third smaller, and it is clear that the work of the Chou and Han Chinese in positional astronomy deserves more adequate recognition than it has had.

The Chinese use of coordinates based on the celestial equator is that used by the modern world, but at the time it seems to have had one disadvantage: it may have been the reason why the Chinese were so late in discovering the precession of the equinoxes. In Greece the discovery was made in the second century B.C. by Hipparchus, who, in compiling his star-catalogue, compared some measurements made 150 years before. His comparison showed that the stars appeared to have changed their positions with reference to the points of the equinoxes, yet if he had been measuring their position along the equator instead of along the ecliptic, this change might not have been so obvious. In China the effect would have been specifically masked by the fact that the Chinese equivalent of right ascension was the position within an individual *hsiu* and not the distance from the equinoctial point. Only the very slow movement of the equinoctial and solstitial points themselves brought Yü Hsi to the parallel discovery in the first half of the fourth century A.D.

But when Chinese astronomers did begin to pay more attention to positions on or near the ecliptic they found something new and interesting. About 725 the monk I-Hsing, who had access to Indian and Greek astronomical ideas (page 81), and the engineer Liang Ling-Tsan constructed measuring instruments which gave values for star positions which differed from those in the ancient star-catalogues and star-maps. In more than 10 of these cases I-Hsing found a north–south movement relative to the ecliptic, and remarked on it. Yet such a north–south change as he observed could not be due to precession, which was by then well known, and meant that some stars had real motions of their own. In the West this movement was not discovered until Edmond Halley detected it in 1718, also by comparing ancient observations with those he had made himself. This I-Hsing seems to have pre-dated Halley's discovery by almost a thousand years. However, the movement concerned is very small and it is now thought that I-Hsing's results may have been due to instrumental errors and not a true effect. Yet it is clear that his mind was entirely open to the possibility of such motions of the 'fixed' stars.

Star recognition

The next question which arises is to what extent there was any similarity between Chinese and European recognition of groups of stars and constellations. As will appear, the answer is that there was very little. The same groups of stars were not seen in the same patterns, and frequently a single European constellation appears on the Chinese planisphere as several small star groups or asterisms. For example, our Hydra comprises the three *hsiu* Chang, Hsing and Liu, together with eight other star groups having no similarity with the European constellation. And if we compare the constellations of the West with the Chinese groupings (Table 26) we see that only three of our zodiacal constellations, and seven outside the zodiacal band, show any similarity. A list is given in Table 27, and it is clear that the correspondences are not at all impressive, and strongly suggest that the recognition and naming of the Chinese constellations grew up in almost complete independence of the West. Indeed, it is worth noting that maritime names are totally lacking in the Chinese heavens; there is nothing corresponding to Cetus (the Whale), Delphinus (the Dolphin), Cancer (the Crab), etc. And, on the other hand, the overwhelmingly agrarian and bureaucratic nature of ancient Chinese civilisation led to a multitude of star names in which the hierarchy of earthly officials found their counterparts. Not only, then, was Chinese star nomenclature independent of the West; it also represented a society which grew up in comparative isolation and independence. This still remains true, even though there seems to have been a transmission to China of Babylonian astrological lore before the sixth century B.C.; nor does it militate against the belief that certain basic ideas, such as 'roads' of the planisphere which gave rise to the *hsiu*, the use of the gnomon, and the recognition of the position of the celestial pole and equinoxes, were transmitted about a thousand years earlier.

In Table 26 the last two lines record constellations unrecognised and unnamed by Europeans until the seventeenth and eighteenth centuries. These groups are all in the southern hemisphere and include the southern circumpolar stars. It has often been supposed that they were not known to the Chinese until the coming of the Jesuits, but knowledge of these southern hemisphere constellations had been growing long before, as this example from the tenth-century *Chin Thang Shu* (Old History of the Thang Dynasty) shows:

> In the 8th month of the twelfth year of the Khai-Yuan reign period (A.D. 724) (an expedition was sent to the) south seas to observe Canopus (Lao jen) at high altitudes and all the stars still further south, which, though large, brilliant and numerous, had never in

Table 26. *Relation between occidental constellations and Chinese star groups*

	Occidental constel-lations, total no.	No. con-taining *hsiu*	No. con-taining only Chinese stars and star groups other than *hsiu*	No. of *hsiu* con-tained	No. of other important Chinese stars and star groups contained	Symbolic similarity of constellation names: None	Very doubt-ful	Doubt-ful	Admissible if adjacent constel-lation fields allowed	Positive
Second century A.D. Ptolemaic:										
zodiacal	12	11	1	17½	146	5	1	3	0	3
extra-zodiacal	36	7	29	10½	290	21	1	3	4	7
				28						
Fifth century A.D.	1	0	1	0	6	1	—	—	—	—
					442					
Seventeenth century A.D.	23	0	23	0	50	23	—	—	—	—
Eighteenth century A.D.	14	0	7	0	7	14	—	—	—	—

Table 27. *Some occidental constellations and Chinese star groups*

	Occidental	Chinese
Zodiacal	Capricorn	Niu (ox)
	Leo	Hsüan yuan (dragon backbone – connected with water-raising chain pumps)
	Scorpio	Fang (room)
		Hsin (heart)
		Wei (tail)
Non-zodiacal	Auriga	We chhê (The Five Chariots)
	Boötes	Hsüan ko (The Sombre Axe, and including nearby stars with military names)
	Canis Major	Thien lang (Sirius)
	Corona Australis	Pi (the tortoise)
	Corona Borealis	Kuan so (a coiled thing)
	Orion	Shen (constellation with parallel figure symbolism)
	Ursa Major	Pei tou (the Northern Dipper)

former times been named and charted. These were all observed to about 20° from the south (celestial) pole. This is the region which the astronomers of old considered was always hidden and below the horizon.

It is sad that the results of these investigations failed to survive, but the expedition remains a remarkable one for its time, indeed unique in the early Middle Ages.

Star-maps

Although there is no doubt that star charts were being constructed as early as the third century A.D., and probably also in the Han, none has come down to us from those times. However, astronomical manuscripts from the second century B.C. have recently been discovered and published. We know from Han carvings and reliefs that the system of representing star groups as patterns of dots or circles connected by lines (the 'ball-and-link' convention) goes back at least as far as that period (Fig. 72). Later on it spread widely into Arabic manuscripts and even Latin translations. In our own time this manner of depicting stars could be seen on astronomical flags of Taoist temples (Fig. 73).

Rather crude star charts have been found in tomb cupolas of the Thang period (seventh to ninth century) both in China and Japan. A manuscript star-map from Tunhuang, dating from A.D. 940 (Figs. 74 and

Fig. 72. Han stone carving showing, on the left, the Weaving Girl constellation (Chih nü), with the stars α (Vega), η and γ Lyrae above her head; in the centre is the sun, marked with its symbolical crow; and on the right, another constellation, probably one of the 'bird' *hsiu*, Fang, Hsin or Wei. All the asterisms are represented in the standard 'ball-and-link' convention. The whole composition would thus depict the sun at an hour-angle of about 260°. From S.W. Bushell, 'The Early History of Tibet', *Journal of the Royal Asiatic Society*, 1880, N.S., vol. 12, 435.

Fig. 73. Constellations on a Taoist flag, outside the Lao Chün Tung temple on the South Bank, Chungking; the main asterism depicted is probably Thien chhün in Cetus. (Original photograph 1946.)

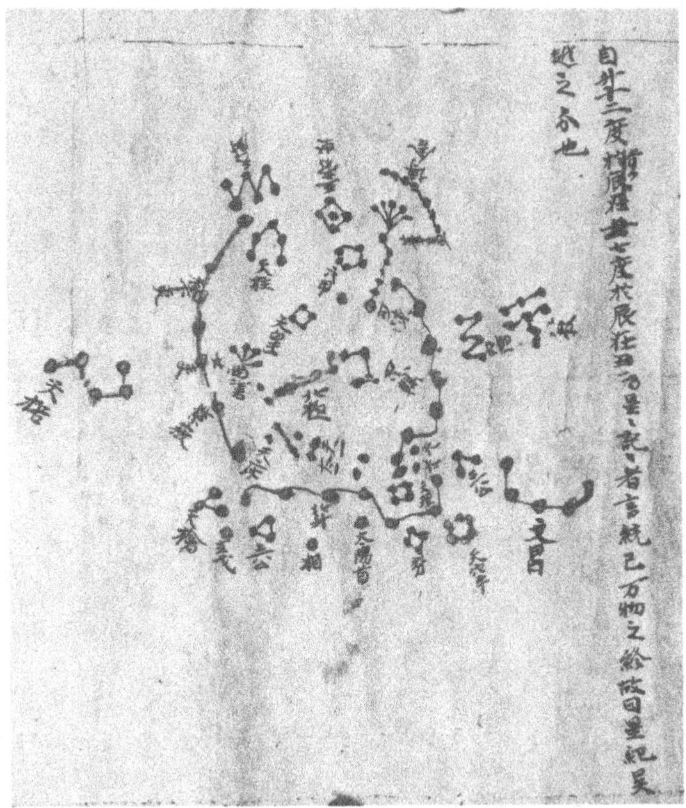

Fig. 74. The Tunhuang manuscript star-map of *c.* A.D. 940 (British Museum Stein no. 3326). To the left is a polar projection showing the Purple Palace and the Great Bear. To the right, on 'Mercator's' projection, an hour-angle segment from 12° in Tou *hsiu* to 7° in Nü *hsiu*, including constellations in Sagittarius and Capricornus. The stars are drawn in three colours, white, black and yellow, to correspond with the three ancient schools of positional astronomers.

75), is almost certainly the oldest extant star chart from any civilisation, if we exclude of course the highly stylised carvings and fresco paintings of antiquity (e.g. Fig. 63). Again, the maps incorporated in the *Hsin I Hsiang Fa Yao* (New Description of an Armillary Clock) by Su Sung must be the oldest printed star charts which we possess. The maps of the book, begun in A.D. 1088 and finished in 1092, are remarkable in several ways. Two use a 'Mercator' method of charting (Fig. 76), already employed a century earlier, although not so accurately; while one of the charts of the polar regions makes use of the latest observations by, it seems, Shen Kua. Other wood block star-map prints of 1005 and 1116 have also been discovered recently.

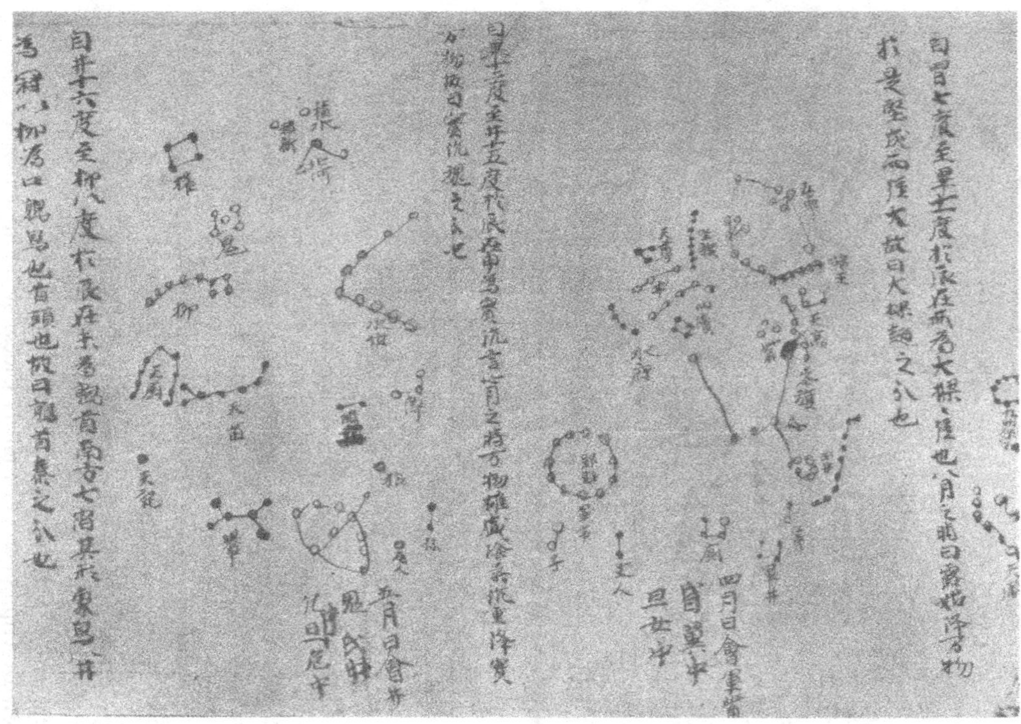

Fig. 75. The Tunhuang star-map of Fig. 74 showing two hour-angle segments on 'Mercator's' projection. To the right a segment from 12° in Pi *hsiu* to 15° in Ching *hsiu*, which includes Orion, Canis Major, and Lepus; to the left another from 16° in Ching *hsiu* to 8° in Liu *hsiu*, which includes Canis Minor, Cancer and Hydra.

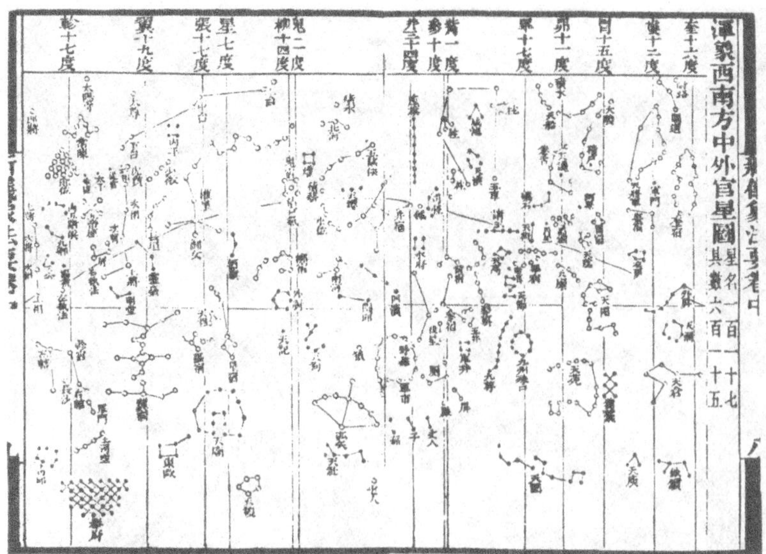

Fig. 76. The star-maps for the celestial globe in the *Hsin I Hsiang Fa Yao* of A.D. 1092; 14 *hsiu* on 'Mercator's' projection. The celestial equator will be recognised as the horizontal straight line; the ecliptic curves above it. Note the unequal breadth of the *hsiu*.

The most famous of all Chinese planispheres is that which was prepared in 1193 for the instruction of a young man who was to reign as Ning Tsung from 1195 to 1224 (Fig. 77). The map was engraved on stone in 1247, and still exists as a stele in the Confucian temple at Suchow, Chiangsu. The inscription which accompanies it is one of the shortest and most authentic expositions of the Chinese astronomical system. After an introduction drawing on Neo-Confucian philosophy, the text describes the celestial sphere, with its 'red' and 'yellow' roads (the celestial equator and the ecliptic). The 'Red Road', it says, 'encircles the heart of Heaven, and is used to record the degrees of the twenty-eight *hsiu*'; if this forthright statement had been known to modern scholars a great deal of nineteenth-century controversy would have been avoided. The 'white road' of the moon is described, with a correct account of lunar and solar eclipses. It is noted that there are 1565 named stars. An interesting section refers to the role of the Plough or Dipper as a seasonal indicator, and shows that the ancient system of keying circumpolars to the *hsiu* had not been forgotten. The treatment of the planets, however, is astrological, correlating regions of the sky with Chinese provinces and cities.

Other planispheres are known. A Korean one, engraved in 1395, gives prominence to the stars lying along the Milky Way, and there is an attempt to depict star brightnesses or 'magnitudes' by variations in the size of the

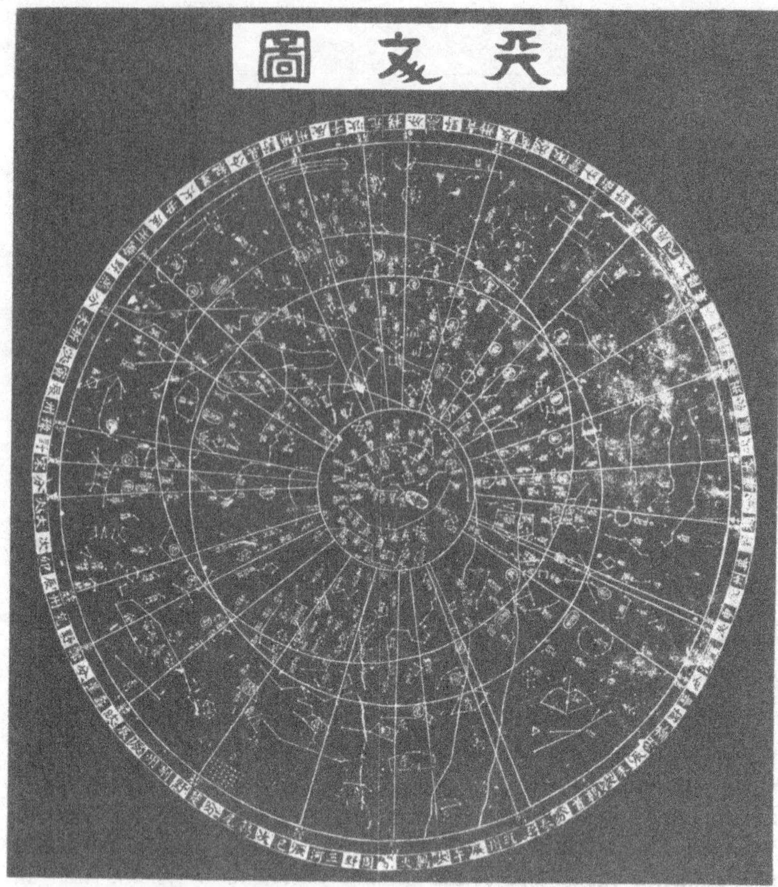

Fig. 77. The Suchow planisphere of A.D. 1193; note the excentric ecliptic and the curved course of the Milky Way. The planisphere with its explanatory text was prepared by the geographer and imperial tutor Huang Shang, and committed to stone by Wang Chih-Yuan in A.D. 1247.

dots. The inscription accompanying it includes a review of the ancient Chinese theories of the universe, and tables of the stars culminating at dawn and dusk for the 24 divisions of the year. Yet another, in the form of a bronze bowl some 34 cm in diameter, shows the stars as small raised dots on the bowl's surface; although this is probably late, and not likely to be earlier than the seventeenth century, it is nevertheless of fully traditional Chinese type.

The planispheres just described all agree in maintaining Thien shu as the Pole Star, thus perpetuating the system of 350 B.C. They also differ from Su Sung's best map in their placing of the points of the equinoxes, using those corresponding to A.D. 200. It is tempting to connect these retrograde steps with contemporary social and political trends. Su Sung and Shen Kua worked at a time when the reform movement associated with the great minister Wang An-shih was in its prime. The Suchow planisphere of a century later was made when conservative traditionalism had once again got the upper hand. What we know of astronomical map-making in other parts of the world, however, suggests that we must not undervalue this whole tradition of Chinese star charts from the Han time down to the Yuan and Ming. Europe had little or nothing to show before the Renaissance comparable with the Chinese tradition of celestial chart-making.

Star legend and folklore

As might be expected, the Chinese had their own star lore, and there were legends, many associated with the person of I-Hsing, the Buddhist astronomer of the Thang. One is in the *Ming Huang Tsa Lu* (Miscellaneous Records of the Brightness of the Imperial Court (of Thang Hsüan Tsung)) of A.D. 855. A quaint and amusing example, it runs as follows:

> I-Hsing in his youth had belonged to a very poor family, and had had as a neighbour one Wang Lao, who often helped him. I-Hsing had tried to make some return, especially during the Khai-Yuan reign-period when he was in high favour with the emperor. Eventually Wang Lao, having killed someone, was imprisoned, and called upon I-Hsing for help. I-Hsing went to see him and said, 'If you want gold and silver I can give you all you want, but as for the law, I cannot change it.' Wang Lao reproached him, saying, 'What good was it to me that I ever knew you', and so they parted.
>
> Later I-Hsing was in the Hun-Thien (Armillary Sphere) Temple where there were several hundred workers. He ordered some of them to move a huge pot into an empty room. Then he said to two servants 'In a certain place there is a ruined garden. Do you hide there secretly tomorrow, from noon to midnight. Something will come – if it is seven in number, put them in the pot and cover them up, and if you lose one I shall give you a great beating.' About six o'clock in the evening, sure enough, a herd of seven pigs appeared, and they caught them all and put them in the pot; then they ran off and told I-Hsing, who was very pleased. Covering the pot with a wooden cover and matting, he wrote certain Sanskrit words upon it in red, the meaning of which his students did not understand.
>
> Before long I-Hsing received a message to go urgently to a certain palace, where the emperor met him and said, 'The Head of the Astronomical Bureau has just informed me that the Great Bear has disappeared. What can it mean?' I-Hsing replied, 'This sort of thing has happened before. In the Later Wei dynasty they even lost

> the planet Mars. But there are no previous records of the disappearance of the Great Bear. Heaven must be giving you an important warning, perhaps of frost or drought. But your Majesty, with your great virtue, can influence the stars. What would most affect them would be a decision on your part in favour of life rather than death. So do we Buddhists preach forgiveness to all.' The emperor agreed, and issued a general amnesty.
>
> Later the seven stars of the Great Bear reappeared in their places in the heavens. And when the pot into which the pigs had been put was opened, it was found to be empty.

This mastery of I-Hsing over the stars of the Great Bear becomes perhaps more comprehensible when we remember that he did indeed write certain tracts or books about their astrological significance and astronomical relations.

THE DEVELOPMENT OF ASTRONOMICAL INSTRUMENTS

The gnomon and the Gnomon Shadow Template

The most ancient of all astronomical instruments, at least in China, was the simple vertical pole. With this one could measure the length of the sun's shadow by day to determine the solstices, when the noonday shadow is either at its longest (winter solstice) or shortest (summer solstice). The gnomon could also be used at night to observe the slow annual revolution of the celestial sphere, and thus to determine the sidereal year. It was called *pei* (碑) or *piao* (表). *Pei* can be written with the bone radical (髀) or the wood radical (椑), in which case it means a shaft or handle. Ancient oracle-bone forms of the phonetic component show a hand holding what seems to be a pole with the sun behind it at the top ([illegible]), so that although this component alone came to mean 'low' in general, it may perhaps have referred originally to the gnomon itself. This is after all an object low on the ground in comparison with the sun, and shows the long shadow of the winter sun at solstice, the moment which the Chinese always took as the beginning of the tropical year (i.e., the year determined by the annual passage of the sun through a particular solstice or equinox).

Other evidence that the Shang people were conscious of the casting of shadows may be derived from the now purely literary word *tsê* (仄), which means the late afternoon when the sun is setting. Ancient bone forms of this ([illegible] and [illegible]) show the sun and a man's shadow at different angles.

Probably the earliest literary reference to solstice observations which has come down to us is the passage in the *Tso Chuan* bearing the date 654 B.C., and it is therefore worth quoting.

> In the fifth year of Duke Hsi, in the spring, in the first month (December), on a *hsin-hai* day, the first of the month, the sun (reached its) furthest south point. The duke (of Lu), having caused the new moon to be announced in the ancestral temple, ascended to the observation tower in order to view (the shadow), and (the astronomers) noted down (its length) according to custom.

From the words inserted into the translation it will be seen that the annalist was vague as to what exactly the prince did when he joined the party of astronomers, but this was a winter solstice, and there can be no doubt they were measuring the length of a shadow. So important was this reading that the attendance of the ruler of the State in person is not surprising.

Although the *Huai Nan Tzu* (Book of (the Prince of) Huai-Nan) preserves a tradition that gnomons of 3 m (10 feet) length were anciently used (which would fit in with the strong evidence for decimal measurement in the Chou already mentioned in the section on Chinese mathematics), this was early abandoned, presumably because it did not readily lend itself to simple calculations about the sides of right-angled triangles. A length of 2·4 m (8 feet) is generally mentioned in ancient and medieval texts, and even in the Yuan period, when the interests of accuracy demanded something larger, a multiple, 12·2 m (40 feet), was used.

The earliest measurements of the shadow's length were no doubt made with the foot-rules of the time, but eventually it was realised that these varied according to bureaucratic prescription and local custom, so that a standard jade tablet, which may be called the Gnomon Shadow Template, was made for this purpose only. It is mentioned in the *Chou Li*, and actual specimens, made of terra cotta, one dated A.D. 164, are extant. The moment of solstice could thus be determined by placing the calibrated template due north at the base of the post for several days around the expected time, and taking noon of the day when the shadow most nearly coincided with it. In practice, the winter solstice was determined indirectly by observing the summer solstice, partly because a much shorter template was then required. Another reason was that a sidereal mark-point could more easily be found at this time, since for several centuries around 450 B.C., the determinative star of the Niu *hsiu* (β Capricorni) was exactly opposite on the celestial sphere to the sun at summer solstice.

The template system was an attempt to overcome the chaos of primitive systems of measurement, and did not persist, although it does seem to anticipate, in some sense, modern devices such as the standard platinum metre. In A.D. 500 Tsu Kêng-Chih made bronze instruments in which the gnomon and a horizontal measuring scale were combined. Some 50 years earlier Ho Chhêng-Thien had proceeded to more careful measurements of the winter solstice shadow, and it was not until this had been done that the inequality of the seasons could be discovered.

Study of the shadow lengths at solstices furnishes exact knowledge of the obliquity of the ecliptic (figure 56). This was probably known by the fourth century B.C. as soon as Shih Shen and Kan Tê began to measure right ascensions and declinations of celestial bodies. No precise figure for the sun's declination at the solstices is available before the *Hou Han Shu* (History of the Later Han Dynasty), where Chia Khuei, in his exposition of A.D. 89, gave definite figures. Many further determinations were made in subsequent centuries by Chinese astronomers. Fig. 79 compares various measurements of the obliquity (ε) made in China and elsewhere, taking account of its gradual change with time. This gives some indication of the accuracy of early values, but there are so many uncertainties – the date, the place, the height of the gnomon, and the nature of the units of length – that the very precise Chinese value of Liu Hung and Tshai Yung may well be somewhat fortuitous. Nevertheless it is hard to reject entirely the figures given in the *Chou Li*, and probably the best solution is to take them as traditional measurements which had been made somewhere during the Chou period and which had afterwards been carefully handed down through uninstructed intermediary hands. The margin of error seems great enough to justify us in choosing a date for the observations, probable on quite other historical grounds, as between the ninth and third centuries B.C. It is worth remarking, though, that the whole series of Chinese observations, culminating in the very accurate ones of Kuo Shou-Ching, proved of value to eighteenth-century astronomers in their discussions about the gradual change which the obliquity undergoes.

Another error arising from primitive computations was the long-standing Chinese idea that the shadow length increased by 2·5 cm for every 1,000 *li* (nearly 440 km) north of the 'earth's centre' at Yang-chhêng, and decreased by the same amount for every 1,000 *li* south. Yet numerous records of experiments disproving this remain. In A.D. 445 Ho Chhêng-Thien had simultaneous measurements made at Chiaochow (modern Hanoi, 5,000 *li* south of Yang-chhêng) and at Lin-I in Indo-China, obtaining 9 cm per 1,000 *li*, while similar observations were made in 508 and 600. The most complete set of figures was that obtained by expeditions under the direction of Nankung Yueh and I-Hsing in 721 and 725, using nine stations on a meridian line of 7,973 *li* (just over 3,500 km), where simultaneous measurements of summer and winter solstice shadow lengths were made with standard 2·4-m gnomons. A value close to 10 cm for every 1,000 *li* was obtained. This work must surely be regarded as the most remarkable piece of organised field research carried out anywhere in the early Middle Ages.

Fig. 78. A late Chhing representation of the measurement of the sun's shadow at the summer solstice with a gnomon and a gnomon shadow template by Hsi Shu (the youngest of the Hsi brothers) in legendary antiquity. From *Shu Ching Thu Shuo*.

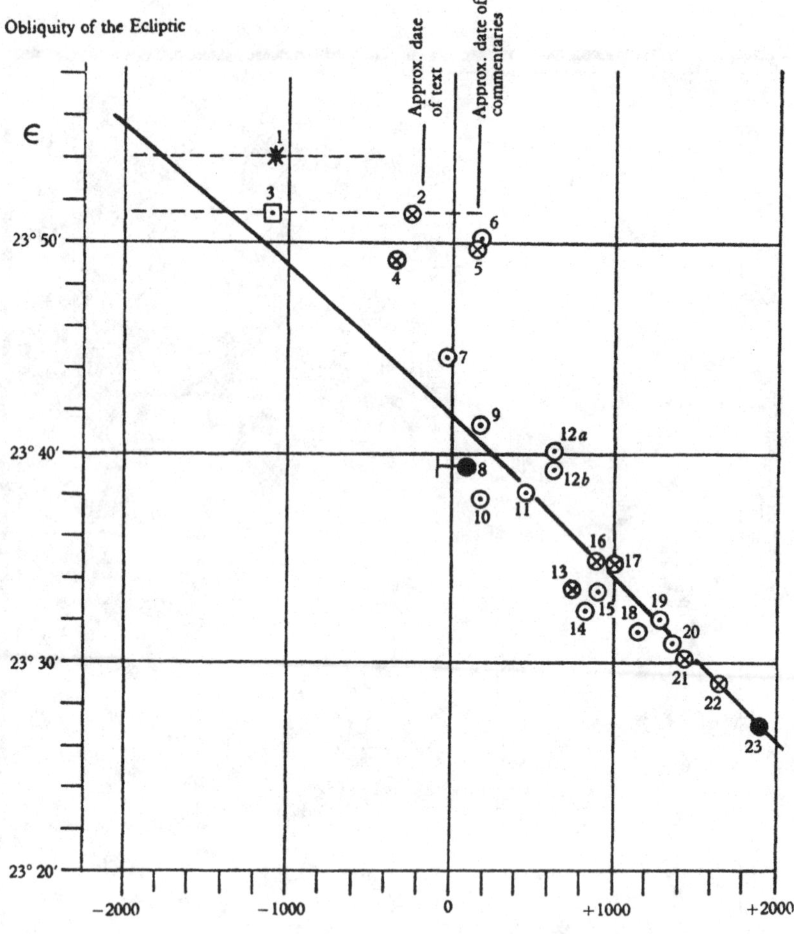

Fig. 79. Plot of ancient and medieval measurements of the obliquity of the ecliptic.

1 *Chou Li* figure, subject to uncertainty as to place, length of units, and height of gnomon.

2–3 (dotted line) *Chou Li* figure recalculated by Hartner, assuming Yang-chhêng as the place.

2 Eratosthenes.

3 Laplace's theoretical value.

4 Pytheas.

5 Ptolemy.

6 Liu Hung & Tshai Yung (taking $67\frac{1}{8}$° Ch. value only; Hartner).

7 Liu Hsiang.

8 Chia Khuei, with backward extension to Lohsia Hung, the range representing also values given in the *Chou Pei Suan Ching*.
9 Liu Hung & Tshai Yung.
10 Liu Hung & Tshai Yung (both solstices, *Hou Han Shu* figures).
11 Tsu Chhung-Chih.
12*a*, *b* Li Shun-Fêng.
13 Value from al-Ma'mūn's observatory (Sind ibn 'Ali and Yaḥya ibn abi Manṣūr).
14 Hsü Ang.
15 Pien Kang.
16 al-Battānī.
17 Ibn Yūnus.
18 Liu Hsiao-Jung.
19 Kuo Shou-Ching.
20 'Ali ibn al-Shāṭir.
21 Ulūgh Beg.
22 Cassini.
23 Contemporary value.

In connection with measurements of this kind, it is worth noting an even earlier record, that of the observation of a military commander, Kuan Sui, who in A.D. 349 led a successful expedition which took him far into Vietnam in pursuit of a vanquished army. Here he set up a gnomon but 'found that the sun was to the north of it, casting a shadow to the south ... Hence the inhabitants of that country' continues the account, 'have the doors of their houses facing north to turn to the sunshine'. Thus the Chinese found what the Greeks at Alexandria had also known, namely, that south of the Tropic of Cancer (which is just north of Canton) the sun will cast shadows at midday to the south during part of the year.

While the observed lengths of the shadows could serve for astronomical calculations, some of which were doomed to inaccuracy owing to the absence of other information (e.g. the curvature of the earth's surface), the most important observation of all was really one of null-points, namely, the moments when the shadow reached exactly its maximum and minimum lengths. These were essential for the construction of calendars. A study of the astronomy and calendrical science of the oracle-bones has shown that the inscriptions make references to the number 548; one such text is datable accurately to 1210 B.C. This has much significance, since $1\frac{1}{2}$ times 365·25 is 547·875; the meaning therefore is that 'the summer solstice recurs at 548 days after the winter solstice'. It can hardly be doubted, therefore, that the Shang people in the thirteenth and fourteenth centuries B.C. were using gnomons to determine the solstices in the traditional calendar, handed down as the 'Old Quarter-Remainder Calendar'. Obviously regular and long-continuing observations were made, covering at least four tropical

Table 28. *Evaluation of the fraction of the tropic and the sidereal year*

	Tropic year fraction (days)	Sidereal year fraction (days)
True value	—	0·25637
True value (computed):		
Han A.D. 200	0·242305	—
Thang A.D. 750	0·242270	—
Yuan A.D. 1250	0·242240	—
Deduced from the oracle-bones (13th century B.C.)	—	0·25
Liu Hsin (San Thung cal. 7 B.C.)	0·250162	
Tsu Chhung-Chih (Ta Ming cal. A.D. 463)	0·242815	—
Kuo Shou-Ching (Shou Shih Shih cal. A.D. 1281)	0·242500	—
Han I (Huang Chhu cal. A.D. 220)	—	0·255989
Liu Chhuo (Huang Chi cal. A.D. 604)	—	0·257610
I-Hsing (Ta Yen cal. A.D. 724)	—	0·256250
Kuo Shou-Ching (Shou Shih cal. A.D. 1281)		0·257500

years, the shortest space of time in which the day-count could come to a whole number. Attempts at exact evaluation of the odd fraction of the tropical year continued throughout Chinese history; the nearest approaches, some quite early, are given in Table 28.

Giant instruments in masonry

One of the most interesting chapters in the history of astronomy concerns a phase of development in which the search for accuracy led to the construction of instruments of extremely large size. The idea behind it was that errors in marking scales would be proportionately less on a large instrument than on a smaller one, though the great advances in practical technology which improved accuracy of graduation in the European Renaissance, and soon spread to other regions, abolished the need for very large instruments in masonry. The movement first declared itself in Chinese and Arabic astronomy, could then be seen in some of the instruments used by the outstanding late-sixteenth-century Danish observer Tycho Brahe, and came to a very late climax in the observatories of eighteenth-century India.

In A.D. 995 Abū'l-Wafā'al-Būzjanī had used a quadrant (for measuring angles) with a radius of almost $6\frac{3}{4}$ m, and a sextant of a contemporary, Ḥāmid Ibn al-Khiḍr al-Khujandi, had a radius of as much as 17 m. The

Fig. 80. The Tower of Chou Kung for the measurement of the sun's shadow lengths at the solstices at Kao-chhêng (formerly Yang-chhêng), some 80 km south-east of Loyang, considered by Chinese astronomers in ancient times to be the centre of the world. (Photograph, Tung Tso-Pin *et al.*, *Chou Kung Tshê Ching Thai Thais Chha Pao Kao* (Report of an Investigation for the Measurement of the Sun's (Solstitial) Shadow), Chhangsha, 1939.) In its present form the structure is a Ming renovation of the building erected by Kuo Shou-Ching about A.D. 1276 for use with the 12-m (40-foot) gnomon. This stood upright in the niche, and the shadow was measured along the horizontal graduated stone scale (see Figs. 81 and 82). One of the rooms on the platform housed a water-clock (perhaps a water-driven mechanical clock), the other probably an armillary sphere.

famous Ulūgh Beg, whose observatory at Samarkand began work in 1420, had a sextant whose height is said to have been as high as the dome of St Sophia in Istanbul, i.e. almost 55 m. In China a giant gnomon was set up during the Yuan dynasty in 1276 by Kuo Shou-Ching. Although markedly independent in character, it was done in the presence of astronomers of the Arab tradition, and after the Marāghah Observatory in Persia had sent models or diagrams as a guide. It seems, then, that this natural development of Chinese astronomy was stimulated by the Arabic trend to large instruments.

Fig. 81. The Tower of Chou Kung, looking along the shadow scale with its water-level channels. (Photograph, Tung Tso-Pin *et al.*, reference as in Fig. 80)

At Kao-chhêng (the old Yang-chhêng), 80 km south-east of Loyang, there stands a remarkable structure known as Chou Kung's Tower for the Measurement of the Sun's Shadow. The tower (Fig. 80) is a truncated pyramid, the sides of which measure some 15 m at the bottom and $7\frac{1}{2}$ m at the top. Two stairways lead from ground level to a platform, on the north side of which stands a single-storey building with three rooms, the centre one having a wide opening to the north giving a good view of the 12-m (40-foot) gnomon (now no longer there) and the shadow which it once cast. The top level was known as the Star Observation Platform, and probably once possessed a thin vertical rod for observations of meridian transits, while the records show that one of the rooms was equipped with a large clepsydra or water-clock. Below the tower, on the north side, lying on the ground and extending for $36\frac{1}{2}$ m, is the Sky Measuring Scale (Figs. 81 and 82). This carries, besides graduations, two parallel troughs connected at the ends and forming a water-level, and it continues into the body of the pyramid, which is cut away to receive it, so that its base is directly below the wall of the central room of the Star Observation Platform. The gnomon was almost certainly an independent pole sunk into a socket at the near end of the horizontal scale. The height of the platform above the scale was about $8\frac{1}{2}$ m, and the whole instrument was built of brick. Tall 12-m gnomons were authorised at Peking, Shangtu (the Xanadu of the English poet Coleridge,

Fig. 82. The Tower of Chou Kung, looking down at the shadow scale. (Photograph, Tung Tso-Pin *et al.*, reference as in Fig. 80.)

and the Mongol imperial summer capital), Nanhai in Kuangtung, and Yang-chhêng, but only at Peking and Yang-chhêng were they set up. Yang-chhêng alone ranked so high as to have a tower as well, for this was the place where the official astronomers, at least as early as the Han, made the 'standard' solstice measurements.

Although it is not certain whether there was any tower approaching in size the one at Yang-chhêng in times earlier than the Yuan, the Confucian temple surrounding it contains a $2\frac{1}{2}$-m stele set up by Nankung Yüeh in A.D. 723. It is arranged so that the sun at summer solstice casts no shadow beyond its truncated pyramidal base. In the *Yuan Shih* (History of the Yuan (Mongol) Dynasty) there is an account of the gnomon and of a device, the Shadow-Definer, which seems to have been an ingenious new invention.

> The shadow definer is made of a leaf of copper 5 cm wide and 10 cm long, in the middle of which is pierced a pin-hole. It has a square supporting framework, and is mounted on a pivot so that it can be turned at any angle, such as high to the north and low to the south (i.e. at right-angles to the incident shadow-edge). This instrument is moved back and forth until it reaches the middle of the (shadow of the) cross-bar, which is not too well defined, and when the pin-hole is first seen to meet the light, one receives an image no bigger than a rice-grain in which the cross-beam can be noted indistinctly in the middle. On the old methods, using the simple summit of the gnomon, what was projected was the upper edge of the solar disc. But with this

> method one can obtain, by means of the cross-bar, the rays from the centre of the disc without any error.
>
> (In A.D. 1279) on the 30th May, I observed the summer solstice shadow as 3·7702 metres in length; and on the 11th Dec. of the same year found that the winter solstice shadow was 23·3903 metres long.

This passage was long misunderstood, many thinking that the device was mounted on top of the gnomon whereas, on the contrary, it was moved along the horizontal scale, and had the effect of focusing, like a lens, the image of the cross-bar. That Kuo Shou-Ching should have made use of the pin-hole is not at all surprising since it had been familiar to Chinese scientists at least three centuries earlier; indeed the camera obscura may have passed to them from the Arabs. And certainly, the thirteenth-century work with the 12-m gnomon seems to have been the most accurate which had ever been done on solstice shadows.

The sundial (solar time indicator)

The determination of time by most sundials depends on noting the direction of the shadow, not the length. Perhaps because of its universality and familiarity, clear references to the sundial or sun-clock are rare in Chinese literature: the term *kuei piao* is usually understood to refer to it. An early reference is found in connection with an assembly of calendar experts which Ssuma Chhien advised should be called together in 104 B.C. Then, in the *Sui Shu* (History of the Sui Dynasty), there are numerous references to the same kind of instrument in the sixth century A.D. as was used by Ssuma Chhien's experts. Here we come up against a difficulty, and one which we shall constantly meet when discussing Chinese technology; the problem of determining the exact nature of an ancient instrument from the names loosely used for it. In this case everything depends on whether the gnomon was placed vertically, pointing towards the zenith (as it was for the measurements discussed in the previous sub-section), or whether it was inclined at an angle depending on the latitude, and thus pointing to the celestial pole – only in the latter case are measurements of equal intervals of time possible.

Further study of this question beings up what was long one of the standard enigmas of Chinese archaeology, namely, the significance of the so-called 'TLV-mirrors' of the Han dynasty. This name is given to the design which is frequently to be found on the backs of Han metal mirrors, and Fig. 83*a* shows the typical pattern, while Fig. 84 is a photograph of such a mirror dating from the Hsin (A.D. 9 to 23). The earliest of such mirrors still in existence is dated about 250 B.C.

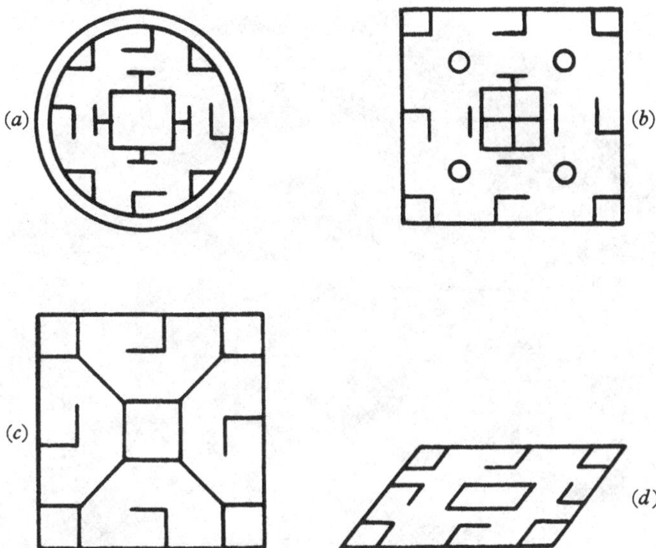

Fig. 83. TLV designs on mirrors and divination boards.

Similar markings have been noticed in some of the reliefs of Han times, and a scene from the Wu Liang tomb-shrines (A.D. 147), illustrated in Fig. 85, shows what was at first thought to be merely a banquet, but was later seen to depict magical operations. In the background a board which bears the TLV markings is apparently hanging on the wall, while on the ground, carefully drawn so as to indicate that it is horizontal, is a small table which may plausibly be identified as the diviner's board or *shih* (volume 1 of this abridgement, page 200). A diagram of the TLV-board is given in Fig. 83*b*. Another stone relief from the Hsiao Thang shrine (A.D. 129) shows two men playing a game at a table the side of which again has a board with these symbols (Figs. 83*c* and 86). We thus come upon an association of divination with games which later (in the next volume of this abridgement) we shall see to have been of great significance in relation to the discovery of magnetic polarity.

In view of all this, it is extremely striking to find that the only existing objects which must have been Han sundials both bear TLV markings. One of these is in jade, the other in grey limestone about 28 cm square (Fig. 87), and a drawing of its inscriptions is given in Fig. 88. The graduations are each one-hundredth of the circumference, and where the lines meet the outer circle there is a series of small sockets numbered 1 to 69 in a clockwise direction. In interpreting this, everything depends on whether they were

Fig. 84. Bronze TLV mirror from the Hsin dynasty (A.D. 9 to 23). This example, from the Cull Collection of Chinese Bronzes, is inscribed with a poem beginning at the pattern of five dots indicated by the arrow at 3.0 p.m.

The Hsin have excellent copper mined at Tan-yang,
Refined and alloyed with silver and tin, it is pure and bright.
This imperial mirror from the Shang-fang (State workshops) is wholly flawless:
Dragon on the east and Tiger on the west ward off ill-luck;
Scarlet Bird and Sombre Warrior accord with Yin and Yang.
May descendants in ample line occupy the centre.
May your parents long be preserved, may you enjoy wealth and distinction,
May your longevity endure like metal and stone
May your lot match that of nobles and kings.

Fig. 85. A scene from the Wu Liang tomb-shrines (*c.* A.D. 147) depicting magician-technicians at work. In the background, a TLV board hanging on the wall, in the centre a *shih* (diviner's board) on the floor. Redrawn by W.P. Yetts.

Fig. 86. A scene from the Hsiao-Thang Shan tomb-shrines (*c.* A.D. 129) showing two men playing a game (perhaps *liu-po*) at a table, the side of which has a TLV board. (Photograph, White & Millman.)

Fig. 87. Plane sundial of the Former Han period, grey limestone, about 28 cm square. (Collections of the Royal Ontario Museum, photograph, White & Millman.)

used in the plane of the horizon or the equator. Remembering the basically polar and equatorial character of Chinese astronomy, it seems more than likely that at an early time the discovery was made that if the base-plate was inclined in the equatorial plane, and the gnomon made to point at the imperial pole of the heavens, a solar time-keeper would result. For this and other reasons it seems likely that the instruments here were also tilted over parallel to the plane of the celestial equator.

Careful investigations make it seem clear, though, that these two devices were not time-keepers but, instead, were used as regulators for clepsydras, the indicator rods which were inserted in the small sockets being placed in the appropriate sockets for the length of the day during which the clepsydras were being checked. As far as the central gnomon is concerned, probably the most likely suggestion is that it was not a gnomon

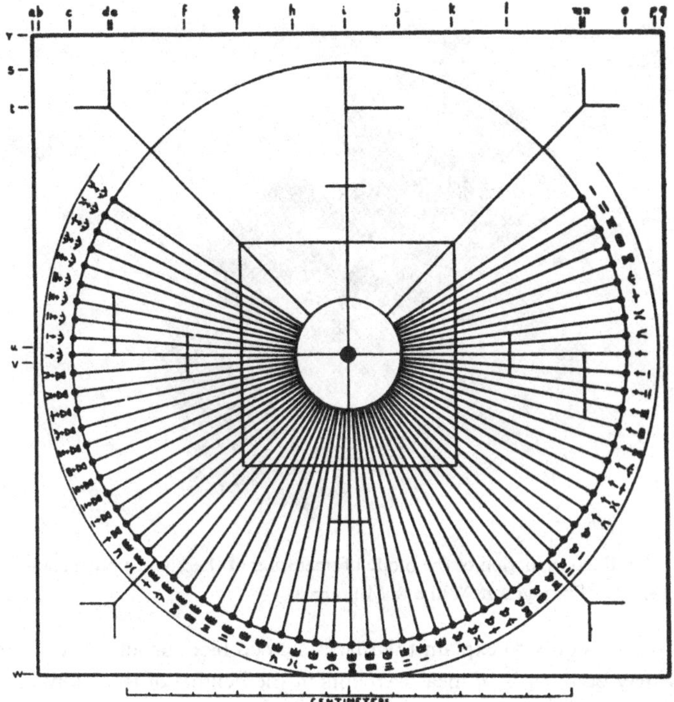

Fig. 88. Inscriptions on the Former Han sundial shown in Fig. 87. After White & Millman.

in the usual sense, but rather a rectangular bronze plate connected with a T-shaped gnomon at the periphery by a bronze bridge. Time would then be shown by that position in which the shadow of the T-shaped gnomon fell on the upright plate in line with the bridge, and the season would be shown by the height of the cross-bar above the bridge, as in the annexed diagram. A reconstructed dial of this is shown in Fig. 89.

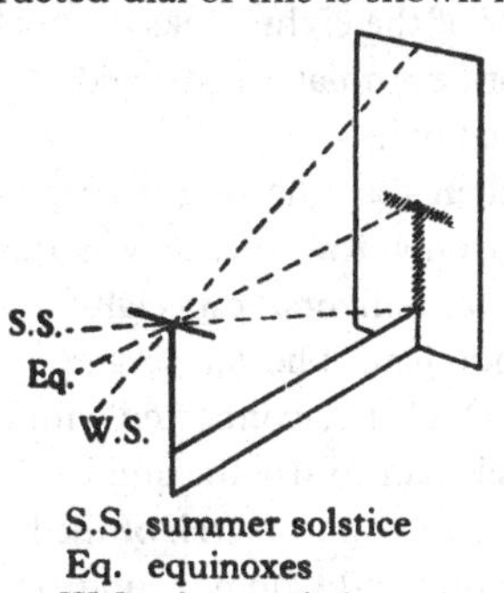

Fig. 89. Reconstruction of the mode of operation of the Han Sundial shown in Fig. 87. (Photograph, White & Millman.)

It remains only to explain the marks shaped like our letters T, L and V. They may be records of measurements of the heights of the gnomons or of the shadows they would be expected to throw at the winter and summer solstices respectively. However, it may well be that the height of the T-shaped gnomon was fixed by the distance from the circumference of the outer circle to the cross-bar of the T, while that of the central plate was fixed by the distance from the centre hole to the cross-bar of the L. At least, this arrangement would give a system which actually works in practice. As for the V marks, it has been suggested from some cryptic passages in ancient texts that the most primitive sundial had been a square of ox-hide or cloth stretched flat and placed in the plane of the celestial equator. The V marks would be a reminiscence of four of the eight hooks originally used to make this arrangement stay in position, eventually restricted to four directions mid-way between the four cardinal points.

Provisionally we may assume then that the original purpose of the TLV markings was a practical and astronomical one. It was quite natural they should have been reproduced on mirrors, especially those which embodied elaborate cosmological symbolism. The *liu-po* game board may have been an independent derivation. Doubtless connected with divination, nothing would have been more natural than to use the sundial face for the game board, since the sundial marks the form and march of the heavens.

In numerous examples of late (Ming and Chhing) sundials still found in China, the dial is graduated on both sides, and the gnomon extends right through so as to stick out below as well as above. This is because such a sundial receives the sun's rays on its upper face for only half the year, when the sun is north of the equator. When this change came in is uncertain, but if not in the Han, then the following passage, not before noted, may throw some light on the matter, for the twelfth-century astronomer Tsêng Nan-Chung evidently thought he was on to something new. In the *Tu Hsing Tsa Chih* (Miscellaneous Records of the Lone Watcher) written by his son or grandson, Tsêng Min-Hsing (1156) says:

> Tsêng Nan-Chung used to say that while there was much in the ancient classics about recording the sun's shadow, it all concerned length, and thus was not comparable with telling time by the water-clock. So at Yüchang he made a Diagram of the Sun Shadow and constructed a sundial. A round (wooden plate) was divided into four (equal divisions), and one of these segments was taken away (i.e. not graduated), so that (the graduated part) was shaped like a crescent moon. Hours and quarters were marked round the edge. The dial was supported on posts so that it was high towards the south and low towards the north (i.e. in the plane of the equator). The gnomon pierced the dial at the centre, one end pointing to the north pole and the other to the south pole. After the spring equinox one had to look for the shadow on the side facing the north pole, and after the autumn equinox one found it on the other (the under) side. This instrument agreed more or less with the water-clock. Tsêng Nan-Chung was very proud of this device, and thought that he had got something which had not been achieved in ancient times.

It would be impossible to have a clearer description of a typical equatorial sundial in which two faces are used according to the time of year. But further evidence would be needed to substantiate the attribution of the system solely to Tsêng Nan-Chung about A.D. 1130. Nevertheless, if we are right in tracing the history of sun-clocks in China to the discovery of the pole-pointing gnomon some time about the fourth century B.C., their development in Asia might be said to parallel that in the West.

Portable equinoctial dial-compasses. In late times portable sundials which incorporated small magnetic compasses for orientation were manufactured in great numbers in China, and are now commonly found in collections of scientific instruments. It does not seem to have been noted, however, that they fall into two quite distinct types, which we may call *A* and *B* (Figs. 90 and 91). For the gnomon, type *A* has a cord which becomes taut when the dial is opened. A reading is taken on the base-plate, and the device has the

Fig. 90. Late Chinese portable sundial of Type *A*; the gnomon is formed by the stretched string and a compass needle is provided for orientation. This type is probably not older than the end of the sixteenth century. From the collection of Mr W.G. Carey.

Fig. 91. Late Chinese portable sundial of Type *B*; the plate, which is aligned with the plane of the equator and can be adjusted for latitude, carries a pole-pointing gnomon. A compass needle is provided for orientation. From the collection of Mr W.G. Carey.

characteristic Western form with the gnomon inclined. It certainly seems to have been unknown in China before the coming of the Jesuits, and could not have been produced before the Ming dynasty: indeed its Chinese name is not only *phing mien jih kuei* but also *yang kuei* (the foreign sundial).

Type *B*, however, is entirely different. Here also there is a plate carrying a magnetic compass, but the dial is graduated on a separate plate which may be raised or lowered so that the gnomon, at right angles to the dial, may point at the pole, whatever the latitude of the observer. The curious thing is that in every specimen we have seen, the ratchet scale (which holds the prop that supports the graduated plate) is marked not with latitudes or names of cities, but with the 24 fortnightly periods or *chhi*. In China each locality must, therefore, have had its local *chhi*. The type *B* dial is wholly in the Chinese tradition, and there is no reason whatever for regarding it as a Jesuit introduction – Shen Kua (eleventh century A.D.) or even Chen Luan (sixth century A.D.) may well have been familiar with it. And indeed there is a mention of a portable sundial in the *Shan Chü Hsin Hua* (Conversations on recent Events in the Mountain Retreat) of 1360.

During the seventeenth century, dials of type *B* became common in Europe, and it may be legitimate to suggest that while the Jesuits introduced into China the horizontal dial, they (or the Portuguese travellers who preceded them) brought back to Europe the simpler plan of the equatorial dial with its pole-pointing gnomon, which they found there. But perhaps it is more likely that a transmission occurred earlier through Arabic or Jewish intermediaries. In any case, the Western type *A* never displaced the old Chinese type *B* for sundials permanently set up, as in palaces, gardens or temples in China.

Besides the two types so far mentioned, there is also found, more rarely, a third *C*. An example is given in Fig. 92. This consists of a series of ivory plates, each one ruled for a different latitude, and bearing a complex pattern of lines from the celestial sphere drawn on a flat surface, using a particular projection. The gnomon, about $2\frac{1}{2}$ cm long, is dismountable for portability. The tip of its shadow indicates not only time but also the period of the year. There is reason to suppose that projects for dials of some such kind were brought to China from the Arabic world in 1267.

The clepsydra

The sundial measures true or apparent solar time, but due to the fact that the earth's orbit round the sun is not a circle but an ellipse, and also to the tilt of the earth's axis with respect to its orbit, apparent solar time varies in rate. In brief, the sun does not appear to move across the sky at the same rate throughout the year. If we want an unchanging rate, a 'mean' solar time

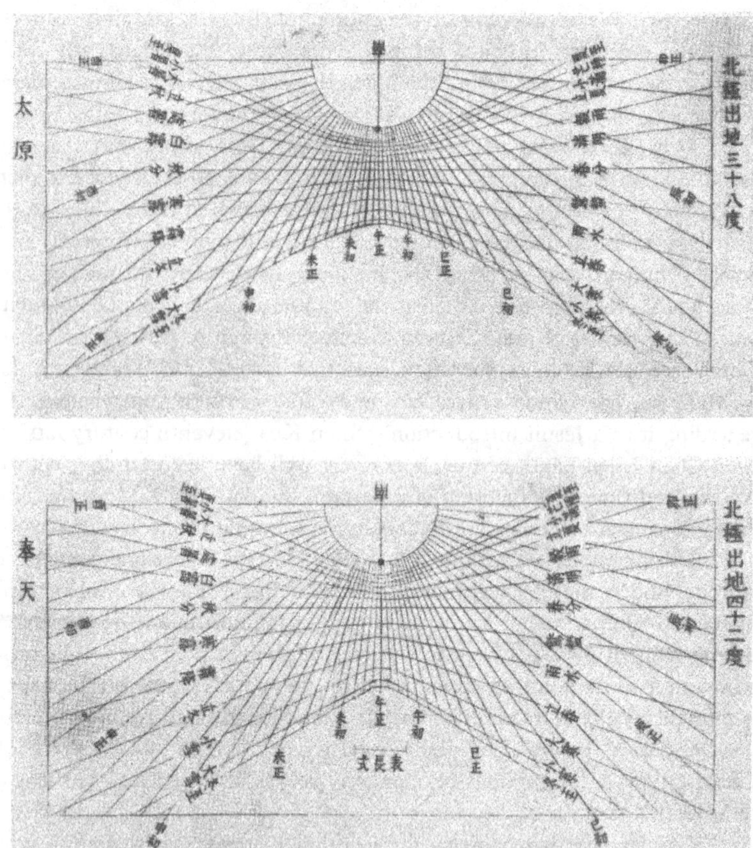

Fig. 92. Late Chinese portable sundial of Type *C*; one only of the series of plates for different latitudes is shown. From the collection of Dr R. Clay.

must be used, and so from very early ages, methods of measuring time other than by the sun were employed. While in Europe reliance was placed mainly on the sundial until the development of mechanical clocks in the fourteenth century A.D., the Chinese paid great attention to the clepsydra, and in their culture it was brought to its highest perfection.

It is nevertheless certain that the clepsydra was not invented in China. Both in Babylonia and in Egypt, as we know from cuneiform tablets and from actual objects and models recovered from Egyptian tombs, it had already been used for centuries by the time of the early Shang period (about 1500 B.C.). There were two obvious ways of arranging to measure time by the drip of water, either by noting the time taken for a vessel of particular shape to empty itself ('outflow type') or by seeing how long a vessel with no lower opening would take to fill ('inflow type'). The Babylonian clepsydras seem to have been mainly of the outflow type, but the Egyptians used both forms, although the inflow type is later and scarcer. Not unnaturally, in subsequent times (about the third century B.C. onwards) the clepsydra attracted the attention of Alexandrian scientists and engineers, who sought to combine the drip of water through narrow orifices with mechanical devices such as gear-wheels. Ctesibius (about the middle of the third century B.C. – a contemporary of Chhin Shih Huang Ti) seems to have been the first to introduce a float into an inflow vessel; this could, for instance, carry an indicator rod which showed the time by its height above the vessel's lid. However, it is to be doubted whether the Greeks knew of the clepsydra much before the third century B.C.

In China the clepsydra was known as the 'drip vessel' (*lou hu*) or the 'graduated leaker' (*kho lou*). The problem of its first appearance there is not solved, but it may have been as early as the seventh century B.C., in view of the antiquity of the clepsydra in Babylonian civilisation.

Clepsydra types: from water-clock to mechanical clock. The general development of clepsydra technique in China may readily be summarised. The most archaic type was no doubt the outflow clock, received in ancient times from Egypt and Mesopotamia, but the Chinese also knew of another archaic device, a floating bowl with a hole in its bottom, so adjusted that it took a specific time to sink. This was the inverse of the outflow clepsydra. Yet from the beginning of the Han onwards the inflow clepsydra with an indicator rod fitted to a float, came into general use. At first there was only a single reservoir for the water, but it was soon understood that the falling head of pressure in the reservoir greatly slowed down the rate at which the water dripped out, and so impaired the time-keeping.

Throughout the centuries two principal methods were used to avoid this difficulty (Fig. 93). One simple and ingenious plan (Type A) was to use one or more tanks between the reservoir and the 'inflow vessel' receiving the water. Here each successive tank increased the compensation for the falling head of pressure, and as many as six tanks above the inflow vessel are known to have been used. The other principal method (Type B) involved the insertion of an overflow or constant-level tank in the series between reservoir and inflow vessel.

Some Chinese clepsydras incorporated a steelyard, a type which has hitherto been somewhat overlooked. These seem to include at least two kinds, one in which the typical Chinese steelyard (a balance with unequal arms) was applied to the inflow vessel (Type C), and another in which it weighed the amount of water in the lowest compensating tank (Type D).

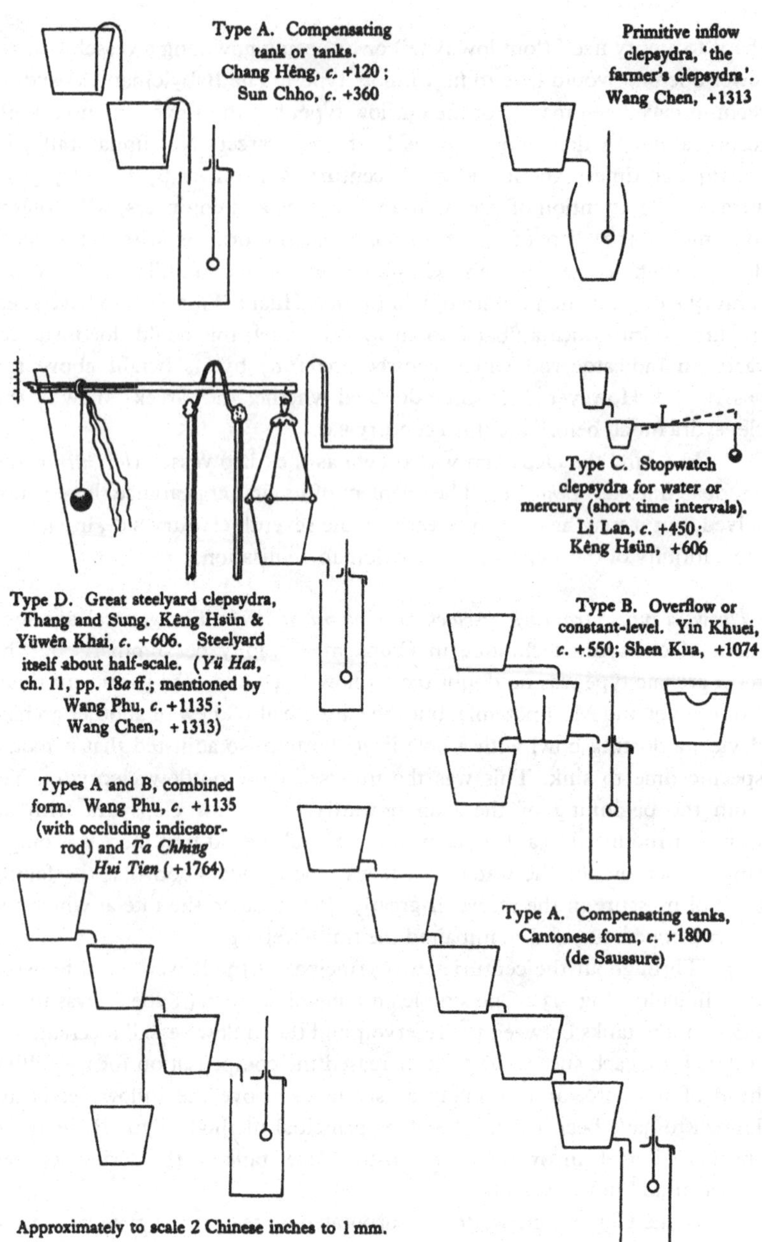
Type A. Compensating tank or tanks. Chang Hêng, c. +120; Sun Chho, c. +360
Primitive inflow clepsydra, 'the farmer's clepsydra'. Wang Chen, +1313
Type C. Stopwatch clepsydra for water or mercury (short time intervals). Li Lan, c. +450; Kêng Hsün, +606
Type D. Great steelyard clepsydra, Thang and Sung. Kêng Hsün & Yüwên Khai, c. +606. Steelyard itself about half-scale. (Yü Hai, ch. 11, pp. 18a ff.; mentioned by Wang Phu, c. +1135; Wang Chen, +1313)
Type B. Overflow or constant-level. Yin Khuei, c. +550; Shen Kua, +1074
Types A and B, combined form. Wang Phu, c. +1135 (with occluding indicator-rod) and Ta Chhing Hui Tien (+1764)
Type A. Compensating tanks, Cantonese form, c. +1800 (de Saussure)
Approximately to scale 2 Chinese inches to 1 mm.

Fig. 93. Types of Chinese clepsydras or water-clocks.

Type C obviously dispensed with the float and indicator rod, and was usually made small and portable. Here mercury was sometimes used, with reservoir, delivery pipe and receiver all made of a chemically resistant material, jade. Such instruments were well adapted for the measurement of small time intervals, as by astronomers studying eclipses, or by others timing races, and we may reasonably call them 'stopwatch' clepsydras. The larger apparatus of Type D was used for public and palace clocks throughout the Thang and Sung. It made possible the seasonal adjustment of the head of pressure in the compensating tank by having standard positions for the counterweight graduated on the beam; and hence controlled the rate of flow for the night watches which depended on the different lengths of day and night. It also of course avoided the necessity of an overflow tank and warned the attendants when the clepsydra needed refilling. But the whole matter is now under discussion and for Type D there are at least two competing reconstructions, one of which envisages an inner and an outer vessel connected by a siphon working in reverse directions alternately.

Besides these main types, there were less common ones. Most interesting were the so-called 'wheel clocks' (Type E). More will be said of these when we discuss Chinese mechanical engineering (the next volume of this abridgement), but it is worth noting now that by Sung times the escapement essential for clockwork had been invented, so that the machinery could be used to drive armillary spheres (page 171), celestial globes and other equipment. The motive power was a water-wheel with buckets on the rim fed by a constant-level clepsydra flow. This hydro-mechanical clockwork was the precursor of all mechanical clocks.

Clepsydras in history. The most famous ancient allusion to the clepsydra occurs in the *Chou Li* (Record of the rites of Chou), where it is said:

> The Official in charge of Raising the Vessel (Chhieh Hu Shih) hoists up a vessel to indicate where the well (of the encamping army) is . . . On any occasion of army service, he elevates the (clepsydra) vessel so as to let the sentries know how many strokes they are to sound (during the hours of the night). On any occasion of funeral rites, he elevates the (clepsydra) vessel so as to organise the relays of weepers. Invariably he keeps watch with fire and water, dividing the day and the night. In winter he heats the water in a cauldron and so fills the vessel and lets it drip.

It is probable that we have here to do with a simple outflow type of clepsydra, and the text may easily belong to the Warring States period, perhaps the fourth century B.C. Another reference is to be found in the *Chhien Han Shu* (History of the Former Han Dynasty), which records the assembly of calendar experts convoked imperially in 104 B.C. at the suggestion of Ssuma Chhien (page 138).

The most primitive form of inflow clepsydra, consisting only of reservoir and receiver, persisted as a rough time-measure in the countryside down to the fourteenth century A.D. But the first sure evidence for floats and indicator rods is reached with Chang Hêng and others about A.D. 85, and also possibly for siphons, since the term *ya chhiu* appears, which certainly meant this in later times. From that point on it is possible to follow the development of type A. Archaeological evidence shows us, however, that such a clock can by no means have been new at the end of the first century A.D. Two Former Han clepsydras survived, at least down to the Sung, at which time a drawing was made and a description given of one of them. It was a cylindrical bronze vessel with a pipe inserted in the bottom, and a hole in the lid for the indicator rod to slide through. The date of the inscription is doubtful; it may be 75 B.C. at the latest and 201 B.C. at the earliest.

The relation between the Han float-and-pointer inflow clock with that of the Alexandrians is therefore rather a difficult matter to determine. There was contact between China and Roman Syria at the end of the first century A.D., though the series of 'embassies' from the West does not really begin until A.D. 120. There might have been earlier ones not recorded, but the possibility that a clepsydra of this type was in use much earlier, in fact at the beginning of the Han (200 B.C.) makes one hesitate to conclude that it was essentially an Alexandrian invention. For the present the problem is unsolved, although it seems likely that the invention spread in both directions from Egypt and Mesopotamia.

The quotation from the *Chou Li* (page 151) shows that the Han people were well aware that water becomes more viscous at lower temperatures and made rough efforts to keep its temperature constant in winter, or at least to prevent it from freezing. Huan Than (40 B.C. to A.D. 30) tells us:

> Formerly, when I was a Secretary at the Court I was in charge of the clepsydras (and found that) the differences in degrees varied according to the dryness and the humidity, the cold and the warmth (of the surroundings). (One had to adjust them) at dusk and dawn, by day and by night, comparing them with the shadow of the sun, and the divisions of the starry *hsiu*. In the end one can get them to run correctly.

He thus noticed variations in evaporation rate as well as viscosity. The text is interesting, too, in that he speaks of the comparison of time by the stars (sidereal time) with clock time.

From at least the Later Chou period onwards the day and night was

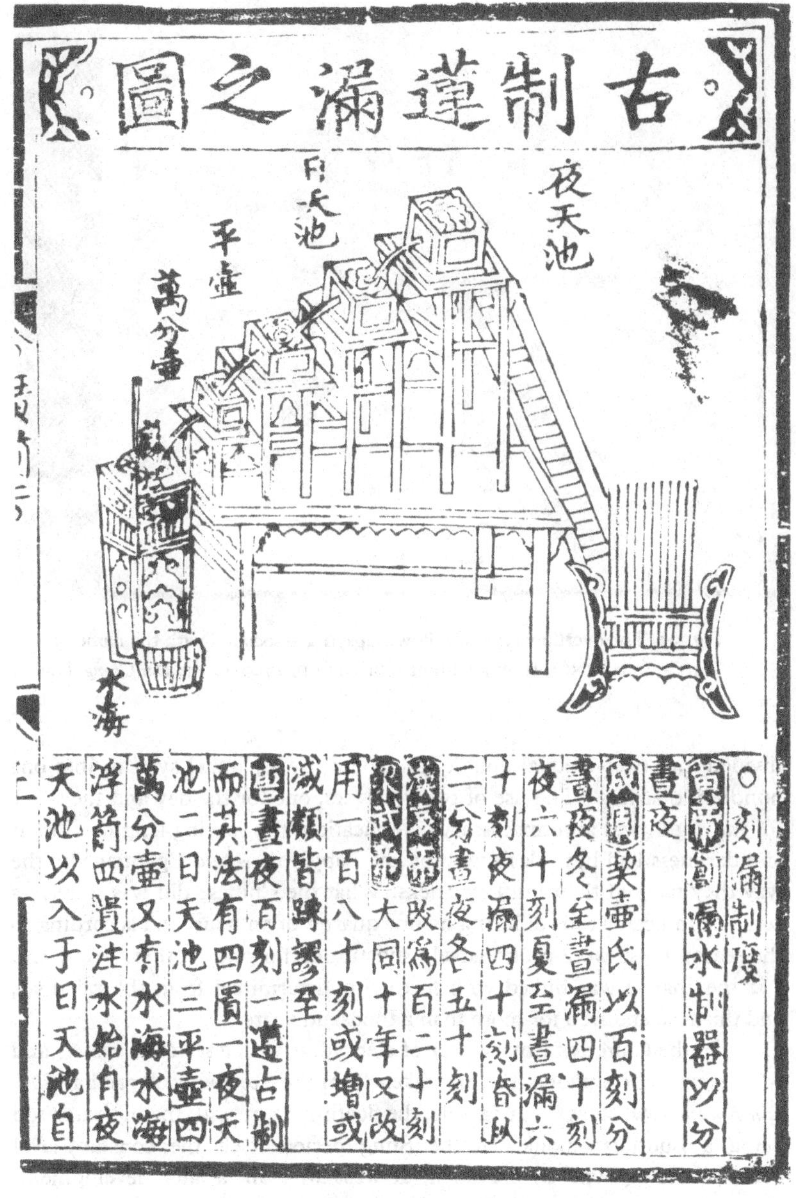

○刻漏制度
黃帝創漏水制器以分
晝夜
成周挈壺氏以百刻分
晝夜冬至晝漏四十刻
夜六十刻夏至晝漏六
十刻夜漏四十刻春秋
二分晝夜各五十刻
漢哀帝改爲百二十刻
梁武帝大同十年又改
用一百八十刻或增或
減頗皆踳謬至
唐晝夜百刻一遵古制
而其法有四匱一夜天
池二日天池三平壺四
萬分壺又有水海水海
浮箭四匱注水始自夜
天池以入于日天池自

Fig. 94. Inflow clepsydra with four compensating water tanks. Associated with the name of Lü Tshai (*d.* A.D. 665), this illustration is from a copy of the *Shih Lin Kuang Chi* encyclopaedia printed in A.D. 1478 and in the Cambridge University Library. The text below refers to the legendary invention of the clepsydra by Huang Ti and other matters concerned with time-keeping.

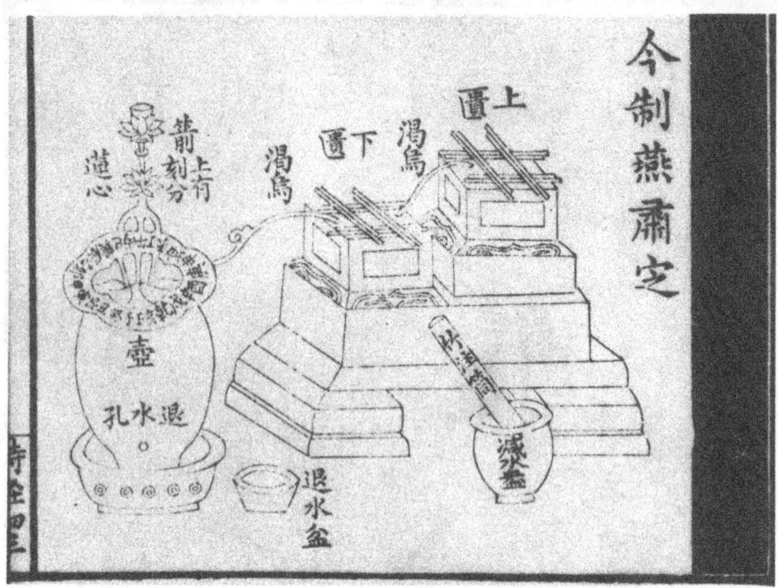

Fig. 95. The overflow type of inflow clepsydra associated with the name of Yen Su (A.D. 1030); from a Chhing edition (A.D. 1740) of the *Liu Ching Thu.*

divided into twelve equal double-hours and, by a parallel system, into one hundred 'quarters'. The use of two receivers, one for the day and the other for the night, was an unnecessary complication; the varying lengths of light and darkness had been dealt with by the Babylonians and Egyptians by the way they marked their outflow clocks. What the Chinese did was to have a series of indicator rods differing by one quarter of an hour, and according to the season these were progressively substituted for one another. From A.D. 102 the change was linked with fortnights determined from the calendar, and this rule endured for more than a thousand years.

In the Chin dynasty we hear of clepsydras with a series of vessels and what seem to be siphons. Fig. 94, taken from the Sung encyclopaedia, *Shih Lin Kuang Chi*, shows four vessels, the floating figure with the indicator rod being a Buddhist monk. In the Sung period itself the constant-level overflow tank (Type B) prevailed. It was, after all, a later development, although the oldest description of this kind of clepsydra comes from the *Lou Kho Fa* (Clepsydra Technique) of Yin Khuei, about A.D. 540. Here one tank only intervened between the reservoir and the receiver; it contained a partition where the water 'hesitated'. Water issuing from the reservoir was filtered, but the arrangements in this clepsydra may date from somewhat earlier, perhaps from the late fifth century.

Early in the Northern Sung dynasty, the overflow principle was adopted in the 'lotus clepsydra' of Yen Su, so called because the top of the receiver was shaped in the form of this Buddhist emblem. He first presented it in A.D. 1030, and after many tests it was officially adopted six years later. Indeed it was taken as a prototype for centuries afterwards (Fig. 95) and the term 'lotus-clock' became proverbial. In the course of time, the two types A and B became (perhaps somewhat illogically) combined into one, so that the system consisted of a reservoir, a compensating tank, an overflow tank, an inflow receiver with a rod-float, and an overflow receiver (fig. 96).

Something must finally be said about types C and D which measured time not by the rise of an indicator rod on a float, but by the actual weighing of the vessel. The simplest plan – attaching the vessel to a balance, usually a steelyard – started either in the Han or Three Kingdoms (San Kuo) periods. The oldest text we have is by the Taoist Li Lan; it dates from about A.D. 450, and he speaks as if the device were already well known. As reproduced in the *Chhu Hsüeh Chi* (Entry into Learning) encyclopaedia of A.D. 700 it runs:

> Water is placed in a vessel (whence it issues by) a siphon of bronze shaped like a curved hook. Thus the water is conducted to a silver dragon's mouth which delivers it to the balance vessel. One *sheng* of water dripping out weighs one *chin*, and the time which was elapsed is one quarter (-hour).

This is the small steelyard clepsydra. Li Lan then goes on to say:

> (It is by means of) jade vessels, jade pipes and liquid pearls (that the) rapid stopwatch portable clepsydra (is constructed). 'Liquid pearls' is only another name for mercury.

Thus he indicates that the short-interval stopwatch clepsydra which could easily be carried about for use in the field needed chemically resistant parts to allow of the use of mercury, but worked on the same principle as all other ordinary steelyard clepsydras. It is hard to say how widely these forms were used, but by the Sui period (A.D. 605 to 616) a carriage bearing these portable time-keepers was considered an indispensable part of any imperial procession.

The clepsydras of type D were more complicated. They seem to have orininated as the invention of two eminent Sui technicians, Kêng Hsün and Yüwên Khai. The new departure made at this time, which produced a large type of palace and public clock, constantly in use until the fourteenth century, can best be appreciated from the diagram of it in Fig. 93.

Combustion clocks and the 'Equation of Time'. There has been much to say about sundials and clocks, but as we have mentioned and most people

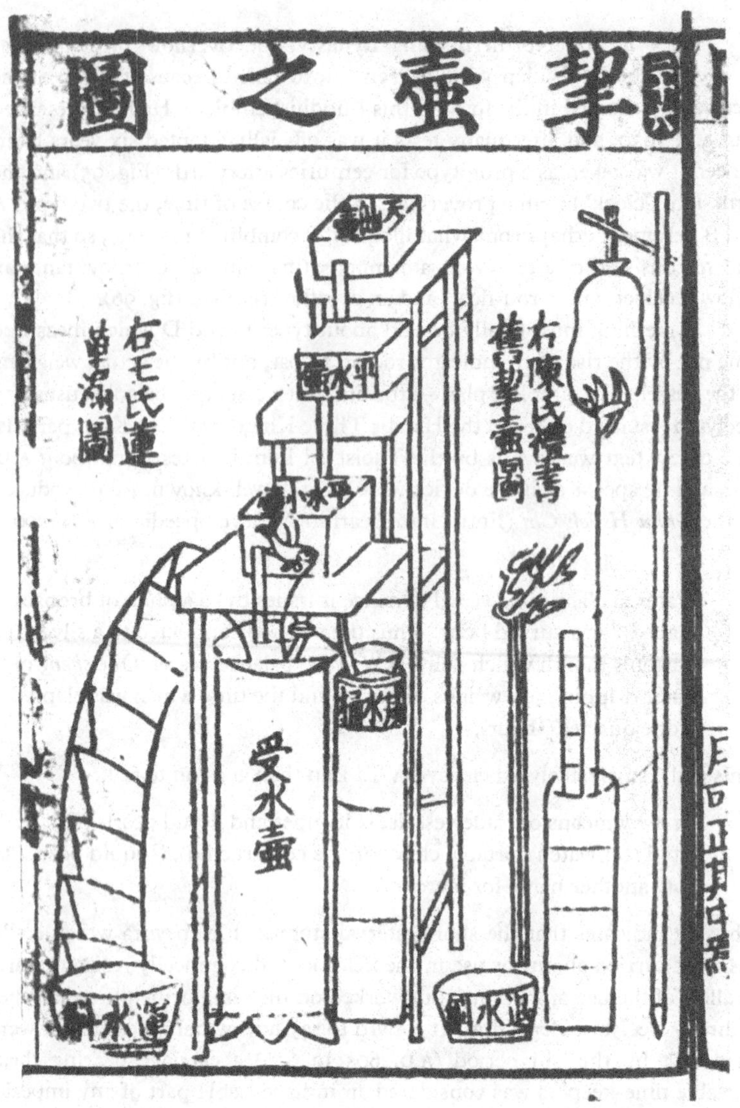

Fig. 96. The oldest printed picture of a clepsydra; an illustration from a Sung edition of Yang Chia's *Liu Ching Thu* (A.D. 1155). It shows the combined multiple tank and overflow type described by Wang Phu in his *Kuan Shu Kho Lou Thu* of *c.* 1135, which incorporated an automatic cut-off device acting when the indicator rod rose to its fullest extent. On the right of the picture Yang Chia showed the oldest and simplest type of inflow clepsydra to illustrate the description in the *Chou Li*.

Fig. 97. An undated metal incense clock; the burning point of the tindery powder winds its way through the strokes of stylised characters. The maze of the tray at the back on the right is in the form of the word *shou* (longevity). *Shuang hsi* (double happiness) can be made out on the cover in the right foreground. (Photograph, Science Museum, London.)

know, the time which they keep is not always the same. Since the rate of apparent motion of the sun across the sky varies with the season, there is a difference between this apparent solar time indicated by a sundial, and the regular average passage of time indicated by a clock; this difference is known as the Equation of Time. Apparent solar time can lead clock time (mean time) by as much as $14\frac{1}{2}$ minutes (this occurs in February) and mean time can lead apparent solar time by up to $16\frac{1}{2}$ minutes (in November). A discrepancy like this – about one Chinese quarter – might have been more easily detectable by a clepsydra than by the mechanical clocks of eleventh-century China or fourteenth-century Europe.

Though the future lay with mechanical clocks, there may have been other methods of time measurement which might, under certain circumstances, surpass water-clocks in accuracy. Hsüeh Chi-Hsüan, a Sung scholar of the mid-twelfth century A.D., said that apart from the water-clock and the sundial there was also the 'incense-stick' method of measuring time (Fig. 97). It would seem very improbable that these were, or ever could have been, made so as to burn with any high degree of regularity. But they have always been widely used in China, and made a great impression on one of the seventeenth-century Jesuits, Gabriel Magalhaens, who wrote:

> The Chinese have also found out, for the regulating and dividing the Quarters of the Night, an Invention becoming the wonderful Industry of that Nation. They beat to Powder a certain Wood, after they have peel'd and rasp'd it, of which they make into a kind of Past, which they rowl into Ropes and Pastils of several shapes. Some they make of

> more costly Materials, as Saunders, Eagle, and other odiferous Woods, about a fingers length, which the wealthy sort, and the Men of Learning, burn in their Chambers ... They make these Ropes of powder's Wood of an equal Circumference, by the means of a Mould made on purpose. Then they wind them round at the bottom, lessening the circle till they come to be of a Conick figure, which enlarges itself at every Turn, to one, two and three handsbreadths in Diameter, and sometimes more; and this lasts one, two and three days together, according to the bigness they allow it ... These Weeks (wicks) resemble a Fisher's Net, or a String wound about a Cone; which they hang up by the Middle, and light the lower end, from which the Fire winds slowly and insensibly, according to the windings of the string of powder'd Wood, upon which there are generally five marks to distinguish the five parts of the Watch or Night. Which manner of measuring Time is so just and certain, that you shall never observe any considerable Mistake. The Learned Men, Travellers, and all Persons that would rise at a precise hour about Business, hang a little weight at the Mark, which shews the Hour when they design to rise; which when the Fire is come to that point, certainly falls into a Copper Bason, that is plac'd underneath, and wakes them with the noise of the fall...

It may be asked whether the sand hour-glass was known in medieval China, and certain phrases of the eleventh-century poet Su Tung-Pho have been thought to imply its use in the Sung. More probably they refer to clepsydras, though we do know that from the beginning of the Ming sand replaced water in certain types of mechanical clocks with driving-wheels.

None of these methods could have demonstrated the Equation of Time. Yet conceivably a lamp in a secluded place, if fed with oil of regular composition and quality, and if well designed, might act as a time-keeper sufficiently good to detect the sundial and clock discrepancy. For this reason there is much interest in something which Yang Yü tells us in his *Shan Chü Hsin Hua* of A.D. 1360, about the 'perpetual lights' which were kept burning in a certain temple.

> Fan Shun-Chhen was a Khaifêng man,... particularly skilled in astronomy. In the Chih-Chun reign-period (1330 to 1333) he was head of the Yung-Fu Buildings Upkeep Office. He once told me that the perpetual lights in the prayer-halls had each a reservoir which used twenty-seven *ko* of oil annually. During the Chih-Yuan reign-period (1264 to 1294) the issue had originally been fixed at thirteen *chin* to the *ko*, so that the total amount was 351 *chin*. In the course of

a year, it was found, when a test was made, that there was an excess of fifty-two *chin*. So there was clearly a discrepancy between this and the time as measured by the sundial...

If this story is to the point it must be garbled, for the discrepancy cannot be seen on a yearly basis; but the fact that Yang Yü says distinctly that a difference between lamp-clock time and sundial time was noticed, suggests at least that Fan Shun-Chhen had been keeping careful comparative records for certain periods within the year.

The sighting-tube and the circumpolar constellation template

Leaving gnomons and time-keepers, we now approach the instruments which the Chinese astronomers used for direct observations of the heavens. We have already met with the sighting-tube as a means of observing the north celestial pole (page 115) and there is no doubt that such tubes (not, of course, equipped with lenses) were commonly employed in ancient Chinese astronomy, and were also used by surveyors, as shown in the Sung *Ying Tsao Fa Shih* (Treatise on Architectural Methods) of A.D. 1103 (Fig. 98). The old Babylonian astronomers seem to have used sighting-tubes which could certainly be effective in rendering dim stars more readily visible by cutting out extraneous light. For instance, while no stars dimmer than sixth magnitude are ordinarily visible with the unaided eye, stars of the eighth magnitude (i.e. some six times dimmer) may be seen at night through a 6-mm hole in a black screen placed some $4\frac{1}{2}$ m from the eye; whether ancient sighting-tubes were narrow enough to have so pronounced an effect is uncertain, but they probably made for some improvement. But although the sighting-tube was thus known in European as well as in Asian culture – it is to be seen illustrated in Western manuscripts (Fig. 99) – only in China was it standard practice to incorporate it into armillary spheres (page 164), and much the most important ancient text on the sighting-tube is contained in the *Shu Ching* (Historical Classic). Its date is very uncertain, but might be within a couple of centuries more or less of the sixth century B.C.

For reasons which will become clear shortly, we turn now to consider two of the most renowned objects of ancient Chinese jade work (of the Shang and Chou periods) – the 'jade images and symbols of the deities of Heaven and Earth'. The 'jade images' are flat discs with a large central hole known as *pi*; the 'symbols', known as *tshung*, are tubes or hollow cylinders whose outsides are square except for a small length of round tube which projects at each end, and are carved out of the solid. Fig. 100 illustrates these two ritual objects, which were doubtless carried in the imperial temple ceremonies. For centuries speculation has centred on their significance. The

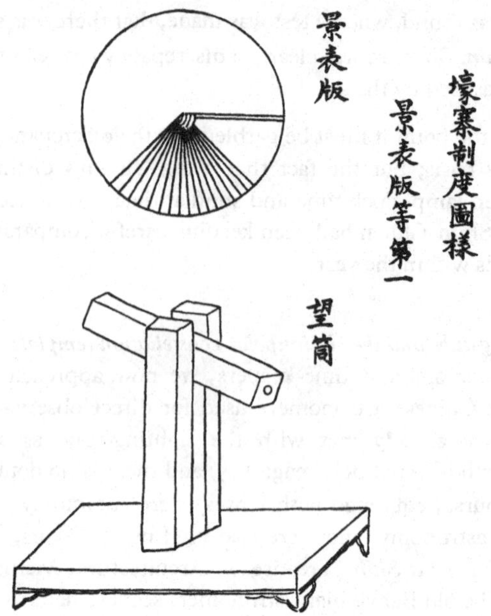

Fig. 98. Sighting-tube and quadrant from the *Ying Tsao Fa Shih* of A.D. 1103.

tshung was used by attendants of the empress, and has possible status as a womb-symbol. The *pi* had male status, and examples exist showing it with one, two or four projections shaped in phallic symbolism or as square 'handles'; it appears to have been used in sacrifices to heaven, the sun, the moon and the stars. Yet there is more to them than this.

For there is another kind of *pi*. In a good many examples, the outer edge is very curiously carved (one might almost say graduated), being divided into three sections of equal length, and continuing with a series of teeth of variable shape until a smooth edge intervenes before the next set of graduations. Furthermore, on one side these discs bear two cross-lines almost at right-angles which because of their constancy and regularity, cannot be dismissed as chance marks of the jade-cutter's saw. The graduations and lines at once become explicable when it is realised that this type of *pi* – known as a *hsüan-chi* – is an astronomical instrument. Fig. 101 shows its use: when the stars *α* and *δ* Ursae Majoris are made to occupy one of the three chief indentations, the present Pole Star, *α* Ursae Minoris, will occupy the second, and the stars of the Tung Fang constellation (stars in our Draco and Cepheus constellations) will fit tolerably well along the third set of teeth. About 600 B.C. this would mean that the centre of the *hsüan chi* would

Fig. 99. A sighting-tube in use by a medieval Western astronomer; from a tenth-century codex of St Gall.

lie on the north celestial pole, while β Ursae Minoris would be seen revolving in the field of view. If now the *tshung* is fitted into the *pi* so that one of its flat edges follows the double line engraved on the back of the *hsüan-chi*, there will be four times in the year, for a given hour, at which the flat surface of the sighting-tube is either parallel to the horizon or at right-angles to it.

What then was the meaning of the single line at right-angles to the double one? Probably it represented the direction of the solstice point, for it must be remembered that about 1250 B.C. in the Shang, *a* Ursae Majoris had a right ascension of 90° (equal to the right ascension of the solstice), instead of 163°, as at the present time. Indeed it should be possible to date

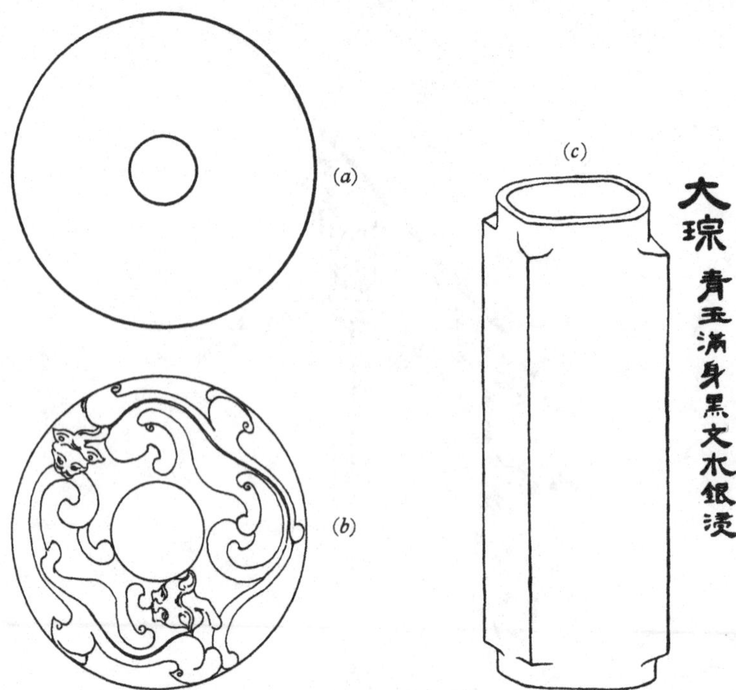

Fig. 100. The ancient ritual objects of jade, *pi* and *tshung*. (*a*) is a plain *pi*; (*b*) is a *pi* ornamented with dragons; (*c*) is a *tshung*. After Laufer, *Chinese Pottery* . . . and Michel.

such *pis* by noting the angle the single line makes with α Ursae Majoris, and two notable examples appear, therefore, to date from about 600 B.C. and 1000 B.C.

The question arises as to the value of the instrument. What could it do which simple inspection of the stars could not? One answer is that it was primarily used for determining the position of the true celestial pole and hence the direction of the solstice. The *Chou Pei Suan Ching* (Arithmetical Classic of the Gnomon and Circular Paths (of Heaven)) makes this clear:

> The true celestial pole is in the centre of the *hsüan-chi* . . . Wait and observe the large star at the middle of the north pole (area) . . . At the summer solstice, at midnight, the southern excursion of this is at its furthest point . . . At the winter solstice the northern excursion of this star is at its furthest point . . . On the day of the winter solstice, at the hour Yu (5–7 p.m.) the western excursion of this star is at its furthest point. On the same day, at the hour of Mao (5–7 a.m.) the eastern

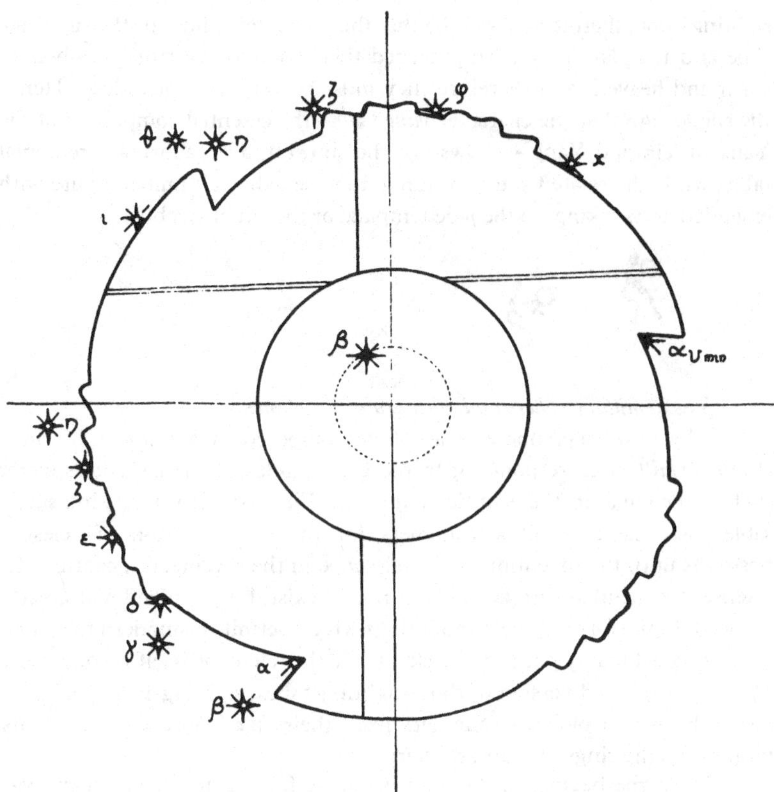

Fig. 101. Diagram to illustrate the use of the circumpolar constellation template. After Henri Michel.

> excursion of this star is at its furthest point. These are the four excursions of the north polar star (as seen through the) hsüan-chi.

There can be no doubt that what was here being observed was the rotation around the pole of β Ursae Minoris. It has also been suggested that the four stars which on many Chinese star-maps surround the polar point in a rough square (see, for instance, Fig. 77) were not meant to be real stars but the four positions of β Ursae Minoris. Apart from this, the *hsüan-chi* could have served as a very rudimentary kind of orienting instrument for the study of constellations closer to the celestial equator. The various teeth on its circumference would give a series of right ascensions, as we should say, and this would be quite in keeping with the principle of 'keying' circumpolar to equatorial constellations, so characteristic of Chinese astronomy.

If we consider the *tshung*, the hollow cylinder, as an idealised sighting-tube, then it may well be that the astronomical use of the sighting-tube and template may have preceded their adoption as ritual symbols of earth and heaven, and therefore they must be very ancient indeed. Hence the suggestion that the character *yang* (昜) – the essential component of the Yang of Yin and Yang – shows not the sun, but a *pi*, above a ceremonial table, while the related yang, meaning 'to raise', shows a human figure with extended arms taking up the jade template or the ritual symbol.

The armillary sphere and other major instruments

The armillary sphere is a sphere composed of a number of rings (Latin *armillae*) corresponding to the great circles of the celestial sphere, such as the equator, the ecliptic, and so on (Fig. 102). It was the indispensable instrument of all astronomers for the determination of celestial positions until the invention of the telescope in the seventeenth century. In a sense, the armillary rings still continue to exist, but modified and greatly reduced, in the form of those mountings which permit the modern telescope to be directed to any point in the sky; while the telescope itself is conversely a greatly magnified version of those sighting-tubes, sighting-marks or other means by which older astronomers fixed their star before reading off its position on the rings of their armillary sphere.

From the beginning the Chinese name for the armillary sphere was *hun i*, the celestial-sphere instrument. It was used in China from early times not only as an observing apparatus, but also as a kind of orrery or model to demonstrate the celestial sphere and so assist in making calendar computations. For this purpose it was moved by water-power, and the movements of the heavenly bodies were reproduced with varying degrees of complexity and accuracy.

It may be said at the outset that there was a fundamental difference between the armillary spheres of China and those of the Europeans and Arabs, owing to the fact that the Chinese remained persistently faithful to what have become the 'modern' celestial coordinates (right ascension and declination) while the West used the ecliptic (see page 68). This means that although most Chinese instruments had a ring for the ecliptic, and although they took much trouble to determine equivalent degrees on the ecliptic as well as on the celestial equator, the ecliptic ring was just as secondary to their main function as the ring representing the celestial equator was on European instruments. Later we shall see how this led the Chinese in the thirteenth century A.D. to the invention of an equatorial mounting, essentially identical with that used for most telescopes today.

Fig. 102. Equatorial armillary sphere in the Chinese tradition set up in Peking Observatory by Ignatius Kögler and his collaborators in A.D. 1744. (Photograph, Thomson, reproduced Bushell.)

The general development of armillary spheres. In all probability the most primitive form of armillary sphere was the simple single ring, with some kind of reference line or sights, which could be set up in the meridian or the plane of the equator as desired. Measurement one way gave the north polar distance (the Chinese form of declination), measurement the other way gave the *hsiu* (the Chinese form of right ascension). This was very likely all that Shih Shen and Kan Tê (fourth century B.C.) had at their disposal. From the middle of the first century B.C., however, developments occurred rapidly. Kêng Shou-Chhang introduced the first permanently fixed equatorial ring in 52 B.C., Fu An and Chia Khuei added the ecliptic ring in A.D. 84, and with Chang Hêng about A.D. 125 the sphere was complete with rings for the horizon and the meridian. Chang Hêng was also able to make a demonstrational type, with a model earth at the centre, and its rings driven by water-power, as well as the observational type, and be combined their use, as we shall see.

After this, spheres of both types continued to be made without much change for several centuries. But while it was convenient enough for

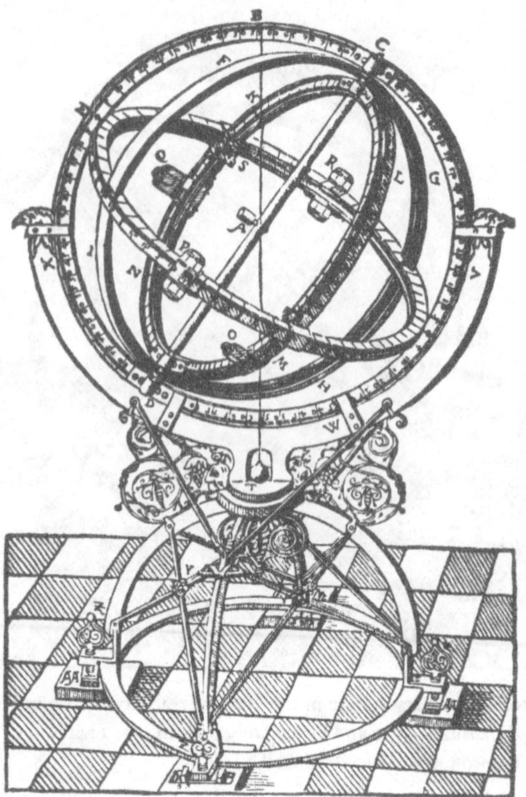

Fig. 103. Tycho Brahe's smaller equatorial armillary sphere (A.D. 1598).

studying annual motion to have the rings for the equator and the ecliptic permanently fixed together, it was inconvenient from the point of view of diurnal motion. In A.D. 323, therefore, Khung Thing so arranged his sphere that the ecliptic ring could be pegged on to the equator at any point desired. This apparatus was the forerunner of all the most important later spheres. Then, three centuries later, Li Shun-Fêng made a radical change; instead of building two nests of rings, he constructed three. The innermost was the declination ring complete with sighting-tube, the new intermediate position had rings for the ecliptic and equator, together with a third ring for the moon's path (though this was dropped after 1050), while the outer rings included a ring for the equator and the rest of the usual complement found in the pre-Thang spheres. Li Shun-Fêng worked in the early part of the seventh century, when Buddhist influence was strong and Greek astronomical ideas were finding their way into China through Indian channels. He proposed, but perhaps the monk I-Hsing was the first to execute,

Fig. 104. Equatorial armillary sphere of Kuo Shou-Ching (*c.* A.D. 1276), made for the latitude of Phin-Yang in Shansi, but later kept at Peking, where this photograph was taken, and at Nanking. The instrument shown here may be one of the exact replicas made by Huangfu Chung-Ho in A.D. 1437.

the plan of having the sighting-tube mounted on the ecliptic ring in order that observations of celestial latitudes (so important in Greek astronomy) could more easily be made. But the original attitudes of Chinese astronomy were never easily modified, and with the exception of an instrument built in 1050, no more mountings of sighting-tubes on the ecliptic were employed until the time of the Jesuits.

It is interesting to compare the final form of the Chinese equatorial armillary sphere with those which issued from the Western tradition. Tycho Brahe's smaller armillary of 1598 (Fig. 103) differed mainly from those of Kuo Shou-Ching (about 1276) (Fig. 104) and that of Su Sung (1088) chiefly in being simpler, for it dispensed with the equator and horizon rings in the outer nest, and with the ecliptic in the inner. This was undoubtedly because Tycho's equipment included other ecliptic armillary spheres, and because he was not trying to make a 'compendium instrument' in the Chinese style.

Perhaps we should consider the Sung (or rather specifically the Northern Sung) as the high point of Chinese armillary sphere making, for we often hear in later times of the four famous armillary spheres then constructed (between 995 and 1092). But the fall of the capital to the Chin

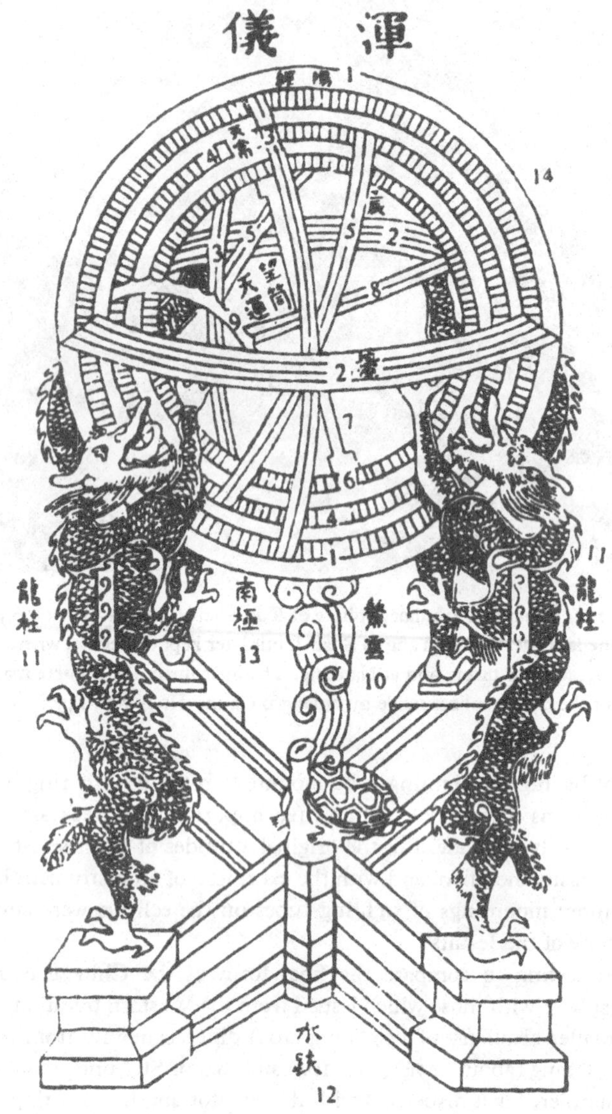

Fig. 105. Su Sung's armillary sphere (*hun i*) of A.D. 1090, described in the *Hsin I Hsiang Fa Yao* (redrawn and labelled by H. Maspero); the first to be provided with a clock drive.

Outer Nest:

1 meridian circle.

2 horizon circle.

3 outer equator circle.

Middle Nest:
4 circle through celestial pole and solstitial points.
5 ecliptic circle.
9 diurnal motion gear-ring, connecting with the power-drive.

Inner Nest:
6 polar-mounted declination ring or hour-angle circle, with
7 sighting-tube attached to it and strengthened by a
8 diametral brace.

Other Parts:
10 vertical column, concealing the transmission shaft.
11 supporting columns in the form of dragons.
12 cross-piece of the base, incorporating water-levels.
13 south polar pivot.
14 north polar pivot.

Tartars in 1126 proved a heavy blow to astronomy, and in spite of the efforts of devoted men, the Southern Sung could never equal the instruments which had been lost. Although the armillary sphere of Su Sung, the last of the four, was (as we shall shortly see) unique in an important respect, in its general construction it well typifies the 'great tradition' of Chinese armillary art. This armillary is illustrated in Fig. 105 where its parts are enumerated, although their proper delineation was beyond the eleventh-century artist who prepared the illustration for Su Sung's *Hsin I Hsiang Fa Yao* (New Design for an Armillary Clock) of 1094.

Armillary instruments in and before the Han. The greatest uncertainty, of course, concerns the earliest beginnings. The chief text comprises a few words in Yang Hsiung's *Fa Yen* (Admonitory Sayings) of A.D. 5. To someone who asked him about the armillary sphere, he replied that Lohsia Hung had constructed it, Hsienyü Wang-Jen had made calculations for it, and Kêng Shou-Chhang had checked it with actual observations. We seem therefore compelled to the opinion that Lohsia Hung did in fact set up armillary apparatus (about 104 B.C.). A description of such instruments was given by Tshai Yung in a memorial already referred to (page 81), which he presented to the emperor about A.D. 180.

> The bronze instruments used by the imperial astronomers are based upon (the Hung Thien theory). A sphere ... represents the shape of the heavens and the earth. By means of it the ecliptic (graduations) are checked. The rising and setting of the heavenly bodies are observed, the movements of the sun and moon are followed, and the paths of the five planets traced. (Such instruments have been found to yield) wonderful results...

The oldest text which we have on the armillary sphere is the remaining fragment (about A.D. 125) of Chang Hêng's *Hun I* (On the Armillary Sphere). It is significant, and opens as follows:

> The equatorial ring goes round the belly of the armillary sphere $91\frac{5}{19}°$ (Chinese degrees, i.e. 90° Western degrees) from the pole. The circle of the ecliptic also goes round the belly of the instrument at an angle of 24° (Chinese degrees) with the equator. Thus at the summer solstice the ecliptic is 67° and a fraction away from the pole, while at the winter solstice it is 115° and a fraction away.
>
> Hence (the points) where the ecliptic and the equator intersect should give the north polar distances of the spring and summer equinoxes. But now (it has been recorded that) the spring equinox is $90\frac{1}{4}°$ away from the pole, and the autumn equinox is $92\frac{1}{4}°$ away. The former figure is adopted only because it agrees with the (results obtained by the) method of measuring solstitial sun shadows...

Both these passages give us a glimpse of bronze armillary spheres of the Han, but the second is interesting for another reason. We have already seen (page 81) that about A.D. 320 Yü Hsi discovered the precession of the equinoxes, yet here in Chang Hêng's text we see that Chinese astronomers were already becoming uneasy about such matters at a much earlier date. An apparent inequality between the north polar distances of the equinox points could only have come about by the use of traditional values for the shadow lengths cast by the sun at the solstices, and continued reliance on a traditional pole-star at a time when a new determination ought to have been made of shadow lengths, and a change of pole-star recognised.

Chang Hêng's text also goes on to describe graduation of the ecliptic by direct measurement from the equator ring of the armillary. He describes a practical method:

> ... One therefore takes a small armillary sphere and graduates both equator and ecliptic in $365\frac{1}{4}°$ starting from the winter solstice point ... Next one fixes a thin ruler of flexible bamboo to each of the two poles, thus spanning exactly half the circumference of the sphere ... One then moves the ruler round, starting from the winter solstice point, and noting how many degrees less or more on the ecliptic correspond to those on the equator ... Moreover, by counting the number of degrees marked on the (semicircular) bamboo ruler, one obtains the north polar distance, at any point...

This procedure remained in use for fifteen hundred years, until the coming of Arabic and Western influences in the Yuan and the Ming. It was essentially a means of expressing movements along the ecliptic in equatorial coordinates; traditional Chinese astronomy never extended the degrees of the ecliptic to form segments of celestial longitude radiating from the pole of the ecliptic as they thought of the *hsiu* as segments of what we should call right ascension.

In discussing such instruments we get very few glimpses of the human side of the story. A particularly interesting one comes from the writings of Huan Than (40 B.C. to A.D. 30);

> Yang Hsiung was devoted to astronomy ... He made an armillary sphere himself. An old artisan once said to him: 'When I was young I was able to make such things following the method of divisions (graduations) to scale (literally 'feet and inches') without really understanding their meaning. But afterwards I understood more and more. Now I am seventy years old, and feel that I am only just beginning to understand it all, and yet soon I must die. I have a son also, who likes to learn how to make (these instruments); he will repeat the years of my experience, and some day I suppose he will in his turn understand, but by that time he too will be ready to die.' How sad, and at the same time how comical, were his words!

The text teaches us several things. In the time of Yang Hsiung (53 B.C. to A.D. 18) there was clearly a tradition of instrument-making; a fact which strengthens the case for the use of armillary rings by Lohsia Hung. Moreover, one can sense the lack of technical education, and the paucity of those who could explain the celestial sphere; perhaps this is to be expected, because it was regarded as a State secret, or at least 'restricted information'.

The invention of the clock-drive. The first armillary sphere said to have been slowly rotated by means of a water-wheel powered by a constant head of pressure from water in a clepsydra was one of those built by Chang Hêng about A.D. 132. The seventh-century *Chin Shu* (History of the Chin Dynasty) says:

> ... Chang Heng made his bronze armillary sphere and set it up in a closed chamber, where it rotated by (the force of) flowing water. Then the order having been given for the doors to be shut, the observer in charge of it would call out to the watcher on the platform, saying the sphere showed that such and such a star was rising, or another star just culminating, or yet another star just setting. Everything was found to correspond (with the phenomena) like (the two halves of) a tally.

The general picture is thus clear. The procedure was for the observers to compare the indications of the mechanised sphere in a closed laboratory with the celestial phenomena actually occurring overhead.

All this was long before the invention of the first clock escapement mechanism – the essence of all mechanical time-keepers – by I-Hsing and Liang Ling-Tsan about A.D. 723 (see the next volume of this abridgement), after which armillary spheres and celestial globes for demonstration continued to revolve by means of a power-drive down to the coming of the Jesuits. From the beginning of the eighth century A.D. these instruments were thus nothing more nor less than great astronomical clocks, antedating the first mechanical clocks of the fourteenth century in Europe. Before the eighth century the mechanism would not have been driven at a precise enough rate for it to be more than a device for demonstration and computation, designed to show a rough approximation to time-keeping. In fact, throughout this long period of Chinese developments the form was often that of an observational astronomical instrument, but the essence was that of a clock. For the aim of the astronomer-engineers, from the time of Chang Hêng onwards, was to persuade a driving-wheel to rotate sufficiently slowly to keep pace with the apparent daily rotation of the heavens. After the beginning of the eighth century that problem was substantially solved. How Chang Hêng did it is still unknown.

The question naturally presents itself – why did Chang Hêng and his successors want to do this? The *Shang Shu Wei Khao Ling Yao* (Apocryphal Treatise on the *Historical Classic*; the Investigation of the Mysterious Brightnesses), which dates from about the time of Chang Hêng, has a relevant passage of great interest:

> If the (computational armillary) sphere indicates a meridian transit when the star (in question) has not yet made it, (the sun's apparent position being correctly indicated), this is called 'hurrying'. When 'hurrying' occurs, the sun oversteps his degrees, and the moon does not attain the *hsiu* in which it should be. If a star makes its meridian transit when the (computational armillary) sphere has not yet reached that point (the sun's apparent position being correctly indicated), this is called 'dawdling'. When 'dawdling' occurs, the sun does not reach the degree which it ought to have reached, and the moon goes beyond its proper place into the next *hsiu*. But if the stars make their meridian transits at the same moment as the sphere, this is called 'harmony'. Then the wind and rain will come at their proper time, plants and herbs luxuriate, the five cereals give good harvest and all things flourish.

This gives us the explanation of the two spheres, the computational one in a closed room, the observational one on the terrace outside. Han astronomers, like their predecessors and their successors, were deeply concerned with all divergences or discrepancies between the indicated positions of the stars on the one hand and those of the sun and moon on the other. If, as is made highly probable from the many texts available, the system was to have small objects representing sun, moon and planets attached somehow to the sphere or globe, yet freely movable across it, then the computer within would adjust their positions in accordance with the predictions of the calendar. If the star transits and other movements which he announced were not in accord with what the observer outside was seeing, then due corrections could be made in the calendar computations.

In his memorial to the emperor in 1092, Su Sung quoted the above passage from the Han apocrypha, adding:

> From this we may conclude that those who make astronomical observations with instruments are not only organising a correct calendar so that good government may be carried on, but also (in a sense) predicting the good and bad fortune (of the country) and studying (reasons for) the resulting gains and losses.

This was an interesting way of rationalising the power to foretell the future naturally attributed by the common people to the work of all medieval astronomers. We do in a way foretell what is to happen, says Su Sung, because we know that if the calendar is always well adjusted the work of the farmers will keep in perfect time with the seasons, and so (apart from special calamities) bring the best harvest. Nevertheless, it would probably be unwise to deny all astrological significance to the spheres and globes of pre-Thang China, rotating more or less unevenly, without escapements, through nights of storm and days of rain or cloud.

So much for clocks and pre-clocks. But there is another significance in these texts which Sinologists have previously missed. The armillary sphere, driven by water-power and set up on top of Su Sung's clock tower in Khaifêng in A.D. 1090, had a sighting-tube. To an astronomer such provision of a water-powered drive to an observational instrument can mean only one thing, namely that the sighting-tube was being moved round automatically to ensure that it was kept pointed to a particular celestial body, tracking it as it slowly passed across the sky because of the rotation of the earth on its axis. The rate at which the sighting-tube (and armillary) would have to move would be no more than 15° per hour, and this would have meant about 163 ticks of Su Sung's clock. Since the bronze instruments must have weighed 20 tonnes or more, the mechanism would have been intended, in part at least, as a kind of coarse adjustment. The fine adjustment could then be made by levering with wooden bars.

There is, all the same, reason for envisaging a further function of the apparatus. Su Sung himself states that a celestial body could be held automatically in the field of the sighting-tube. He speaks indeed of maintaining the sun in this position by day. Presumably this was intended as a check upon the time-keeping of the whole clock. The movement would simply have brought the sun's rays again directly down the sighting-tube each time the machinery advanced by one tick.

In the West in 1670 Robert Hooke proposed an automatically adjusted clock-driven telescope to follow celestial bodies, although for technical reasons this was not put into practice until 1824 by Joseph Fraunhofer. We may conclude, then, that the mechanisation of the armillary spheres of medieval China, though generally applied to instruments used for computing or demonstration, anticipated the invention of the clock drive of the modern telescope. Su Sung, in coupling his clockwork to an observational sphere, fully accomplished this, thereby anticipating Hooke by six centuries and Fraunhofer by seven and a half.

The invention of the equatorial mounting. Of the armillary spheres of the Southern Sung little is known, but some of the instruments of Kuo Shou-Ching of the thirteenth century A.D. are still in existence, kept at the Purple Mountain Observatory of the Academia Sinica, north-east of Nanking. Made under the Yuan dynasty, they were still in use at the time of arrival of the Jesuits in 1600. Matteo Ricci wrote about them, and his description of one particular instrument is of significance here:

> The fourth and last instrument, and the largest of all, was one consisting of, as it were, 3 or 4 huge astrolabes in juxtaposition; ... Of these astrolabes, one having a tilted position in the direction of the south represented the equator; ; a second which stood crosswise on the first, in a north and south plane, the Father took for a meridian, but it could be turned round on its axis; a third stood in the meridian plane with its axis perpendicular, and seemed to represent a vertical circle, but this also could be turned round to show any vertical whatever. Moreover, all these were graduated and the degrees marked by prominent (metal) studs, so that in the night the graduation could be read by touch without any light. This whole compound astrolabe instrument was erected on a marble platform with channels round it for levelling.

The authentic Chinese texts from which we gain information about the re-equipment of the imperial observatory between 1276 and 1279 include, of course, the *Yuan Shih* (History of the Yuan Dynasty). There this instrument is listed simply as *chien i* or 'simplified instrument', illustrated in Figs. 107, 108 and 109. We can regard it as a simplified version of the medieval instrument known as the torquetum, which consisted of a series of discs and circles, not arranged concentrically as in an armillary sphere. Such an

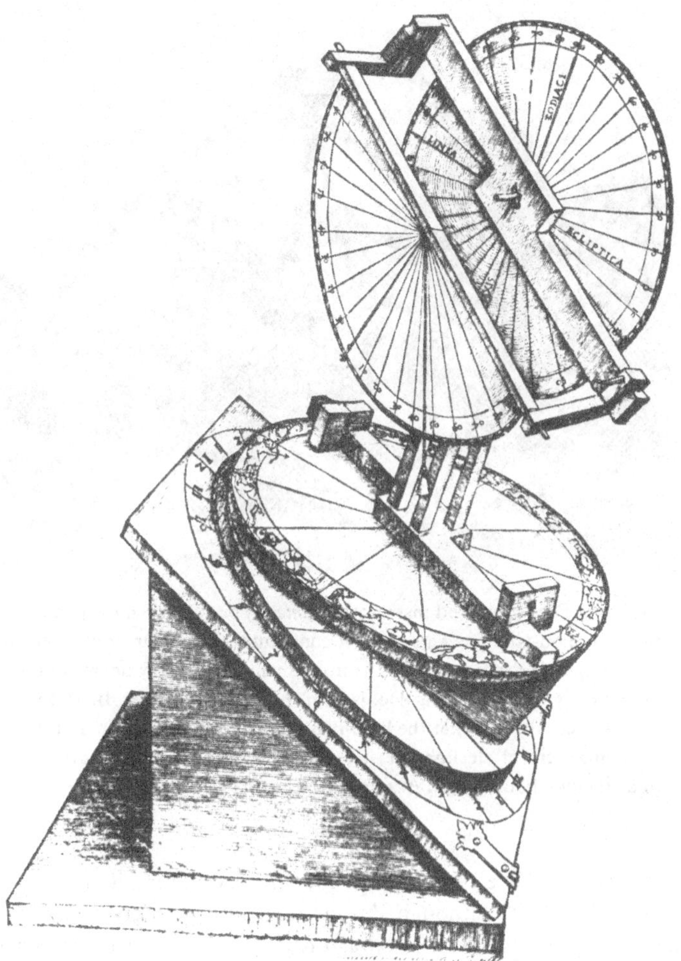

Fig. 106. The torquetum of Petrus Apianus (A.D. 1540).

instrument can be seen in Fig. 106, and it was probably used mainly for computing as it permitted the direct conversion of ecliptic coordinates to equatorial ones, and vice versa. It seems to have been an Arab invention, probably due to the Spanish Muslim Jābir ibn Aflah (born *c.* 1130).

How the Arabic invention arrived in China is not certain, but it seems to have been by the hand of an astronomer from the Marāghah observatory in Persia – possibly Jamāl al Din ibn Muhammad al-Najjārī – who reached China in 1267. With him the astronomer brought seven instruments, all described in the *Yuan Shih*, of which only a terrestrial sphere and an

Fig. 107. The 'simplified instrument' (*chien i*) or equatorial torquetum devised by Kuo Shou-Ching *c*. A.D. 1270. It constitutes the precursor of all equatorial mountings of telescopes. Here the instrument, which may be one of the replicas of Huangfu Chung-Ho, is seen from the south-east. Its size may be gauged from the fact that the base-plate measures 5·5 × 3·8 m and the revolving (central) declination ring which carried the sighting-tube is 1·8 m in diameter. (Photograph, Academia Sinica.)

astrolabe seem to have been new to the Chinese. There is no record of a torquetum (Fig. 108) although the circumstantial evidence seems strong enough to make it likely that it was on this occasion that one arrived in China.

The great point of interest here is that Kuo Shou-Ching's 'simplified instrument', while a 'dissected' armillary sphere and recognisably related to the torquetum, is nevertheless a true equatorial (Fig. 106). Doubtless it was because the ecliptic components had been removed that the instrument received its name 'simplified'. This means that though Arabic influence may have been responsible for suggesting its construction, Kuo adapted it to the specific character of Chinese astronomy, namely equatorial coordinates. And in so doing, he fully anticipated the equatorial mounting so widely used for modern telescopes, because, with such a mounting, a rotation about only one axis (that parallel to the earth's polar axis) is needed to follow the

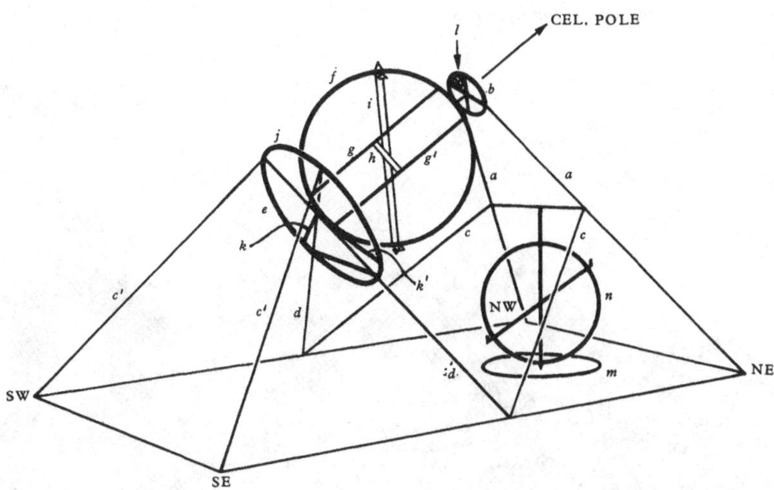

Fig. 108. Diagram of Kuo Shou-Ching's equatorial torquetum, seen from the south-east, for comparison with Fig. 107. *a*, *a*, North pole cloud frame standards; *b*, normal circle, diameter 71 cm (fixed); *c*, *c*, *c'*, *c'*, dragon pillars; *d*, *d*, south pole cloud frame standards; *e*, fixed diurnal circle, graduated in the 12 double-hours and the 100 quarters, each of the latter having 36 subdivisions. This circle has four roller-bearings in its northern face which allows the easy rotation of the equatorial circle (*j*) upon it. Diameter 1·93 m. *f*, Mobile declination ring or meridian circle, 1·83 m in diameter, graduated on both sides in degrees and minutes of arc, and carrying a sighting-tube or arm with open sights (alidade) (*i*) for determining the north polar distance; *g*, *g'*, stretchers; *h*, double brace to prevent deformation; *j*, mobile equatorial circle, 1·83 m in diameter, graduated in degrees and minutes of arc, and marked with the boundaries of the 28 *hsui*. It is strengthened by cross-stretchers like the declination ring. *k*, *k'*, Independently movable radial pointers with pointed ends. It is not clear whether these carried sighting-vanes. Their name 'boundary bars' (*chieh hêng*) indicates that they were used to mark off the boundaries of the *hsiu*. *l*, Pole-determining circle, of diameter equivalent to 6°, attached to the upper part of the normal circle (it can be seen only in Fig. 109). The circle has a cross-piece inside it, with a central hole, and seems to have been used for determining the moment of culmination of the Pole Star itself. Observation was made through a small hole in a bronze plate attached to the south pole cloud frame standards. The main polar axis through the centre of the diurnal and equatorial circles and that of the normal circle was also provided with holes which constituted a polar sighting-tube. *m*, Fixed terrestrial coordinate azimuth circle; *n*, revolving vertical circle with arms with open sights (alidades), for measurement of altitudes.

Fig. 109. Kuo Shou-Ching's equatorial torquetum seen from the south, so as to show the 'pole determining circle' at the top, with its cross-bars and central hole, attached to the 'normal circle'. (Photograph, Academia Sinica.)

curved path of a star across the sky (Fig. 110). Here again we see the faithfulness of the Chinese to what afterwards became the coordinates universally used, a faithfulness which also exhibits a simplicity due to practised skill in economy of effort.

The Chinese do not seem to have taken up the terrestrial globe at that time, nor do they appear to have adopted the astrolabe, of which not a single example is known from China. However, the Chinese did possess one piece of apparatus which in the West gave rise to the astrolabe, and this was the anaphoric clock, a form of clepsydra not uncommon in Hellenistic times. It consisted of a bronze dial made to rotate by the rise and fall of a float attached to its drum by a counterweight. A peg representing the sun was plugged in to an appropriate hole in the disc which rotated behind a network of wires representing the meridian, celestial equator, etc. This was just the reverse of the astrolabe where it is the stars that are shown on a metal fretwork, with the dial engraved with its various circles rotating beneath. There is evidence (to be discussed in the next volume of this abridgement) that the dial of the anaphoric clock, the first of all clock faces, was known and used in some Chinese clepsydras.

Fig. 110. A twentieth-century equatorial telescope mounting, the 100-inch (2·5-m) reflector at Mount Wilson, California. (Photograph, Hale Observatories.)

The celestial globe

The idea of representing the constellations and stars on the surface of a globe goes back rather far in Greek antiquity, and may just possibly have been derived from Babylonian practice. Although it has long been generally accepted that the classical technical term for a celestial globe was *hun hsiang*, the history of the thing itself has never been properly written. Since the star-position measurements of Shih Shen and Kan Tê were made during the fourth century B.C., there is no reason why 'images of the spherical heavens' in the form of solid globes should not have been made in the Chhin or early Han. This is the time from which we have the oldest literary references to the celestial sphere theory and to armillary rings, and the problem really depends on the interpretation of certain technical terms.

Traditionally the celestial globe (*hun hsiang*) has been contrasted with the observational armillary sphere (*hun i*) but account must be taken of three, not two, types of apparatus. Besides the two just mentioned there was also the armillary sphere used for demonstration, with or without a model earth supported on a pin at its centre, and with or without arrangements for rotating continuously the nest of rings. The conclusion seems to be that before the fourth century A.D., the term *hun hsiang* often, if not indeed always, meant the demonstrational armillary sphere. The existence of such instruments with earth models is proved by a statement of Wang Fan about A.D. 250 in which he explained his plan of removing the model from the centre of the sphere and replacing it by a horizontal box-top outside to symbolise the earth. It would seem that the demonstrational armillary sphere half sunk in a box-like casing persisted at least until the time of I-Hsing and Liang Ling-Tsan in the early eighth century. On the other hand the true solid celestial globe seems not to have originated before the time of Chhien Lo-Chih (A.D. 435), and it is quite significant that this occurred only after Chhen Cho had constructed in about A.D. 310 a standard series of star-maps based on the *Hsing Ching* catalogue which had been begun in the fourth century B.C.

These remarks may be illustrated by a quotation from the astronomical chapter of the *Sui Shu* (History of the Sui Dynasty) written about A.D. 654 by Li Shun-Fêng:

> *Uranographic Models:*
> The characteristic of uranographic models (models of the heavens) is that they have the round parts, but not the straight part.
>
> At the end of the Liang dynasty (*c.* A.D. 550) there was one in the Secret Treasury. It was made of wood, as round as a ball, several arm spans in circumference, and pivoted on the north and south poles, while round the body of it were shown the twenty-eight *hsiu*,

> as also the stars of (each of) the Three Masters (i.e. of Shih Shen, Kan Tê and Wu Hsien), the ecliptic, the equator, the Milky Way, etc. There was also an external horizontal circle surrounding it, at a height which could be adjusted to represent the earth. The southern extremity of the polar axis penetrated below this to represent the south (celestial) pole, while the northern extremity rose above it to represent the north (celestial) pole. When the globe rotated from east to west, the stars which made their meridian transits morning and evening correspond exactly with their degrees, and the equinoctial and solstitial points, as well as the fortnightly periods, all also checked – there was absolutely no difference from the heavens.
>
> This was not at all like the (observational) armillary sphere which must have a sighting-tube for measuring and computing the motions of the sun and moon and the positions of the stars in degrees.

This, together with further remarks and some words of Wang Fan, brings us to the following picture. When the first fully developed armillary spheres were made by Chang Hêng about A.D. 125 he made them in two forms, one for observation, the other for demonstration and computation. It does not seem likely that he placed an earth model at the centre. Nor did Liu Chih in 274. This appears to have been the contribution of Ko Hêng in the middle of the third century, and almost immediately afterwards Wang Fan adopted the alternative of sinking the sphere in a box, the top of which represented the earthly horizon.

Meanwhile, at the beginning of the fifth century A.D., the true solid celestial globe had started on its career. Chhien Lo-Chih made at least two instruments in 435, one of the traditional armillary type with an earth model, and one true celestial globe. The limiting factor was probably the preparation of standard star-maps based on the *Hsing Ching*. Chhen Ho, who had been Astronomer-Royal of the Wu State, compiled the first standard maps in 310, and we are expressly told that Chhien Lo-Chih's globe was based on these. After this there was a long interval, and the next great celestial globe of which we can be sure is that placed in his astronomical clock-tower by Su Sung in 1090 (Fig. 111), to be followed by the globe of Kuo Shou-Ching (1276). This tradition was inherited by the Jesuits. It is very regrettable, though, that none of the earlier medieval Chinese celestial globes has survived.

Compared with European developments the Chinese were in some respects tardy, in others advanced. For although the early fifth century A.D. was rather late for the first appearance of the solid celestial globe, the practice of setting a model earth at the centre of a demonstrational armillary sphere from at least the middle of the third century A.D. was extremely

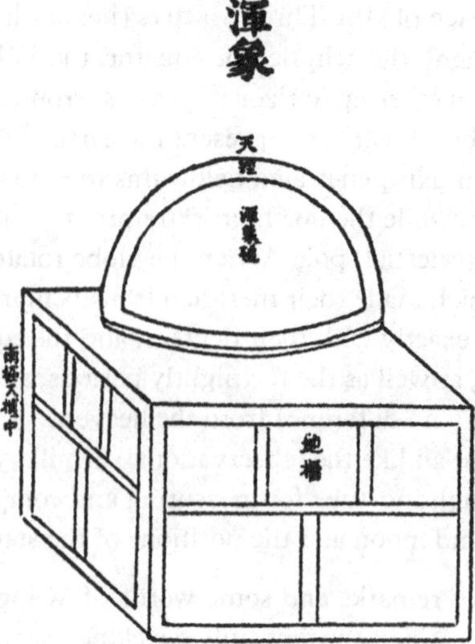

Fig. 111. Su Sung's celestial globe, sunk in a casing and rotated by a water-driven mechanical clock; from the *Hsin I Hsiang Fa Yao* of A.D. 1092.

enlightened. Europeans did not come to this until the end of the fifteenth century.

ASTRONOMY OF THE CALENDAR AND THE PLANETS

Although there is a large literature on the Chinese calendar, its emphasis has sometimes been more archaeological and historical than scientific. But, as we saw above (page 79), the long series of 'calendars' in Chinese history, a hundred or so during twenty-two centuries, was really a series of ephemerides or astronomical tables continually gaining in accuracy. A calendar is only a method of combining days into periods suitable for civil life and cultural or religious observances. Some of its elements are based on those of astronomical cycles which have obvious importance for man, such as the day, the month and the year; others are artificial, such as the week and the subdivisions of the day. The complexity of calendars is due simply to the fact that cycles based on the sun are not exactly divisible by those determined from the motion of the moon. The length of the synodic month (one lunation or new moon to new moon) is at present 29·53050 days, while that of the tropical year (vernal equinox to vernal equinox) is 365·24219 days, and one figure is not an exact multiple of the other. A calendar based on the moon, depending only on lunations, makes

Table 29. *Values for the length of the lunation*

	Lunation (days)
True value	29.530588
Deduced from the oracle-bones (13th century B.C.)	29·53
Yang Wei (Ching Chhu cal. A.D. 237)	29·530598
Ho Chhêng-Thien (Yuan Chia cal. A.D. 443)	29·530585
Tsu Chhung-Chih (Ta Ming cal. A.D. 463)	29·530591

the seasons unpredictable, while calendars based on the tropical year cannot predict full moons, the importance of which in ages before the introduction of widespread artificial lighting was considerable. Thus we have in Western civilisation, the movable date of Easter on the solar calendar, because it is a festival determined by a lunar calendar. The whole history of civil calendar-making, therefore, is that of successive attempts to reconcile the irreconcilable. We shall therefore treat the subject only briefly, the real scientific interest lying in the accuracy of the estimates of the periods of revolution of the planets which were formed in ancient and medieval times.

Perhaps the most important source for all Chinese calendrical questions is the *Ku Chin Lü Li Khao* (Investigation of the (Chinese) Calendars, New and Old) of Hsing Yün-Lu (about 1600); but many reliable works, both in Chinese and Japanese, have been written since then. In Table 28 (page 134) we gave some estimates of accuracy with which the Chinese ascertained the 'year fraction' of the tropical and sidereal years. From the fifth century A.D. both were rather precisely known. Similar information is available for the lunation and this is given above, Table 29.

Motions of the sun, moon and planets

Indigenous Chinese astronomy never attempted an advanced geometrical analysis of the moon's motions. One of the earliest mentions of them is in the *I Chou Shu* (Lost Books of the Chou) dating perhaps from some time before 300 B.C. The *Huai Nan Tzu* (Book of (the Prince of) Huai-Nan) of about 120 B.C. gives an account of them, stating that the eastward motion of the moon among the stars is 13 Chinese degrees, a figure long afterwards accepted (modern value 13° or 12·8 Chinese degrees). Shih Shen (fourth century B.C.) knew that the rate of motion varied, and that the moon's path diverged from the ecliptic north and south. The first mention of the 'Nine Roads of the Moon' – a succession of paths due to the gradual rotation of the moon's orbit about the earth (Fig. 112) – is in the *Hung Fan Wu Hsing Chuan* (Discourse on the Hung Fan chapter of the *Shu Ching* in relation to the Five Elements) of Liu Hsiang, written about 10 B.C., where a detailed description is given. The roads were traditionally assigned colours, but white predominated.

The unequal length of the seasons was a remarkably late discovery in China. Though the fact that the equinoxes are not placed at exactly equal intervals between the solstices may have been implied by the figures available to the Han astronomers (and had already led the Greek astronomer Hipparchus in the second century B.C. to his theory of the eccentric motion of the sun), it was not recognised in China until the time of Chang Tzu-Hsin, and his pupil Chang Mêng-Pin in the Northern Chhi dynasty, about A.D. 570. Perhaps this was due to the need for the development in China of suitable methods of algebra (Hipparchus of course used a geometrical approach based on the Greek geometrical model of planetary paths). As we saw earlier in the section on mathematics (page 50), the Thang mathematician Li Shun-Fêng knew of the Method of Finite Differences, an important algebraic procedure for finding the constant terms in an equation representing a process such as the sun's angular motion across the sky. As this method is believed to go back to Tsu Chheng-Chih and Liu Chhuo in the fifth and sixth centuries A.D., it would therefore have been available to the two Changs, but not before, even though its origins may lie way back in Babylonia. Parallels between Chinese and Babylonian ephemerides (tables of computed positions of celestial bodies on specific dates) for lunar and solar motion have indeed long been suspected.

Sexagenary cycles

The most ancient day-count in Chinese culture did not depend on the sun and moon at all. It was that sexagesimal cyclical system already referred to (volume 1 of this abridgement, page 198), a system counting in groups of 60; it contained a series of 12 characters (the so-called 'earthly branches'; *chih*) combined alternately with a series of 10 (the so-called 'heavenly stems'; *kan*) so as to make 60 combinations, at the end of which the cycle started all over again. These characters are among the commonest on the oracle-bones of the mid-second millennium B.C., and in the Shang period (*c.* 1520 to 1030 B.C.) they were used strictly as a day-count. The practice of using them for the years did not come in until the first century B.C., but from then on both uses continued down to modern times.

These 22 characters or cyclical signs run through all Chinese culture, being used in many different ways. Besides the counting of days and years they were employed in numeration systems for books, volumes and chap-

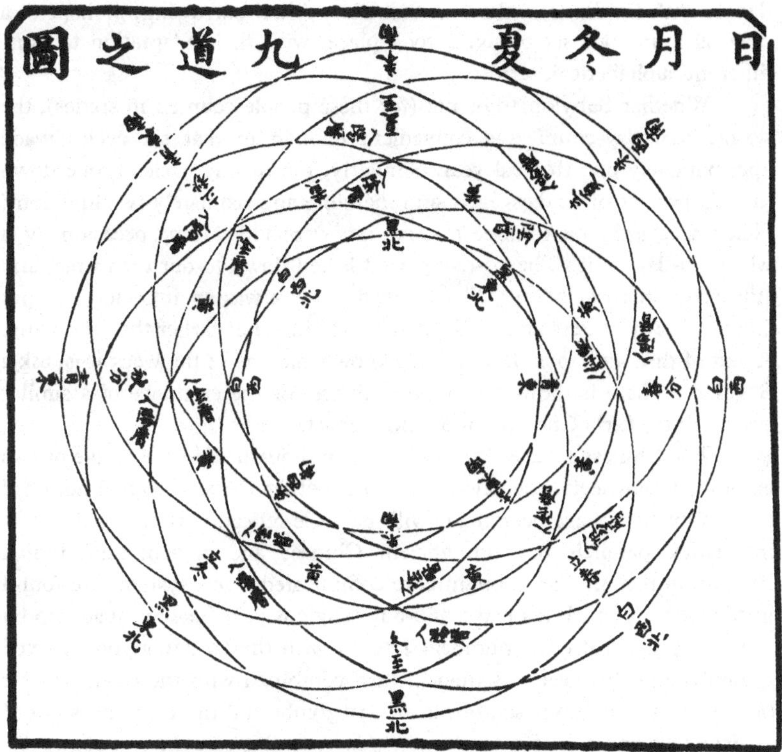

Fig. 112. The Nine Roads of the Moon, a late Chhing representation. The diagram shows the progressive forward motion of the major axis of the moon's orbit. Eight different positions of apogee (the moon's most distant orbital point from the earth) are represented by the outermost bulges on the diagram, being passed through in eight to nine years (actually 3232·575 days). The 'road' should of course be drawn as one single interweaving line, not as nine separate lines, but that was the old Han tradition.

ters, for keying explanations to diagrams, and many other serial uses; along with another set of a thousand characters traditionally attributed to Chou Hsing-Ssu (died A.D. 521). The original meanings and etymologies of the 22 ancient cyclical characters are still obscure, but the 10 *kan* at least may have been names for sacrifices appropriate to each day of the 10-day week. Recently there has been the fascinating suggestion that the 22 letters of the earliest Semitic alphabets (such as that of Ras Shamra), from which the Greek and Latin alphabets derived, show such close phonetic correspondences with the *kan* and the *chih* that trans-Asian borrowing, perhaps of a magical-numinous set of signs, took place, with their adaptation to quite different, alphabetical, scripts.

Whether Babylonian or not (for these people counted in sixties), the sexagesimal day-count was convenient enough in that six cycles made approximately one tropical year. Similarly, the 60-day cycles broke down into six periods of 10 days each, and therefore approximately two lunations. When necessary, one to three 10-day units were added in or, occasionally, a whole 60-day cycle. The 10-day period lasted down to our own time, and still exists in rural places. The seven-day week was an introduction into China, by way of Persian and Sogdian merchants, not earlier than the Sung. A text of that time says 'If you don't know what day of the week it is, ask a Sogdian'. There is, however, some evidence for the existence of a similar system in the early Chou but it did not persist.

The Chinese sexagesimal cycle can be thought of in the image of two enmeshed cogwheels, one having 12 and the other 10 teeth, so that not until 60 combinations have been made will the cycle repeat. It is one of the most interesting parallels between ancient Chinese and ancient Amerindian civilisation that similar, though more complicated, cycle systems are found in the latter. The Mayas had a 260-day religious year (*tzolkin*) made up by combining cyclically the numbers 1 to 13 with the 20 day glyphs (carved symbols) and this year was then in turn combined with the calendar year (*haab*) which was composed of the 20 day glyphs and the 19 months (18 of 20 days each and one of 5 days), or 365 days. In this case the two cogwheels were of 260 and 365 teeth respectively, giving a total period of 18,980 days, or 52 years. The cyclical period so formed was adopted by the Aztecs and other Central American cultures. But as its multiples were awkward to handle arithmetically, the Mayas developed a parallel system based on a vigesimal principle (a count of 20) which gave them an extremely precise chronology.

The usual view about the 10 'stems' or *kan* was that they had been developed by combining the Five Elements with the Yin–Yang dualism, but this cannot be true, because they long antedate these late Chou theories. It is much more probable that they were the names of the days of the 10-day week, and by the early Han they had come to be associated with 10 astrological names of obscure origin. The 12 'branches' or *chih* were already in very ancient times applied to the lunations (months) of the tropical year, but also served in other ways, particularly as compass points and as names for the double-hours of each sidereal day (i.e. the day timed by the stars instead of the sun, and about four minutes shorter than a mean solar day). Some think that the 12 *chih* derive from the rites proper for each lunation. The Chinese also had a parallel list of astrological names which seem to have been borrowed from a cycle of years connected with the planet Jupiter; planetary cycles thus demand consideration next.

Planetary revolutions

The *Khai-Yuan Chan Ching* (The Khai-Yuan reign period Treatise on Astrology and Astronomy) gives the essential technical terms which came down from the astronomical schools of the fourth century B.C. All the Chinese astronomical literature confirms them, and recently hitherto unknown manuscripts on planetary motions have been discovered, notably in the Ma-wang-tin tomb of about 168 B.C. Each planet was associated with one of the Five Elements and one of the cardinal points of the compass. As is well known, because of changes in the relative motions of the earth and

(1) Jupiter	Sui hsing (the Year-star)	Wood	East
(2) Mars	Ying huo ('Fitful Glitterer')	Fire	South
(3) Saturn	Chen hsing (the Exorcist)	Earth	Centre
(4) Venus	Thai pai (the 'Great White One')	Metal	West
(5) Mercury	Chhen hsing (the Hour-star)	Water	North

the other planets, when the planets are observed in the night sky they appear most of the time to move eastwards across the background of the stars but, on occasions, they seem to stand still, then to move westwards (in backwards or retrograde motion), then to stand still once more before moving eastwards (in direct motion) again. The Chinese had a number of technical terms to cover all the apparent movements.

According to the means at their disposal the observers of the Warring States period must have mapped the planetary motions; in the *Shih Chi*, Ssuma Chien has much to say about retrograde motion. Here reproduced (Fig. 113) is a diagram of such motion for Mercury from a very late source, but illustrating facts which must have been familiar in ancient China. By comparing the figures for the estimates of the periods of revolution dating from various times between 400 B.C. and A.D. 100, we can obtain a clear indication of astronomical refinement during that time (Table 30). It will thus be seen that before the end of the first century A.D. the estimates of the synodic periods (periods of one apparent circuit of the sky) had become quite reliable. This was the case a little later in the West, for about A.D. 175 Cleomedes the Stoic gave figures of about the same degree of precision. Naturally Mercury caused the most difficulty as it is the hardest of the planets to follow with the naked eye.

In spite of so much accurate observation, Chinese study of planetary motion remained purely numerical. Unlike that of the Greeks in which the

Table 30. *Estimates of periods of planetary revolutions*

	Mercury	Venus	Mars	Jupiter		Saturn	
	Days	Days	Days	Days	Years	Days	Years
True values:							
Sidereal Period[a]	87·969	224·7	686·98	4333	11·86	10759	29·46
Synodic Period[b]	115·877	583·921	779·936	398·884		378·092	
Shih Shen (4th century B.C.)[c]	—	736	780	—		—	
Kan Tê (4th century B.C.)[c]	68+	585	—	400		—	
Eudoxos of Cnidos (408 B.C. to 355 B.C.)	110	570	260	390		390	28
Huai Nan Tzu[d] (2nd century B.C.)	—	635	—	—		—	
The author of the *Lo Shu Wei Chen Yao Tu* (perhaps 2nd century B.C.)	74+	—	—	—		—	
Ssuma Chhien (1st century B.C.) (*Shih Chi* figures)	—	626	—	395·7	12	360	28
Liu Hsin (end 1st century B.C.) (*Chhien Han Shu* figures)	115·91	584·13	780·52	398·71	11·92	377·93	29·79
Li Fan (85 A.D.) (*Hou Han Shu* figures)	115·881	584·024	779·532	398·846	11·87	378·059	29·51

[a] The sidereal period is the time of the planet's revolution in its orbit about the sun relative to the stars.

[b] The synodic period is the interval between two conjunctions with the earth relative to the sun, i.e. the time during which the planet makes one whole revolution as compared with the line joining the earth to the sun.

[c] Figures preserved in the *Khai-Yuan Chan Ching*. The motions of the planet Mars are also referred to in the probably 4th-century B.C. *Yen Tzu Chhun Chhiu* (Master Yen's Spring and Autumn Annals), ch. I, p. 22*a*.

[d] Ch. 3, p. 4*a*.

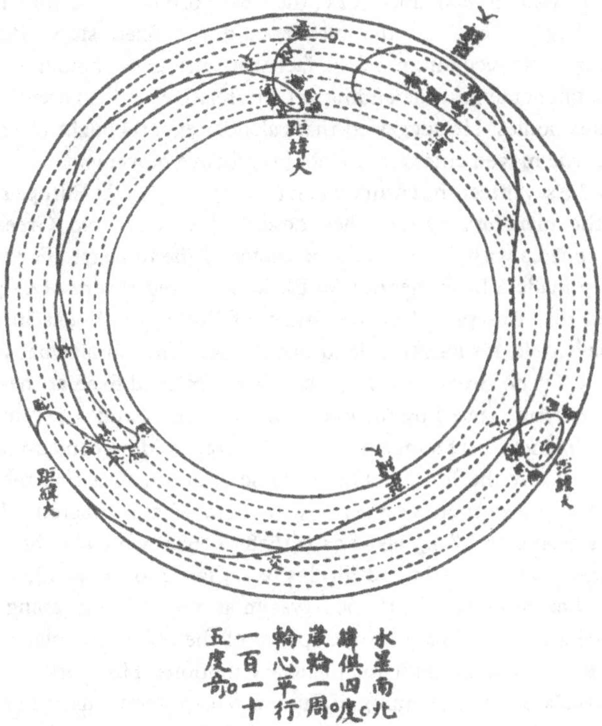

Fig. 113. A plot of the circular area of the ecliptic to show the retrograde (backward) motions of Mercury (from the *Thu Shu Chi Chhêng*, A.D. 1726). Many of the technical terms for the motions inserted in the diagram are ancient, according with their definitions in the *Chin Shu* (A.D. 635).

geometry of curves and circles gave a scheme of planetary motion and an appropriate theory to go with it, the Chinese system perpetuated the algebraic treatment of the Babylonian astronomers such as Naburiannu and Kidinnu, and never sought a geometrical model of planetary motions. If this gave it, at the time of the arrival of the Jesuits, a somewhat archaic character, one must also remember that the attachment of the Greeks to circular motions was almost as great, so that in the West astronomers like Kepler and his successors had to make titanic struggles to break away from them to the idea of elliptical orbits.

Nevertheless, it would be rash to assume that the Chinese astronomers never visualised planetary orbits. In an interesting chapter of the *Chu Tzu Chhüan Shu* (Collected Works of Chu Hsi), which records conversations taking place about A.D. 1190, the philosopher speaks about 'large and small circular tracks', i.e. the small 'orbits' of the sun and moon, and the large 'orbits' of the planets and the fixed stars. Particularly interesting is his recognition of the fact that retrograde motion was only an apparent phenomenon, depending on the relative and different speeds of the various bodies. He suggested that calendar experts ought to realise that all retrograde movements were in fact progressive movements.

In the seventeenth century when the Jesuits brought European ideas to China, the planetary system they described was the old Greek one of celestial spheres with the earth at the centre of the universe. For just a few years they made a brief mention of the sun-centred theory of Copernicus, but when that was placed on the Index of Prohibited Books in Rome in 1616, they could no longer defend and transmit it. An alternative system devised by Tycho Brahe, in which the earth remained fixed at the centre of the universe and orbited by the moon and the sun, round which in turn the rest of the planets were supposed to revolve, received no condemnation and was described by the Jesuits. The outcome of all this was that the Chinese gained an impression of Western views as somewhat incoherent. The great Chinese astronomer Wang Hsi-Shan (1628 to 1682) who, like the rest of his countrymen, was not limited in his acceptance of new ideas by any theological orthodoxy, took Brahe's system and modified it, going so far as to propose a force acting from the sphere of the outermost planet inwards towards the sun, to account for planetary motions. His work, and that of others stimulated by the coming of European astronomy, inspired a Chinese rediscovery and a revival of interest in their own astronomical traditions which had been ignored for some three centuries.

Duodenary series

The cycle of Jupiter's motion must have attracted attention at a very early date on account of the fact that the sidereal period of this planet (the time it takes to complete an orbit round the sun) is almost 12 years (actually 11·86 years), which seemed to fit in some way with the 12 cyclical characters and the 12·37 lunations in the tropical year. The *Chi Ni Tzu* (Book of Master Chi Ni), transmitting a fourth-century B.C. tradition from the south, says:

> When the Great Yin (Thai Yin) is at the position of the element Metal during the first three years, there will be abundant harvest. When it is at Water, there will be damage to crops for three years. When it stands in Wood, for another three years, there will be prosperity; when it stands in Fire, again for three years, there will be drought. So some times are suitable for storing agricultural products while at other times rice may be given away. Accumulations need not be for more than three years...

Here Thai Yin is not, as might be supposed, the moon, but an invisible 'counter-Jupiter' which moved around in the opposite direction to the planet itself. Jupiter (*Sui hsing*), with the other planets, appears to move eastwards through the stars, so a 'shadow-planet' (*Thai sui* or *Sui yen*) was invented to move with them, accompanying the sun in its westward motion. The 12 Jupiter-stations were named *tzhu* and the whole cycle of years a *chi*. The names for the years have survived in two forms, one set astronomical, the other astrological; it was natural therefore that they should sometimes also have been used for the months of a single year and for the double-hours of a single day, as well as for the years of the Jupiter cycle. The astronomical terms were used to designate the positions of Jupiter, the astrological or calendrical terms applied to the positions of the counter-Jupiter.

Month-names, however, were not the most important in the old Chinese calendar. The 12 months (*chhi*) of the tropical year were divided into 24 fortnights, of which 12 were '*chhi*-centres' and 12 were '*chhi*-nodes' (the analogy here being with the nodes of a bamboo). These are given in Table 31. They are undoubtedly very ancient; one of the earliest literary references to them occurs in the *Mu Thien Tzu Chuan* (Account of the Travels of the Emperor Mu) which is of late Chou time. Before then, in the Shang, there were months of 29 days and 30 days; sometimes two of the latter came successively.

At some quite early time the twelve-fold division became associated with a cycle of animals, the ox, sheep, dragon, pig, and so on. There has been great debate among scholars, both Eastern and Western, about the origin of these associations, some maintaining that the Chinese took them over from neighbouring Turkic peoples or from the ancient Middle East, and others labouring to prove that they were essentially Chinese in origin. Most of the arguments need re-examination in the light of modern knowledge about dates and reliability of texts, but in any case, so far as the historian of science is concerned, whoever invented the animal cycle is welcome to it. Its interest seems purely archaeological and ethnological. It is worth noting, however, that the naming of years for animals still continues in numerous Asian cultures, e.g. those of the Mongols and Tibetans as well as the Chinese.

Resonance periods

In view of the impossibility of dividing the tropical year by a whole number of lunations, calendar-makers in all cultures have attached importance to certain 'resonances', i.e. those rather longer lapses of time which end with approximate agreement between the solar and lunar reckonings. Thus 19 tropical years are almost equivalent to 235 lunations. In the West this period is associated with the name of Meton of Athens (*c.* 432 B.C.); in

Table 31. *The 24 fortnightly periods* (chhi)

	Chinese name		Translation	Beginning
1	Li Chhun	立春	Beginning of Spring	5 Feb.
2	Yü Shui	雨水	The Rains	20 Feb.
3	Ching Chê (or Chih)	驚蟄	Awakening of Creatures (from hibernation)	7 March
4	Chhun Fên	春分	Spring Equinox	22 March
5	Chhing Ming	清明	Clear and Bright	6 April
6	Ku Yü	穀雨	Grain Rain	21 April
7	Li Hsia	立夏	Beginning of Summer	6 May
8	Hsiao Man	小滿	Lesser Fullness (of Grain)	22 May
9	Mang Chung	芒種	Grain in Ear	7 June
10	Hsia Chih	夏至	Summer Solstice	22 June
11	Hsiao Shu	小暑	Lesser Heat	8 July
12	Ta Shu	大暑	Greater Heat	24 July
13	Li Chhiu	立秋	Beginning of Autumn	8 Aug.
14	Chhu Shu	處暑	End of Heat	24 Aug.
15	Pai Lu	白露	White Dews	8 Sept.
16	Chhiu Fên	秋分	Autumn Equinox	24 Sept.
17	Han Lu	寒露	Cold Dews	9 Oct.
18	Shuang Chiang	霜降	Descent of Hoar Frost	24 Oct.
19	Li Tung	立冬	Beginning of Winter	8 Nov.
20	Hsiao Hsüeh	小雪	Lesser Snow	23 Nov.
21	Ta Hsüeh	大雪	Greater Snow	7 Dec.
22	Tung Chih	冬至	Winter Solstice	22 Dec.
23	Hsiao Han	小寒	Lesser Cold	6 Jan.
24	Ta Han	大寒	Greater Cold	21 Jan.

Note: Each of these 24 periods corresponds to 15° motion of the sun in longitude on the ecliptic. The average fortnightly period is 15·218 days; the half (lunar) month, 14·765. The names of the periods suggest that the list was first established in, or north of, the Yellow River valley.

China it was known as the *chang*. Similarly, four 'Metonic cycles' were equivalent to exactly 76 years or 27,759 days, and this unit is associated with the name Callipus of Cyzicos (*c.* 370 to 330 B.C., and therefore a contemporary of Shih Shen), who obtained a rather satisfactory calendar if one day was dropped out of each such period. In China the 76-year cycle was known as the *pu*.

Other resonances were found. For example 27 *chang* were equivalent to 47 lunar eclipse periods of about 135 months each, or 513 years, and this was termed a *hui*. Three *hui* (or 81 *chang*) constituted a *thung* (1,539 years), a period less than this not giving a round number of days. The smallest period of agreement of sexagenary cycles, lunations, tropical years and eclipse periods, was found to be three *thung* or 4,617 years. Another period was the *chi* (already met with as a term for Jupiter's sidereal period) or the *sui* or *ta chung*, which was 20 times a *pu*, i.e. 1,520 years or 19 × 487 sexagesimal day cycles. Three *chi* constituted a *ta pei*, *yuan* or *shou*. The *Chou Pei Suan Ching* (Arithmetical Classic of the Gnomon and the Circular Paths of Heaven) of the Han period or earlier goes on to say that seven *shou* equalled one *chi* or 31,920 years, after which time, 'all things come to an end and return to their original state'. This 'ground period' is interesting, for it equals exactly four 'Julian cycles', one Julian cycle (7,980 years) being a period proposed in the West in the late sixteenth century by Joseph Scaliger, some centuries after the Chinese text. The *chi* is the lowest common multiple of the Metonic *chang*, the Callipic *pu*, the 80-year period in which the 60-day cycle recurs, and the 60-year cycle itself, as well as the Western 'sabbatical' cycle of 28 years and the Roman 'indiction' cycle of 15 years.

There are other interesting relationships. Planetary cycle constants were termed *chi mu* and day–month–year cycle constants *thung mu*. With the Han value for the length of the lunation, the smallest number of lunations which would give a round number of days was 81 (i.e. 2,392 days), and when this was combined with the lunar eclipse cycle of 135 months, the former multiplied by 5 and the latter by 3, both give 405 lunations or 11,906 days. This is the shortest period of whole days in which the eclipse cycle could be completed, and is identical with the *tzolkin* of the Mayas.

The parallel between the shorter of these cycles and those known to the Greeks has seemed to some scholars rather strong evidence of transmission from the West to China. However, there are reasons for thinking that the *chang* and the *pu* were known to the people of the Shang dynasty, not indeed by those names, which have not so far been found on the oracle-bones, but because from certain dates, which were evidently considered of great importance by those who inscribed the bones, precisely those intervals can be derived. Two of these are the eclipse dates (1311 B.C. and 1304 B.C.) which have been confirmed by modern computation methods as correct, and the other two are new moon dates (1313 B.C. and 1162 B.C.). The time elapsing is exactly 2 *pu* or *chang*. If this should be confirmed we can hardly suppose that these periods were not known to the Babylonians as well as to the Chinese of the Shang.

It was natural that the Han astronomers should wish to incorporate into these systems the synodic and sidereal periods of the planets. Their thought was that 'at the beginning' (of the world or of a world-cycle) there had been a general conjunction of the planets (so that all appeared together in the sky), and that at the end of the world, this would happen again. There are conjunctions of Jupiter and Saturn at almost the same place in the sky every 59·5779 years – by coincidence very close to the sexagesimal year cycle. Conjunctions of Jupiter, Saturn and Mars recur every 516·33 years, which may account for the 500-year period given in the *Mêng Tzu* (The Book of Master Mêng) of about 290 B.C. for the recurrence of the birth of great sages. Planetary conjunctions and their occultations or 'eclipses' of one another were closely watched. Occasionally there may have been distortion of the date for political reasons, as in the conjunction of 205 B.C., but more often the records, when recalculated today, are found to be quite reliable. The occultations of Mars by the moon in 69 B.C. and of Venus in A.D. 361 are cases in point. We also hear from time to time of predictions which proved to be accurate, such as one by Tou Yen about A.D. 955.

The notion of a 'general conjunction', found also in Greek and Hellenistic literature, seems to derive from Berossus (third century B.C.) and the Babylonians. Now three *thung* periods would be equivalent to nine times the lunar eclipse cycle of 135 lunations (i.e. to 4,617 years), and to 28,106 sexagesimal day cycles; and this was the unit the Han scholars, notably Liu Hsiang and Liu Hsin, combined with the planetary cycles. It was thought that in 138,240 years, all the planets would exactly repeat their motions, so that by combining this 'cogwheel' with the 4,617-year period, a whole 'world-cycle' was taken to be 23,639,040 years. The beginning of it was known as the 'Supreme Ultimate Grand Origin'. This idea continued, in various forms, throughout the later history of Chinese astronomy. It was characteristic of the amplitude and spaciousness of medieval Chinese cosmological thought that when I-Hsing in the early eight century A.D. used indeterminate analysis for calendar calculations (page 50 above) he arrived at a figure of 96·96 million years as the time which had elapsed between the last general conjunction and 724. A thousand years later in the Western world, Archbishop Ussher was reckoning the date of creation as no earlier than 4004 B.C.

RECORDS OF CELESTIAL PHENOMENA

Eclipses

The conception of an eclipse of the sun or moon (Figs. 114, 115) as the gradual eating of the luminary by some celestial dragon manifests itself in the earliest term for eclipse, *shih* (食), which appears on the Anyang oracle-bones. This word simply means 'to eat' and the original pictograph showed a food vessel with a lid ( ). The formation of a more precise technical term *shih* (蝕) by addition of the 'insect' radical, did not take place till comparatively late, after the Han.

How far back in Chinese history eclipses were observed is a question

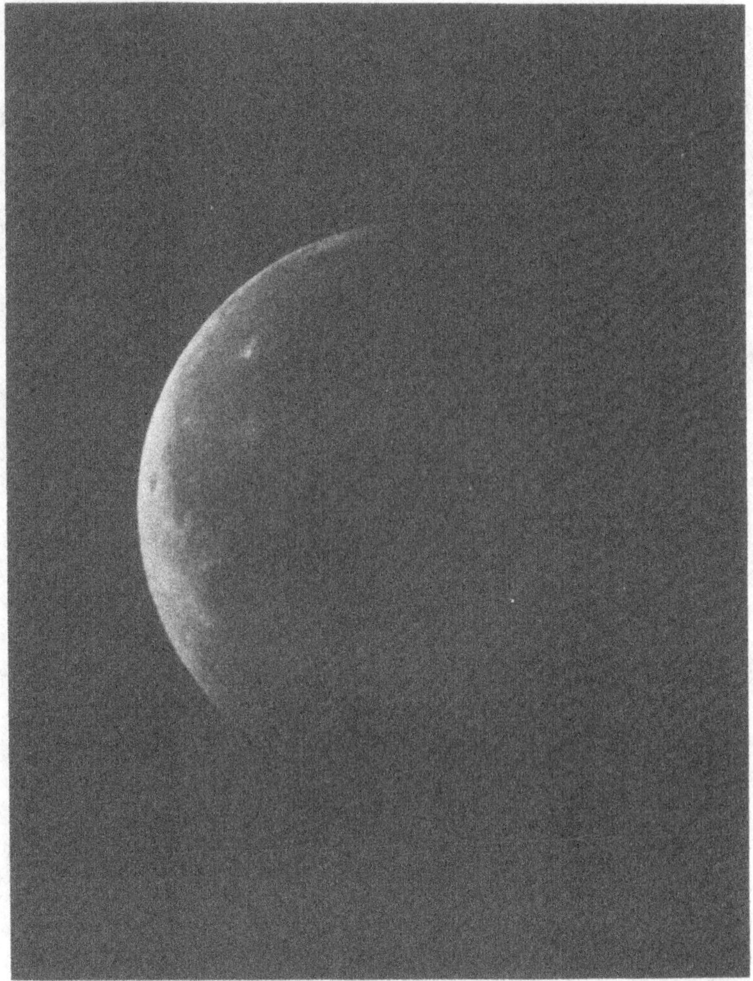

Fig. 114. Eclipse of the moon. Here the bright disc of the full moon gradually becomes very dull and, usually, brownish red in colour with only, at most, a crescent remaining bright.

which has been in dispute for centuries, ever since growing confidence in eclipse cycles induced I-Hsing to make computations about the eclipses of the Hsia and Shang dynasties. Traditionally, the solar eclipse recorded in the *Shu Ching* (Historical Classic) was regarded as of the third millennium B.C. 'On the first day of the month, in the last month of autumn, the sun and moon did not meet (harmoniously) in Fang', so runs the text. Identifications of this have varied from 2165 to 1948 B.C., but since this date falls in

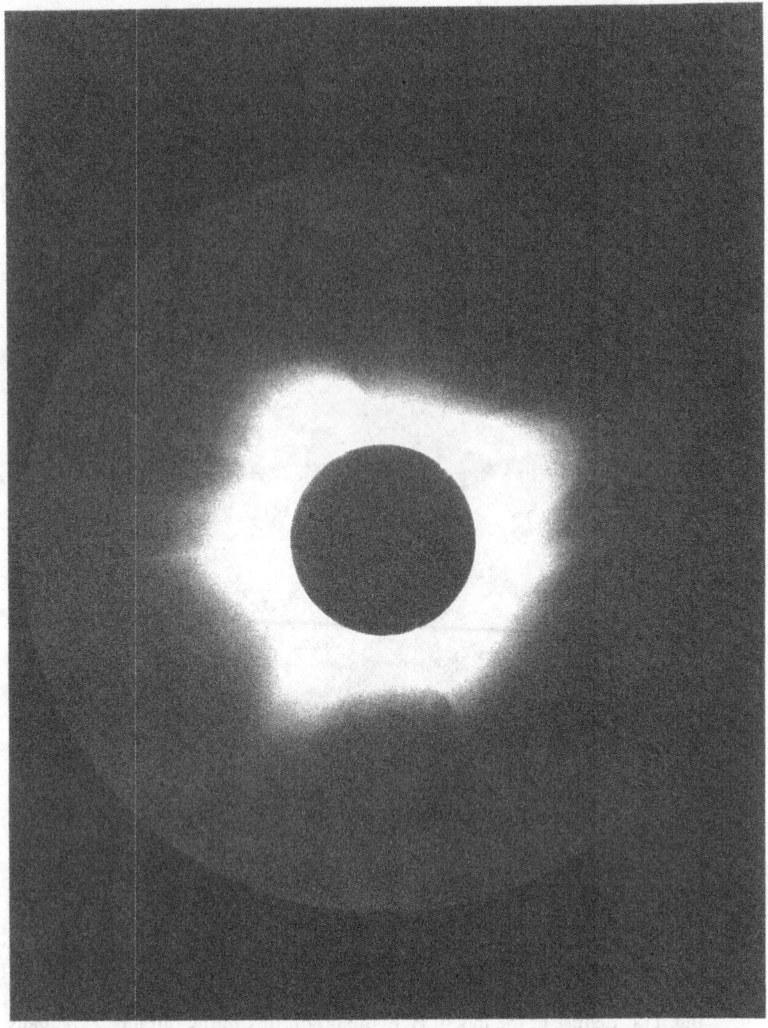

Fig. 115. Total eclipse of the sun. The light of the corona, which surrounds the eclipsed solar disc, is comparatively dim – only about half the brightness of the full moon – and in no way compensates for the sudden and dramatic loss of sunlight.

what might be called a period legendary in character, and since the text is almost certainly of later date, attempts to fix the year exactly have been abandoned. The opinion then grew up that the oldest recorded solar eclipse was that in the *Shih Ching* (Book of Odes), which has been identified as having occurred in 734 B.C. This used to be accepted as the earliest verifiable eclipse in the history of any people, since the Babylonian records were supposedly all lost after the time of Ptolemy. However, copies of some of these, dating from 747 B.C. (Nabonassar's time) or even perhaps 763 B.C., had apparently been sent to Greece by Aristotle's nephew Callisthenes after the capture of Babylon by Alexander the Great in 331 B.C. Nevertheless, recent researches on the oracle-bones have in turn increased the priority of the Chinese for extant records, because six lunar eclipses and one solar eclipse recorded by the Shang people are on the bones which have come down to us. These dates are, for the moon, 1361 B.C., 1342 B.C., 1328 B.C., 1311 B.C., 1304 B.C. and 1217 B.C., and for the sun 1217 B.C. To these may be added a seventh lunar eclipse, that of 1137 B.C., mentioned in the *I Chou Shu*, apparently correctly. More eclipse records have been deciphered but are so far undatable.

An interesting point is that the Shang people were watching lunar eclipses for calendrical rather than astrological reasons, since they can occur only on or very near the date of full moon, and the bone inscriptions indicate the desire of their writers to check as closely as possible the dates of the lunations. We also have a hint of the degree of State organisation at the time, since the first and fifth of the eclipses mentioned above are followed by the word *wên* – i.e. 'this was reported from the provinces, and not actually seen at the capital'. This technical term continued in use in the Han, fifteen hundred years later. It is interesting, too, that the solar eclipse in the *Shih Ching* is termed *chhou* (ugly, abnormal), while lunar eclipses were termed *chhang* (usual), which suggests that the latter were expected while the former were not.

Eclipse theory. By what date did the Chinese arrive at an approximately correct understanding of the nature of eclipses? The question is bound up with the methods used for eclipse prediction. In the West there is a tradition that the Greek philosopher Thales of Miletus foretold the solar eclipse of 585 B.C., but recent research shows that this may just have been more by luck than judgement. The difficulty here was one faced by all the ancients, and simply arises from the fact that the moon's path does not follow the ecliptic exactly; if it did there would be a solar eclipse at every new moon, and a lunar eclipse at every full moon. Hence the need to develop various arbitrary periods permitting approximate predictions of when eclipses would have been likely to occur.

The fact that the moon shines by reflected light has already been mentioned (page 90), but although the Chinese seem to have been observing eclipses as early as anyone in the West, their understanding of the nature of the sun and moon seems to have come rather later than in Greece. Shih Shen in the fourth century B.C. was certainly well aware that the moon had something to do with solar eclipses, because he gave instructions for predicting them based on the relative position of the moon to the sun.

When the moon was in conjunction with the sun, during the dark nights at the beginnings and ends of lunations, such eclipses could, he believed, occur. However, it may be that Shih Shen did not think that the moon was interposed between the sun and the earth, but rather that a Yin influence radiating from it overcame the Yang influence of the sun. Such a view would explain why Kan Tê, perhaps considering sun-spots as incipient eclipses, spoke of them as beginning at the centre of the sun and spreading outwards. It probably also explains why the late Chou and Han astronomers spoke of 'veilings' of the sun which were not full eclipses. As for lunar eclipses, the astronomers of the fourth century B.C. seem to have been less well informed than those of the Shang, since they supposed they could occur at any time of the lunation instead of only at full moon.

In Ssuma Chhien's time (about 100 B.C.) it was thought that eclipses of the moon could be due to influences emanating from almost any heavenly body, including not only the five planets but also the red stars Arcturus and Antares. Ssuma Chhien certainly knew, however, that there were definite periods for lunar eclipses, but he said nothing of this for solar ones, for which he ventured no theory. This theory of a 'radiating influence' was still alive in the first century A.D. for Wang Chhung mentions it. Yet from what he says the correct view had long been known, although he himself preferred the hypothesis that the sun and moon both shrank or faded periodically. His remarks are interesting:

> According to the scholars, solar eclipses are brought about by the moon. It has been observed that they occur at times of new moon, when the moon is in conjunction with the sun, and therefore the moon can eclipse it. In the 'Spring and Autumn' period there were many eclipses, and the *Chhun Chhiu* (Spring and Autumn Annals) says that at such and such a month at new moon there was an eclipse of the sun, but these statements do not imply that the moon did it. Why should (the chroniclers) have made no mention of the moon if they knew that it was really responsible?
>
> Now in such an abnormal event the Yang would have to be weak and the Yin strong, but (this is not in accord with) what happens on earth, where the stronger subdue the weaker. The situation is that at the ends of the months the light of the moon is very weak, and at the beginnings almost extinct; how then could it conquer the sun? If you say that eclipses of the sun are due to the moon consuming it, then what is it that consumes (in a lunar eclipse) the moon? Nothing, the moon fades of itself. Applying the same principle to the sun, the sun also fades of itself.

Wang Chhung then mentions again the problem of how the moon can consume the sun, states the correct view once more and comments:

> ... That the sun and the moon are in conjunction at times of new moon is simply one of the regularities of the heavens.
>
> But that the moon covers the light of the sun in solar eclipses – no, that is not true. How can this be verified? When the sun and the moon are in conjunction and the light of the former is 'covered' by the latter, the edges of the two must meet at the beginning, and when the light reappears they must have changed places. Suppose the sun is in the east and the moon is in the west. The moon falls back quickly eastwards and meets the sun, 'covering' its edge. Soon the moon going on eastwards passes the sun. When the western edge (of the sun), which has been covered first, shines again with its light, the eastern edge which was not 'covered' before should (now) be 'covered'. But in fact we see that during an eclipse of the sun the light of the western edge is extinguished, yet when (the light) comes back the western edge is bright (but the eastern edge is bright also). The moon goes on and covers the eastern (inner) part as well as the western (inner) part. This is called the 'exact intrusion' and 'mutual covering and obscuring'. How can these facts be explained (by astronomers who believe that the moon covers the light of the sun in solar eclipses)?

After this description of an annular, not a total, eclipse of the sun, where an outer ring (annulus) of the sun's disc remains uneclipsed, Wang Chhung goes on to question whether the sun and moon are really spherical. He concludes that the fact that they appear so is an illusion, and in support he quotes the evidence of 'shooting stars' (meteorites) which, on examination, are found to be no more than irregularly shaped pieces of 'stone'.

To summarise, it is clear that the correct theory was already widely held in Wang Chhung's time (about A.D. 80). Yet he preferred the view that the sun and moon had some intrinsic rhythms of their own, an idea almost identical with some of the speculations of the Roman poet Lucretius on the subject (first century B.C.). Wang Chhung supported this with objections drawn, for instance, from observations of annular eclipses, and from rather sophisticated doubts about the shapes of heavenly bodies. But perhaps the real interest of his position lies in the fact that he – a Confucian sceptic – was criticising the theories of practical astronomers not unconnected with Taoism. He was probably not inclined to draw any great distinction between magicians and star-clerks, or between prognosticators and calendar calculators. The fact that the imperial court employed them all merely discredited the court. He, Wang Chhung, was against them all and all their theories – but sometimes the theories happened to be right.

Wang Chhung's pulsation theory is an admirable example of the inhibitions possible from a world-view otherwise good in itself – organic naturalism, the idea that the universe is a vast interdependent organism. This was the perennial philosophy of China. Even the simple conceptions of converging paths and mutual shadowing were too mechanical in flavour for Wang Chhung; he preferred to believe in a rhythm emanating from the intrinsic natures of the celestial organisms in question. And Wang Chhung was no isolated case; Liu Chih, two centuries later, was another.

Yet the correct view is already to be found in Liu Hsiang's *Wu Ching Thung I* (The Fundamental Ideas of the Five Classics), of about 20 B.C. 'When the sun is eclipsed it is because the moon hides him as she moves on her way.' It would seem likely, then, that the right view was reached between the early Warring States period and the middle of the Former Han. After Wang Chhung it was well established.

Extent, reliability and precision of the records. Thirty-seven eclipses, from 720 B.C. onwards, are recorded in the *Tso Chuan* (Master Tsochhiu's Enlargement of the Spring and Autumn Annals), and from the beginning of the Han onwards there are systematic records in all the dynastic histories. By a remarkable coincidence, the list of lunar eclipses given in Alexandria by Ptolemy in his *Almagest* started from 721 B.C. The eclipses of the Han dynasty have been studied with particular care, and attempts made to assess the reliability of the ancient official astronomers in China, as shown in Table 32. The failure to record some eclipses may have been due to bad weather, while some of the unidentified may have been calculated according to some cycle and then inserted in the records. However, eclipses may also have been inserted for political reasons (to apportion blame) or omitted when the rulers were doing better. Thus during the reign of the unpopular and criminal empress Kao Tsu, an eclipse announcement occurred (186 B.C.), though none could have taken place. Yet a very detailed examination leads to the conclusion that falsification of the records was rare; indeed, the case of the empress just mentioned may be an isolated one. But the records, it seems, were often left incomplete, and the extent of their incompleteness tallies with the popularity of the reign. If 'warnings from Heaven' seemed not to be needed, the astronomers may have noted eclipses but refrained from memorialising about them, hence they were not recorded by the historians. Popularity seems here to refer to popularity with the high officials of the court rather than the mass of the people, and we should therefore expect the Han records most likely to be complete in the reigns unsatisfactory to the Confucian bureaucracy.

able 32. *Observations of solar eclipses by ancient Chinese astronomers*

	Chhun Chhiu (Tso Chuan)	Chhien Han Shu
Identifiable and verified by modern computation		
Very striking	21	12
Visible	5	9
Not striking	2	6
Hardly visible	3	6
Partial	1	5
	32	38
Identifiable if slight textual errors are assumed	0	14
Unidentified	3	0
Impossible	2	3
Total number recorded	37	55
Computed by modern methods as striking, but of which no observation was recorded	14	28

Many passages exist, if one can find them, throwing light on the customs and mental processes of members of the Astronomical Bureau in the various dynasties. An example appears in the *Shan Chü Hsin Hua* (New Discussion from the Mountain Cabin) written by Yang Yü in A.D. 1360. It throws a flood of light on what went on. Yang Yü says:

> When I was a Co-signatory Observer in the Bureau of Astronomy, there came a special imperial edict that we were to pay particular attention to celestial presages. On the first day of the seventh month in the sixth year of the Chih-Yuan reign-period (+1340), there came (to my house) one of the Senior Observers, a Mr Chang, who asked me to go to the Observatory as quickly as possible. When we arrived there together, we were met by Commissioner Li, dressed up in state apparel, who said: 'Last night there appeared the Ching Hsing phenomenon. That is a very auspicious omen. I consider that it ought to be memorialised immediately. I suppose we shall be richly rewarded.' So I looked up the files which contained the records of earlier memorials, and came to a very different conclusion. I said, 'Although the phenomenon has occurred on the last day of the month (i.e. at the new moon), its shape was slightly different from what it ought to be. Besides, if the Ching Hsing appears, there ought to be reports coming in of wine-sweet springs, phoenixes, purple herbs,

> and felicitous clouds, in order to corroborate (the celestial omen). But (on the contrary) there are epidemics and catastrophes in Shensi, brigands and robbers in the central provinces, and in Fukien rebels are active. I am afraid it won't do. Why should the Tao of heaven be proclaiming the opposite (to the Tao of earth)?' But Mr Li was most obstinate, and stuck to his opinion. So I said 'Up to now, only the six Observers here have seen the phenomenon. In the unlikely possibility of its having been generally seen by people throughout the country, will they not have taken it as an omen of evil?' Finally he agreed to wait and see if it appeared again (that night), before we memorialised about it. And indeed only nine days later the planet Venus 'crossed the meridian'. All this shows how careful one has to be not to take lightly responsibilities like these.

It is thus fairly clear that before Chinese records can be made full use of by modern astronomers or meteorologists interested in periodic events over long periods, a good deal of further historical analysis and research will be needed. Nevertheless, if the records were not more accurate than would appear from the severest of their critics, it would have been impossible to find known periodicities in them, as has in fact been done in the case of the sun-spot cycle (page 211 below). Again, there is the striking case of the solar eclipse of 96 B.C. This had been among an unidentified group of eclipses, until the discovery of some calendar tablets in a Han watch-tower showed that an intercalary month had been wrongly inserted by scholars; once the necessary correction had been made, the statement about the eclipse in the *Chhien Han Shu* (History of the Former Han Dynasty) was shown to be accurate.

It is interesting to see the gradual advance in precision of the records. Already in the *Chhun Chhiu* (Spring and Autumn Annals) there are three cases where the word *chi* occurs, showing that the eclipse was total. Then with regard to the solar eclipses of 442 B.C., 382 B.C. and 300 B.C., the *Shih Chi* (Historical Records) says that the daylight was so darkened that the stars could be seen. Han records describe some eclipses as nearly total, or as crescent-shaped, as well as those which were total. A partial eclipse of three-tenths is also once mentioned. The degree of partiality is recorded in all subsequent dynasties, and the Thang records have a stock phrase: 'all the great stars could be seen'. The Han records sometimes mention the duration of eclipses and the time of their onset and ending, correct to a quarter of an hour. Records of the Thang and Sung frequently have exact details, though not always. The celestial positions of eclipses were generally noted in the Han and always in the Thang, and a modern check of them shows that most were faithfully recorded.

Eclipse prediction. Chinese astronomers naturally devoted much attention throughout the centuries to the prediction of eclipses, though like all such efforts before the Renaissance, this could only be based on previous experience. We see now that an eclipse period such as the *saros* of 18 years 11 days (223 synodic months), at the end of which eclipses recur in the same relative positions of the sun and moon, depends simply on the periods of revolution of the moon and its nodes (those points where its orbit crosses the ecliptic). But of course this is not easy to determine, because total solar eclipses cannot all be observed from one centre, since each is visible over only a small part of the earth's surface. Although there are more solar eclipses than lunar ones in a *saros*, the difficulty remains. For solar eclipses to recur in approximately the same longitude, a longer period must be used; such is the period of 54 years 33 days used by the Greeks. It does not seem to have occurred to the Han people to recognise either of these periods, but they developed one of their own, the *shuo wang chih hui* (later called the *chiao shih chou*) of 135 months. Whereas the *saros* contained 41 solar eclipses, the Chinese period covered 23. It was much used by Liu Hsin in the San Thung (Three Sequences) calendar of 7 B.C., and it must have been developed during the first century B.C.

By the early years of the third century A.D. the moon's path had been analysed more clearly. Liu Hung's method of eclipse prediction recognised the nodes and assessed the angle of the moon's orbit with the ecliptic as 6° approximately (5° 54′ in Western degrees: modern value 5° 8′), and was used in the Chhien Hsing calendar of A.D. 206. In the same century Yang Wei was able to predict the directions of first and last contact of the moon's limb (edge) with the sun for solar eclipses. This was much refined in the next century by Chiang Chi who could apparently predict the proportion of coverage of a partial eclipse, and it was at this time that attempts were also made to predict the geographical path along which solar eclipses would be visible. In the Sung the prediction of eclipses was sometimes assigned to one bureau, and their observation to another, but the high level of prediction continued until the Ming dynasty when there was a steady decline, and the earlier methods were forgotten. When the Jesuits came, their ability to predict eclipses was one of the most important reasons for the credit which they were able to obtain at the imperial court.

Looking back over the whole story, one can see that an interesting change of ideas about the cause of eclipses went on over the centuries. At the beginning, eclipses were regarded with fear and trembling as omens or 'reprimands from heaven', but then slowly but surely they moved over into the realm of predictable events; thus enlarging the domain of non-frightening science and minimising that of celestial displeasure.

Earth-shine and the corona. Among celestial phenomena allied to eclipses, mention may be made of the effect known as 'earth-shine', seen when the sunlit earth illuminates by reflection the unlighted part of the moon. As the 'Ballad of Sir Patrick Spens' puts it:

> I saw the new moon yestereen
> Wi' the old moon in her arms,
> And if we gang to sea, Master,
> I fear we'll come to harm.

The phenomenon was recognised by the Chinese, and Ssuma Chhien says:

> When the sky is serene, then the Ching-hsing (Resplendent orb) appears. It is also called Te-hsing (Orb of virtue). It has no constant form, but it appears (to the people of) countries which follow the Tao.

This was the tradition which caused premature rejoicings in the Peking Astronomical Observatory in 1340, as reported by Yang Yü in the passage quoted on page 201 above. It is strange that the European interpretation was quite contrary in character.

An examination of the oracle-bones of the second millennium B.C. may contain the first recorded observation of the corona, that area of pearly-white light which sometimes displays streamers, and which can be seen round the sun during a total solar eclipse. The date of the bone fragment must be either 1353 B.C., 1307 B.C., 1302 B.C. or 1281 B.C.; it bears characters which have been deciphered as 'three flames ate up the sun, and a great star was visible'. It seems not unreasonable, therefore, to suppose that this was a record of coronal streamers, or possibly of especially striking solar prominences (flame-like masses of gas which sometimes appear at the sun's limb).

Another text which may be relevant is that in the *Tso Chuan* (Master Tsochhiu's Enlargement of the Spring and Autumn Annals) for 490 B.C., where it is said that 'a cloud like a flock of red crows was seen flying round the sun'. The term *jih erh*, which means a kind of solar halo, may also have been used to refer to the corona. It has also been suggested that observations of the corona may have been at the origin of the 'winged sun' symbol, so characteristic of Assyria and Persia, but not unknown in ancient China.

Novae, supernovae, and variable stars

Eclipses are not the only celestial phenomena for which a wealth of records is available to us in Chinese texts. The total number of stars visible to the naked eye is not constant; we know that from time to time stars rise into visibility while others disappear, and that the brightness or 'magnitude' of some stars changes. On occasion stars hitherto very dim or even invisible undergo a million-fold increase of brightness; such stellar explosions are called 'novae' or new stars, or, if the cataclysm is exceptionally great, 'supernovae'. There are other stars which show periodical changes in brightness; these are 'variables'. All are highly important in modern astronomy and knowledge of their behaviour in the past is vital for current theories of cosmology.

What must certainly be the most ancient extant record of a nova is contained in one of the oracle-bones dating from about 1300 B.C. (Fig. 116). Its inscription says, 'on the seventh day of the month, a *chi-ssu* day, a great new star appeared in company with Antares'. Another bone inscription of the same period says, 'On a *hsin-wei* day the new star dwindled (or disappeared)'. That this refers to the same event seems probable, for the second date is only two days after the first, and some such sudden rise and fall would be expected. The term *hsin hsing* was used for a nova until the middle of the Han period when it was replaced by the better known technical term *kho hsing*, guest-star.

There are other examples, and at the end of the thirteenth century A.D. Ma Tuan-Lin gave in his *Wên Hsien Thung Khao* (Comprehensive Study of (the History of) Civilisation) a list of the extraordinary stars which had appeared since the beginning of the Han. When the list of positions of suspected Chinese novae are plotted on a star chart they show a distribution in space similar to known novae in modern times. The guest-stars were thus true novae. This result has a bearing on the arguments about the reliability of Chinese records, for if guest-stars had been invented to criticise the government, it is highly unlikely that they would all have been placed in the right part of the sky.

The giant stellar explosions which give rise to supernovae are now thought to occur about once every one or two centuries in our own galaxy, and as often in other galaxies. Of the four supernovae in our own galaxy of which there are any historical records, one was the 'New Star' of Tycho Brahe observed in A.D. 1572, the second was that seen by his successor Kepler in 1604, the third that of 1006, and the fourth the now famous supernova of 1054 which was recorded almost only by the Chinese. This last was the origin of what is now known as the Crab Nebula, which on photographs appears as a somewhat shapeless and diffuse bright cloud, but when observed visually through a telescope looks something of the shape of a hermit crab. This gaseous cloud is still expanding from a central star. Since the Chinese records say that the maximum apparent brightness of this guest-star was as bright as Venus, it can readily be calculated that at the time of the explosion the star was several hundred million times as bright as our sun.

Fig. 116. The oldest record of a nova. The inscription on this oracle-bone, dating from about 1300 B.C., reads, in the two central columns of characters: 'On the 7th day of the month, a *chi-ssu* day, a great new star appeared in company with Antares'.

The value of the Chinese records for modern astronomy has been underlined in recent years in research into the life cycles of stars, as well as other topics, and there is now no doubt about the identification of the Crab supernova with the guest-star of 1054. Five texts have been assembled which describe it, but only one, from the *Sung Hui Yao* (History of the Administrative Statutes of the Sung Dynasty), need be quoted:

> In the fifth month of the first year of the Chih-Ho reign period, Yang Wei-Tê (Chief Calendrical Computer) said, 'Prostrating myself, I have observed the appearance of a guest-star; on the star there was a slightly iridescent yellow colour. Respectfully, according to the disposition for emperors, I have prognosticated, and the result said, "The guest-star does not infringe upon Aldebaran; this shows that a Plentiful One is Lord, and that the country has a Great Worthy." I request that this prognostication be given to the Bureau of Historiography to be preserved.'

This was done, and the emperor was congratulated. In the month of April 1056 it was reported that the guest-star had become invisible, which was an omen of the departure of guests.

Europe and the Arab countries tended to the ancient belief that the heavens were changeless. The supernova of 1006 was, it is true, mentioned in the West as well as being recorded in detail in the East because of its quite exceptional brightness, but the 1054 supernova was recorded only in the East. Tycho Brahe, however, noted the appearance of the supernova of 1572, and from his own observations and those of others, proved it to be a truly celestial phenomenon, i.e. one occurring far beyond the moon, and therefore outside the usual sphere of change. This was an event of great importance for European astronomy and, indeed, for all Western science. Previously the failure of medieval Europe to consider such phenomena had been due, not so much to any difficulty in seeing them, but to prejudice and spiritual inertia connected with the groundless belief in celestial perfection. By this the Chinese were not handicapped.

Comets, meteors and meteorites

While there exist a few Babylonian cuneiform records of comets as far back as 1140 B.C., and observations of them were quite frequent in ancient and medieval Europe because of the astrological significance attributed to them, the Chinese records are by far the most complete. The computation of approximate orbits of some 40 comets which made appearances earlier than A.D. 1500 has been based almost entirely on the Chinese observations. As in the case of novae, the first compilation of these events, as noted in the dynastic histories, was made by the Chinese themselves. Ma Tuan-Lin incorporated them in his *Wên Hsien Thung Khao*, dealing with comets down to A.D. 1222. A supplementary section took them down much later and this and other examinations of the evidence brought the total list up to 372 comets between 613 B.C. and A.D. 1621.

To give an idea of the care with which the Chinese astronomers described their comets, we may select the record of the comet of A.D. 1472 studied also in Europe by Johannes Müller of Königsberg (often known as Regiomontanus).

> In the seventh year of the Chheng-Hua reign period (A.D. 1472), in the twelfth month, on a *chia-hsu* day (of the sexagenary cycle) a comet was seen in the star-group Thien thien (σ and τ Virginis). It pointed towards the west. Suddenly it went to the north, touched the star 'Right conductor' (η, ι and τ Boötis), and swept through the Thai Wei Yuan (the 'Enclosure' of stars in Virgo, Coma Berenices and Leo), touching Shang chiang (ν Comae Berenices), Hsin chhen (2629 Comae Berenices), Thai tzu (*E* Leonis), and Tshung kuan (2567 Leonis). Its tail now pointed directly west ... On a *chi-mao* day its tail had greatly lengthened. It extended from east to west across the heavens. The comet then proceeded northwards, covering about 28°, touched Thien chhiang (ι, θ, κ Boötis), swept through the Great Bear (Pei tou), ... It was now perfectly visible in full daylight ...

From such a description the path of the comet is easily traceable. The use of the expression 'swept through' was particularly appropriate, since the technical term for comets from very early times in China was *hui hsing* or *sao hsing*, 'brush-stars', although other equally descriptive terms such as 'long stars', 'candle-flame stars', etc. were used from time to time. Confusion with novae is of course always to be checked, for comets do not necessarily have tails: when a comet comes into line with the earth and the sun its tail is no longer visible and its light may appear nebulous. The Chinese had a special term for a comet in opposition, *po hsing*, clearly distinguishing it, at least theoretically, from a nova. Whether any manuscript drawings of comets still exist in the records of the Astronomical Bureau in Peking we do not know, but Fig. 117*a* shows a late one from Korean records.

There is no doubt that Halley's Comet is the most important in astronomy. This comes not only from the fact that it was the first comet to be established as a periodic one, but also because its history can be traced accurately for over two thousand years. That it can be followed back so far is due to the careful way in which the Chinese observations were recorded. Halley's own observations were made in 1682. He recognised his comet as the same which Apianus had seen in 1531 and Kepler in 1607, and predicted that it would return, as it did, in 1758. The importance of this return can hardly be over-estimated; it proved that some comets at least were part of the solar system, and that Newton's laws fitted their motions as well as those of the planets. After the next reappearance in 1835, astronomers and Sinologists set to work together to produce a complete re-computation of all appearances of the comet. The earliest Chinese observation which may have been Halley's comet is that of 467 B.C., but the data are insufficient for certainty; there seems, however, little doubt that the comet of 240 B.C. was Halley's. The reappearance in 163 B.C. is doubtfully identified, but those of

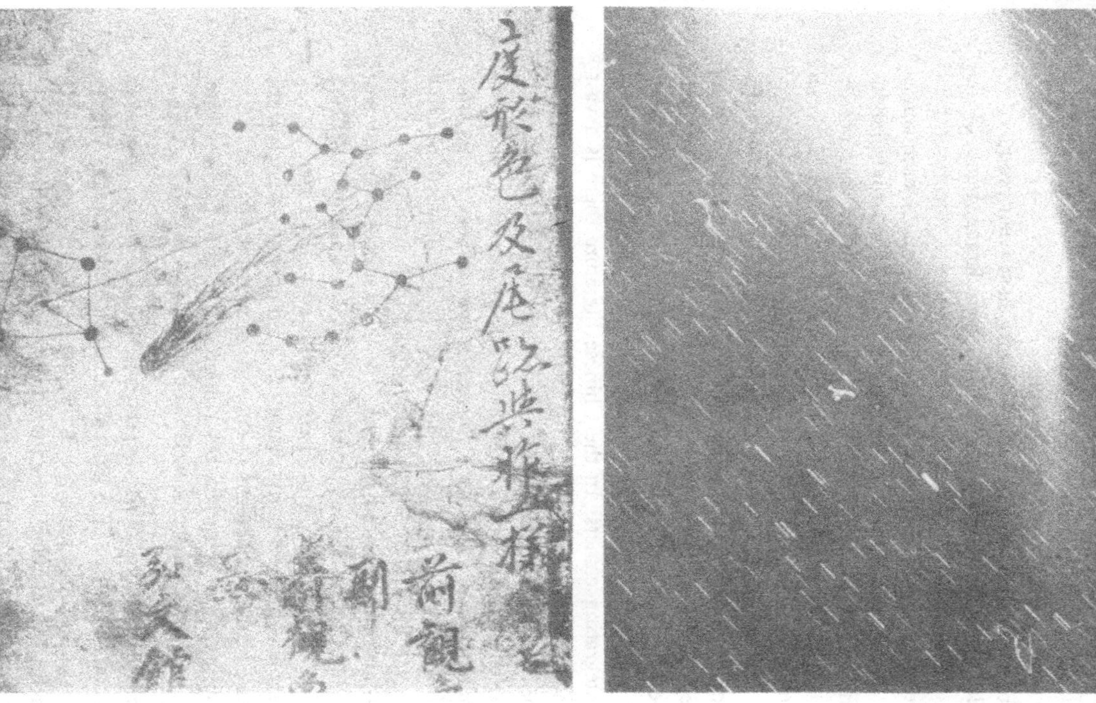

(*a*) (*b*)

Fig. 117. (*a*) Manuscript drawing of a comet passing between the *hsiu* I and Chen, on the night of 28 October 1664, from the records of the Korean Astronomical Bureau. The annotation at the side says that the shape and colour of the comet and its tail were the same as the previous night. The notes below refer to previous observation and the last line on the left signs for the Bureau. (Photograph, W.C. Rufus.) (*b*) Comet Arend-Roland photographed at Cambridge in yellow-red light about mid-night, 28 April 1957. The notable 'spike' pointing away from the tail is clearly shown. Though 'spiked' comets are rare, references to them exist in ancient and medieval Chinese sources. (Photograph, Argue & Wolf.)

87 B.C. and 11 B.C. are quite definite, the latter being observed in detail by the Chinese for nine weeks. Accurate observations were made of its reappearance in A.D. 66 and, after this, every reappearance in the 76-year cycle is found in the Chinese texts, including that of 1066, so familiar in Anglo-Norman history. The Chinese were also the first to observe that the tails of comets always point away from the sun.

Few elaborate theories about comets seem to have been produced in China; naturally some early writers ascribed them to derangements of the Yin and Yang. But as if in premonition of a modern theory, others associated different kinds of comets with the different planets, each comet being thought to originate from a particular planet. Thus the Chinese associated comets with the solar system, the theory being worked out by Ching Fang about 50 B.C.; his original text is lost, but there is a parallel one in another treatise.

Between comets and meteor streams there is a close connection. It is known that the Perseid meteors of August move in the same orbit as Tuttle's Comet, while the November Leonids (which used to give such great showers every 33 years) follow that of Tempel's Comet, and the May Aquarids that of Halley's. They are almost certainly remains of disintegrated or disintegrating comets. In the Chinese literature there is a vast mass of information about meteors, known as *liu hsing*, 'gliding' or 'shooting' stars, or *pên hsing*, 'energetic' stars, and meteor showers (*liu hsing yü*), as well as the actual fall of meteorites on the earth's surface (*hsing yün*). Ma Tuan-Lin summarised all this in the thirteenth century A.D. in his *Wên Hsien Thung Kao*. Many of the observations are highly detailed, and the whole list fills over 200 pages. The material would bulk even larger if the records of local histories were drawn upon. The earliest date from 687 B.C. and 644 B.C. (subject to the usual reservations about the adjustment of Chou texts by Han scholars). A very great Leonid shower was reported in A.D. 931, and so complete are the records that it has been possible to analyse the observations statistically and confirm the periods of recurrence.

Here is an example of careful description of the fall of a meteorite, from Shen Kua's *Mêng Chhi Pi Than* (Dream Pool Essays):

> In the first year of the Chih-Phing reign period (A.D. 1064), there was a tremendous noise like thunder at Chhang-chou about noon. A fiery star as big as the moon appeared in the south-east. In a moment there was a further thunderclap while the star moved to the south-west, and then with more thunder it fell in the garden of the Hsu family in the I-Hsing district. Fire was seen reflected in the sky far and near, and fences in the garden roundabout were all burnt. When they had been extinguished, a bowl-shaped hole was seen in the ground, with

> the meteorite glowing within it for a long time. Even when the glow ceased it was too hot to be approached. Finally the earth was dug up, and a round stone, as big as a fist, still hot, was found, with one side elongated (i.e. pear-shaped). Its colour and weight were just like iron.

Meteorites had many other names in Chinese books besides the *yün* already mentioned, or *yün shih*. Very early on they were called *thien chhuan*, 'hounds of heaven'. Moreover, as in other civilisations, meteorites were sometimes confused with axes of the Neolithic period: in A.D. 660 a meteorite presented to the emperor was called 'the stone axe of the thunder-god', and other names were 'the thunder-god's ink block' or 'thunder lumps'.

Solar phenomena: sun-spots

The outstanding example of a celestial phenomenon which in general escaped the attention of Europeans because of their preconceived idea of the perfection of the heavens, is that of spots on the sun. In Europe their discovery was one of the advances made by Galileo in his use of the telescope. He first saw them towards the end of 1610, but did not publish his results till 1613, by which time the same discovery had been made independently by Harriot in England, Fabricius in Germany, and Scheiner in Holland. Previously, dark spots on the sun seen by the naked eye had been ascribed to the passage of planets across it, and indeed Scheiner thought that the spots might be satellites of the sun. Galileo was able to refute this and showed they must be on or close to the sun's surface.

European records of sun-spots do go back before the seventeenth century: there were observations with the naked eye of spots in 807 and 840, but these were interpreted as transits of Mercury and Venus respectively across the sun's disc; others were seen in 1196 and 1457, but their true nature was not understood then either.

Chinese records are by a long way the most complete we have. They start nearly one thousand years before the first reference in the West, that is, in the time of Liu Hsiang, 28 B.C. Between that time and A.D. 1638 there are 112 descriptions of outstanding sun-spots in the official histories, but there are also numerous mentions in local topographical records, volumes of memoirs, and various other publications. Extensive lists occur in the thirteenth-century *Wên Hsien Thung Khao* and the *Thu Shu Chi Chhêng* encyclopaedia, although there are more modern ones. The black spots are often described as being 'as big as a coin', 'as big as a hen's egg', or as a peach, a plum, etc. The Chinese terms for spots are *hei chhi*, *hei tzu* and *wu*, and the use of *wu*, which means 'crow' as well as 'black', raises the question of whether some references well before 28 B.C. might be based on sun-spot observations. The existence of a crow in the sun (the colleague of the rabbit in the moon) was part of Chinese mythology in Chou and early Han times. Thus we find a passage in the first-century A.D. *Lun Hêng* (Discourses weighed in the Balance) where Wang Chhung says 'The scholars hold that there is a three-legged crow in the sun ...', after which he argues, in his sceptical way, that the thing is impossible. But it may mean that the black spots had been observed perhaps as early as the time of Tsou Yen (late fourth century B.C.).

As in the case of eclipses, it seems likely that from an early date the sun must have been observed through smoky rock-crystal or semi-transparent jade. Indeed, this technique is specifically referred to by the sixteenth-century naturalist Li Shih-Chen, who says that 'books on jade' mention certain kinds which were used for looking at the sun. Haze due to dust-storms from the Gobi would also have permitted the observation of sun-spots.

THE TIME OF THE JESUITS

From hints which have been dropped on the way, it will have been suspected that the coming of the Jesuits was by no means an unmixed blessing for Chinese science. We can draw up a provisional balance-sheet of the merits and demerits of their contribution. In the first place, the new European methods for the prediction of eclipses were greatly superior to the traditional Chinese methods. Secondly, the Jesuits brought a clear exposition of the analysis of planetary motions using geometry, and of course the appropriate mathematics for applying it. This had many other uses, such as (thirdly) the construction of sundials and the geometrical projections necessary for the astrolabe, and in surveying. A fourth contribution was the doctrine of the spherical earth and its division into spaces separated by meridians and parallels. Fifthly, the new European sixteenth-century algebra was made available to the Chinese, with many new computing methods, and ultimately mechanical devices such as the slide-rule. Sixthly, by no means the least valuable transmission was the up-to-date European techniques of instrument-making, the graduation of scales, the cutting of micrometer screws, and the like. The spread of the telescope was the climax of this.

Let us now look at the debit side of the transaction. Although the Jesuit mission was one of the most outstanding historical examples of inter-cultural contact at a lofty intellectual level, the Chinese astronomers quite often had right on their side.

China and the dissolution of the crystalline spheres

The world-picture which the Jesuits brought was that of a closed universe of crystal spheres nesting in one another, with the earth fixed at the centre. Hence they opposed the indigenous Hsüan Yeh doctrine of heavenly bodies floating in infinite space, and the irony was that they did so just at a time when the best minds in Europe were breaking away from the closed system. Hence also (second) they obstructed the spread of the sun-centred universe of Copernicus in China for they could not but be sensitive to the condemnation of Galileo by the theologians of the Church. It followed, thirdly, that they wanted to substitute an erroneous theory of the precession of the equinoxes for the cautious Chinese refusal to form any theory at all about it. Fourthly, they completely failed to appreciate the equatorial and polar character of traditional Chinese astronomy, confusing the *hsiu* divisions with the zodiac, and they equalised the 12 equatorial divisions when there was no need to do so. Fifthly, in spite of the advance to the equatorial coordinates (right ascension and declination) which was just being made by Tycho Brahe in Europe, the Jesuits tried to imposed the less satisfactory Greek ecliptic coordinates on Chinese astronomy.

A fascinating glimpse of this paradoxical situation is seen in letters which Ricci wrote late in 1595, enumerating the 'absurdities', as he called them, of the Chinese. They say, he wrote, that:

(1) The earth is flat and square, and that the sky is a round canopy; they did not conceive the possibility of the antipodes.
(2) There is only one sky (and not ten skies). It is empty (and not solid). The stars move in a void (instead of being attached to the firmament).
(3) As they do not know what air is, where we say there is air (between the spheres) they affirm that there is a void.
(4) By adding metal and wood, and omitting air, they count five elements (instead of four) – metal, wood, fire, water and earth. Still worse they make out that these elements are engendered, the one by the other; and it may be imagined with how little foundation they teach it, but as it is a doctrine handed down from their ancient sages, no one dares to attack it.
(5) For eclipses of the sun, they give a very good reason, namely that the moon, when it is near the sun, diminishes its light.
(6) During the night, the sun hides under a mountain which is situated near the earth.

Here we see the elements of superiority in European science at the turn of the sixteenth and seventeenth centuries imposing a fundamentally wrong world-picture, that of the solid spheres, on the fundamentally right one which had come down from the Hsüan Yeh school, of stars floating in an infinite empty space. But did any stimulus come back in exchange?

It seems that some general ideas about Chinese knowledge of the heavens and their freedom from the restrictions suffered in Western astronomy (such as the perfection and changelessness of celestial bodies), and also the scepticism of the Chinese about the crystal spheres, were not without effect. Especially was this so over the question of life elsewhere in the universe, the 'plurality of worlds' that had been held impossible by scholars in medieval Europe. The seventeenth century saw a proliferation of science fiction about interplanetary travel, and some also had a Chinese theme, like Francis Godwin's *The Man in the Moone; or a Discourse of a Voyage Thither, by Gonsales, the Speedy Messenger* (1638) in which, when the traveller returns to earth, he lands in China, where he meets both mandarins and missionaries and notes that both the inhabitants of the moon and those of China speak a tonal language. In less romantic form the same idea was urged by John Wilkins in his *Discovery of a World in the Moon, tending to prove that 'tis probable that there may be another habitable World in that Planet*, also 1638, and by Christiaan Huygens (1698). The Chinese theme occurs again in an amusing political satire by Daniel Defoe, *The Consolidator; or Memoirs of Sundry Transactions from the World in the Moon* (1705), who mentions both Godwin and Wilkins. Yet a third recurrence comes in Miles Wilson's *History of Israel Jobson, the Wandering Jew*, who visited all the planets and allegedly wrote the account of his travels in Chinese (1757). Indeed, it is hardly possible to take up any book of this period dealing with the plurality of worlds without finding some reference to China.

In a word, it may be that the Chinese idea of an infinite universe which could undergo change, and their complete disbelief in the crystalline spheres of the West, which became evident to the Jesuits as soon as they began to discuss cosmology in China, was one of the elements which combined in breaking up medieval views in Europe, and contributed to the birth of modern astronomy. For such a suggestion there is contemporary evidence in the words of Christopher Scheiner, seeking about 1625 to show that the realm of the stars had a fluid nature:

> The peoples of China [he said] have never taught in any of their innumerable and flourishing academies that the heavens are solid; or so we may conclude from their printed books, dating from all times during the past two millennia. Hence one can see that the theory of a liquid heavens is really very ancient, and could easily be demonstrated; moreover one must not despise the fact that it seems to have

> been given as a natural enlightenment to all peoples. The Chinese are so attached to it that they consider the contrary opinion (a multiplicity of solid celestial spheres) perfectly absurd, as those inform us who have returned from among them.

And we may take as another example the opinion of Johan Nieuwhoff, who accompanied the Dutch ambassadors to Peking in 1656, and who found the idea of the plurality of worlds very Chinese.

In any case the height of irony is reached when we find Wells Williams, in 1848, reproaching the late Chhing popular writers for their belief in the solid celestial spheres, under the impression that this was a primitive Chinese doctrine still persisting.

The imperfect transmission

The sincere (and well-justified) admiration of Ricci himself for the instruments of Kuo Shou-Ching has already been mentioned. But such was the decadence of the late Ming period, and so convinced were the Europeans of their scientific superiority, that the accounts of Chinese astronomy which got through to seventeenth-century Europe were mainly unfavourable. For instance Louis Lecomte could write in 1696:

> They still continue their Observations. Five Mathematicians spend every Night on the Tower in watching what passes overhead; one is gazing towards the Zenith, another to the East, a third to the West, the fourth turns his eyes Southwards, and a fifth Northwards, that nothing of what happens in the four Corners of the World may scape their diligent Observation ... If this had always been practised by able and careful Mathematicians, we should have a great number of curious Remarks; but besides that these Astronomers are very unskilful, they take little care to improve that Science; and provided their Salary be paid as usual, and their Income constant, they are in no great trouble about the Alterations and Changes which happen in the Sky. But if these Phenomena's are very apparent, as when there happens to be an Eclipse, or a Comet appears, they dare not be altogether so negligent.

The two most important features in European astronomy at the time the Jesuits began their work in China were the invention of the telescope and the acceptance of the theory of Copernicus, putting the sun, not the earth, as the centre of the universe. The former they transmitted but the latter, after some hesitations, they held back. Ricci died in Peking in 1610, the same year in which Galileo published his *Sidereus Nuncius* (Sidereal Messenger) announcing his telescopic discoveries. It was in 1616 and in

1632 that the two condemnations of Galileo's Copernican views were made by the Holy Office and these had a considerable effect on the China Mission.

In 1618 the Jesuit Father Terrentius (Johann Schreck) arrived in China; he had been the seventh member of the Cesi Academy, having been elected next after Galileo, and was an astronomer and physicist of great gifts. He brought with him a telescope, which was eventually given to the emperor in 1634, and remained in touch with Galileo, who was not very helpful, and with Kepler, who took more interest. In China the telescope became known as the 'Far-Seeing Optick Glass' (*Yuan ching*).

It is clear that in this early period, especially before the condemnation of Galileo, the missionaries were not at one on Copernicanism. In general it may be said that Chinese books between 1615 and 1635 described telescopic discoveries, but did not mention Copernicanism, then for a short time the heliocentric theory was mentioned, but after news of the condemnation reached China a curtain descended, and a belief in the earth as the centre of the universe (geocentric theory) returned. As we saw above (page 190), the general impression the Chinese got was a garbled one suggesting that the Europeans were not clear in their own minds about the solar system. This did not, of course, affect the Jesuits' abilities in calendar reform, since the heliocentric and geocentric theories are both equivalent from a calendrical point of view.

'Western' science or 'new' science

Between 1629 and 1635, the second generation of missionaries, of which Terrentius was one, produced a monumental compendium of the scientific knowledge of the time. This was entitled, on its presentation in the latter year, the *Chhung-Chên Li Shu* (Chhung-Chên Reign-Period Treatise on Astronomy and Calendrical Science). Ten years later it was re-issued as the *Hsi Yang Hsin Fa Li Shu* (Treatise on Astronomy and Calendrical Science according to the New Western Methods). Eventually it formed the basis for the *Yu-Ting Li Hsiang Khao Chhêng* (Complete Studies on Astronomy and Calendar) printed in 1723, then in 1738, incorporated into the *Thu Shu Chi Chhêng* (Imperial Encyclopaedia) of 1726, and finally improved by the addition of astronomical tables embodying more recent Western observations (1738).

Here we must halt for a moment. The reader will probably have noticed nothing especially significant in the preceding paragraph, seemingly concerned only with the recitation of fact. But actually it raises certain points of extreme importance in these culture-contacts of the two great civilisations. It is vital today that the world should recognise that seventeenth-century Europe did not give rise to essentially 'European' or

'Western' science, but to universally valid world science, that is to say 'modern' science as opposed to the ancient and medieval sciences. Now these last bore indelibly an ethnic image and superscription. Their theories, more or less primitive in type, were rooted in their culture and could find no common form of expression. But when once the basic technique of discovery had itself been discovered, once the full method of scientific investigation of Nature had been mastered, the sciences assumed the absolutely universal nature of mathematics, and in their modern form are at home under any meridian, the common light and inheritance of every race and people. Of arguments about elements or 'humours', Yin and Yang, there could be no end: the disputants could reach no common ground. But the mathematisation of hypotheses led to a universal language, an ecumenical medium of exchange, a reincarnation of the merchants' single-value standard on a plane transcending merchandise. And what this language communicates is a body of incontestable scientific truth acceptable to all men everywhere. The physically unified world of our own time has indeed been brought into being by something that happened historically in Europe, but no man can be restrained from following the path of Galileo and the sixteenth- and seventeenth-century natural philosophers. Meanwhile the period of political dominance which modern technology granted to Europeans is now demonstrably ending.

In their gentle way, the Jesuits were among the first to exercise this dominance, spiritual though in their case it was meant to be. To seek to accomplish their religious mission by bringing to China the best of Renaissance sciences was a highly enlightened proceeding, yet this science was for them only a means to an end. Their aim was naturally to support and commend 'Western' religion by the prestige of the science from the West which accompanied it. This new science might be true, but for the missionaries what mattered even more was that it had originated in Christendom. The Chinese were acute enough to see through all this from the very beginning. The Jesuits might insist that Renaissance natural science was primarily 'Western', but the Chinese clearly understood that it was primarily 'new'.

Already by 1640 the Jesuits noted a strong disinclination on the part of the imperial court to accept the designation *Hsi Yang* (Western) in book titles, but after the Manchu conquest of 1644, they felt that they could get away with it. However, by 1666, when the Khang-Hsi emperor was spending much time himself studying modern science, he insisted that editions of the encyclopaedia of mathematics and astronomy should thenceforward bear the title *Hsin Fa* ('according to the new methods') only. In this way he united himself with that group of men at the other end of the Old World who were gathering together in the Royal Society of London to further the new, or experimental, philosophy. It is doubtful, however, if he ever knew anything much about them, while they would have known barely his imperial name.

The integration of Chinese astronomy into modern science

In 1699 there began a great refitting of the Peking Observatory under the direction of the learned Jesuit Ferdinand Verbiest. The instruments of Yuan or Ming time were taken down from the astronomical platform on the eastern wall of the city, and a new set installed in their place, where they have remained until the present time (Fig. 118). The Jesuit and later instruments are as follows: (1) a simple ecliptic armillary sphere, (2) a simple equatorial armillary sphere, (3) a large celestial globe, (4) an horizon circle for azimuth measurements, (5) a quadrant, (6) a sextant, (7) a quadrant altazimuth for measuring altitudes and azimuths, (8) an elaborate equatorial armillary sphere, and (9) a smaller celestial globe. Items (1) to (6) were established by Verbiest, (7) and (8) were eighteenth century, but (9) is something of a mystery and may have been a thirteenth-century globe of Kuo Shou-Ching given an eighteenth-century housing.

The transmissions of the Jesuits seem to have affected a number of Chinese scholars who were more or less outside their circle. For instance, as we have seen (page 190), the *Wu Hsing Hsing Tu Chieh* (Analysis of the Motions of the five Planets) by Wang Hsi-Shan (1640) proposed what was essentially the planetary system of Tycho Brahe, a compromise made in 1593 in which the sun moves round the earth, but all the rest of the planets move round the sun. He followed it up three years later with a larger work, the *Hsiao-An Hsin Fa* (Wang Hsi-Shan's New (Astronomical) Methods) which was an attempt to synthesise Western and Chinese ideas. His contemporary Hsüeh Fêng-Tsu, who was more closely connected with the Jesuits, though probably a Copernican, wrote in 1650 a conciliation of Western and Chinese astronomy *Thien Hsüeh Hui Thung* (Towards a Thorough Understanding of Astronomical Science) and later a book on eclipses, the first Chinese work to make use of logarithms.

As the eighteenth century went on, Chinese astronomers and mathematicians emancipated themselves more and more from the spell which the Jesuit apparition had woven during the decadent Ming and early Chhing times. At the end of the century, among several important works, there was a treatise on celestial map-making, *Kao Hou Mêng Chhiu* (Investigation of the Dimensions of the Universe), by Hsü Chhao-Chun (1800), and when in 1851 Fêng Kuei-Fen published his *Hsien-Fêng Yuan Nien Chung Hsing Piao* (Table of Meridian Passages of 100 Stars in Right Ascension and Declination for 1851), Chinese astronomical science might be said to have merged at last with that of the world as a whole.

Fig. 118. The Peking Observatory, photographed about 1925 from the north-east corner of the platform. On the right the armillary sphere of Kögler & von Hallerstein (A.D. 1744) and Verbiest's quadrant; in the centre Verbiest's celestial globe. At the back, in the southern row, from right to left, Verbiest's ecliptic armillary sphere and horizon circle, then Stumpf's quadrant altazimuth (1714). The background is formed by roofs and trees of the city of Peking. (Photograph, Whipple Museum Collection.)

Unexpectedly, then, the Jesuit intervention led in due course to a rediscovery on the part of the Chinese of the achievements of their own civilisation before the Ming decadence. All in all, the Jesuit contribution, chequered though it was, had the qualities of a noble adventure. If the bringing of the science and mathematics of Europe was for them a means to an end, it stands for all time nevertheless as an example of cultural relations at the highest level between two civilisations hitherto sundered.

SUMMARY

An epilogue to a long section should have the grace of being short. It will by now be abundantly evident that the Chinese contribution to the development of astronomical science was a very remarkable one (see Table 33). Without running over again all the specific points to which attention

Table 33. *Chart to show the comparative development of astronomy in East and West*

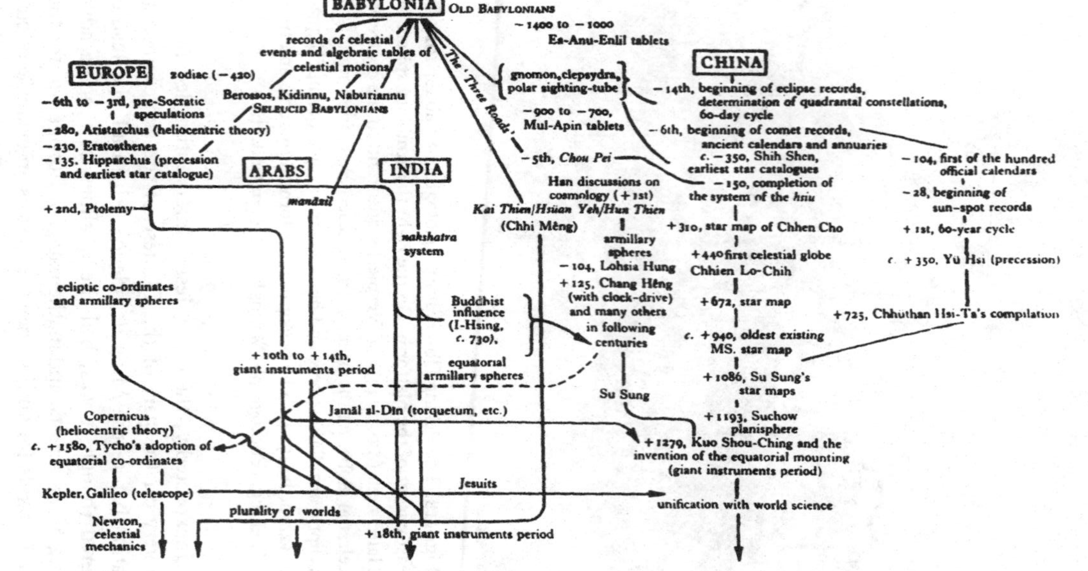

has been drawn, we may mention (*a*) the polar and equatorial system of the heavens, strikingly different from that of the Greek and Hellenistic peoples though equally logical; (*b*) the early conception of an infinite universe, with the stars as bodies floating in empty space; (*c*) the development of quantitative positional astronomy and star-catalogues two centuries before any other civilisation of which comparable works have come down to us; (*d*) the use in these catalogues of equatorial (i.e. essentially modern) coordinates, and faithfulness to them extending over two thousand years; (*e*) the elaboration, in increasing complexity, of astronomical instruments, culminating in the thirteenth-century A.D. invention of the equatorial mounting; (*f*) the invention of the clock drive for that forerunner of the telescope, the sighting-tube, and of a number of ingenious devices to improve astronomical instruments; and (*g*) the maintenance, for longer continuous periods than any other civilisation, of accurate records of celestial phenomena such as eclipses, sun-spots, comets, novae, etc.

The most obvious absences from such a list are just those elements in which Western astronomy was strongest, the Greek geometrical schemes of planetary motion, the Arabic use of geometry for making engraved charts for their astrolabes, and the physical astronomy of the Renaissance. We often hear of 'the Greek genius for inquiry – the desire not only to know the facts but the reasons behind the facts ...', but this is surely a false antithesis. It was not necessary that the reasons should be conceived geometrically or mechanically. The Chinese did not feel the need for these forms of explanation – the component organisms in the universal organism followed their Tao each according to its own nature, and their motions could be dealt with in the essentially 'non-representational' form of algebra. The Chinese were thus free from that obsession of European astronomers for the circle as the most perfect figure. Nor did they experience the medieval prison of the crystalline spheres, those unexpectedly unyielding materialisations of the Greek spirit of geometry. If, like all Chinese science, Chinese astronomy was fundamentally empirical and observational, it was spared the excesses and aberrations, as well as the triumphs of Western theorising. But clearly it requires a much more important place in the history of science as a whole than historians have been wont to give it.

3

The sciences of the heavens: (ii) Meteorology

Meteorology is a word which has undergone much change of meaning since the time when it was born in ancient Greece. For Aristotle (as in his *Meteorologica*) it included the study of many events which are now known to be celestial, such as meteors and meteorites (from which it took its name), comets and the Milky Way, though at that time they were classed as belonging to the 'sublunary' world – the nearer regions of space. In the modern sense, meteorology has become essentially the study of climate, weather and all that goes on in the earth's atmosphere, together with tidal phenomena. There is no work in Chinese equivalent to Aristotle's *Meteorologica*, but that does not mean that the Chinese were not deeply interested in weather matters. And they were long in advance of the West in certain methods of meteorological measurement, keeping records of a more complete nature over a much longer time. As regards the tides, they were also sometimes considerably in advance of Europeans.

CLIMATE IN GENERAL

China's succession of passing weather is the product of the seasonal monsoon circulation, the occasional tropical cyclones, and the procession of continental cyclonic storms, all modified by the relief features of the subcontinent. Though China's weather is affected more by influences from the land-mass to the west than from the Pacific Ocean to the east, yet the arid nature of central Asia contends with the moisture of the south-eastern seas over a perennial Chinese battlefield. During the summer the air masses over inner Asia become heated, expand, rise and overflow towards the encircling oceans. The consequent reduced pressure causes warm moist oceanic air to be drawn along the surface, thus setting up a very large-scale circulation by convection. Winter reverses the process. The resulting winds are the monsoon winds, less regular in China than in India, but equally the basic background of the climate. Hence the fact, familiar to everyone who has lived in China, that the rainfall occurs mainly in a distinct rainy season, occupying usually the three summer months.

The prevailing tendency of the monsoon blowing inwards in summer and outwards in winter is complicated by the eastward movement of high- and low-pressure systems migrating from time to time. The low-pressure areas, or depressions, are most common in spring and early summer and give rise to characteristic unsettled spells of cloudy and showery weather in central and north China. The third element in the Chinese climate is the tropical typhoon, a small but very intense disturbance with extremely low pressure at the centre, steep pressure gradients, and wind velocities up to 265 km an hour. Though the typhoon as a whole moves fast (often several hundred kilometres a day), the total area under the sway of its damaging winds is frequently not more than 160 km in diameter. The typhoon originates in the Pacific, and after travelling westward tends northward as it strikes the Chinese coast, eventually dying out in the interior provinces.

To what extent has there been a long-term change in the Chinese climate? The question has received a good deal of discussion, and the consensus of opinion has been that China (or at least north China) was formerly both warmer and moister than at present. This conclusion has been drawn mainly from climatic and biological evidence contained in ancient texts. For instance, contemporary observations in the Warring States period show that such annual events as the blooming of peach-trees (*Prunus persica*), the commencement of song by the cuckoo (*Cuculus micropterus*), and the first appearance of the house-swallow (*Hirundo rustica*), were all placed by the ancient books a week to a month earlier than the records of the present day. Again, elephants existed at the Yellow River latitude in the Shang and Chou periods, but are not mentioned thereafter, and down to the Sung, but not since, crocodiles were to be found in south China.

As to the growth of weather-lore in China, prediction never advanced beyond the state of peasant proverbs. Nor did it do so in Europe before the Renaissance. Such proverbs form part of the 'omen text' which was one of the main components of the *I Ching* (Book of Changes), and are often met with in ancient books. For example, in the *Tao Tê Ching* (Canon of the Tao and its Virtue) of the fourth century B.C., we read 'A hurricane never lasts a whole morning, and a sudden rainstorm will not go on for the whole day'. In the *I Ching* haloes foretell storms, and eastward travelling clouds (the opposite direction to the monsoon rains) bode well for travellers. Weather observation and prediction must have played a large part in the study of Nature by the Taoists. Yeh Mêng-Tê in the *Pi Shu Lu Hua* (Conversations while Avoiding the Heat of Summer) of A.D. 1156 wrote:

> When I lived in the mountains I often saw that old farmers could predict rain and sunshine, proving right seven or eight times out of

> ten. I asked them their methods, but they said that there was nothing but experience. If you ask those who live in cities, they know nothing. Since at that time I had plenty of leisure, I often rose very early in the morning, and with an empty mind concentrated on the clouds, mountains, river, fields and trees in all their beauty, and found I could predict the weather aright seven or eight times out of ten...

The supposed influence of climate on human beings in health and disease was also discussed, as it was in ancient Greece.

TEMPERATURE

In Han times the 'phenomenalists' interpreted excessive seasonal cold and heat as 'heavenly reprimands' for deficiencies on the part of the emperor or his administration. It was thus essentially for astrological purposes that records were kept of summers especially hot or winters excessively severe, and these found their way into official histories. Temperature was important also for the ancient Chinese on account of the difficulties in making the clepsydra keep accurate time (page 151). Records of excessive cold and heat occupy four chapters in the *Thu Shu Chi Chhêng* (Imperial Encyclopaedia) of A.D. 1726, and from these records long-period pulsations of climate have been inferred. A rough correlation with sun-spot frequency can also be seen (Table 34).

PRECIPITATION

The study of Anyang oracle-bones has shown that as far back as the thirteenth century B.C. rather systematic meteorological records were being kept. An analysis has been made of bone records from 1216 B.C. in which rain, sleet, snow, wind, and direction of rain and wind, are all mentioned for 10-day periods. Many successful predictions are recorded because of the habit of the scribes of writing an additional note on the bone saying that, for example, it did in fact snow, after the divination process had said that it would. As in other civilisations, early meteorology was closely connected with divination.

Of greater scientific interest is the recognition of the meteorological water-cycle in China. Perhaps the oldest indication that this was understood occurs in the *Chi Ni Tzu* (Book of Master Chi Ni), a naturalist work probably of the late fourth century B.C. There it is said that:

> the wind is the *chhi* of heaven, and the rain is the *chhi* of earth. Wind blows according to the seasons and rain falls in response to the wind. We can say that the *chhi* of the heavens comes down and the *chhi* of the earth goes upwards.

Table 34. *Correlation of severe winters with sun-spot frequency*

Century (A.D.)	No. of severe winters per century: Europe	China	Sun-spot frequency (Chinese records)
6th	—	19	7
7th	—	11	0
8th	—	9	0
9th	11	19	8
10th	11	11	1
11th	16	16	3
12th	25	24	16
13th	26	25	6
14th	24	35	9
15th	20	10	0
16th	24	14	2

Later, there are other statements of a similar kind, a book in the Han stating that clouds are water from the Khun-Lun mountains evaporating and rising. This would be about 50 B.C. In the first century A.D., however, the clear distinction between the circulation in the atmosphere and the vast distances of the starry firmament had still not become fully accepted. The Greeks were subject to similar confusions. Wang Chhung, in his *Lun Hêng* (Discourses weighed in the balance) of about A.D. 82 has an interesting passage on the subject:

> The Confucians also maintain that the expression that the rain comes down from heaven means that it actually does fall from the heavens (where the stars are). However, consideration of the subject shows us that rain comes from above the earth, but not down from heaven.
>
> ... How can we demonstrate that the rain originates in the earth and rises from the mountains? Kungyang Kao's commentary on the *Spring and Autumn Annals* says: 'It evaporates upwards through stones one or two inches thick, and gathers. In one day's time it can spread over the whole empire, but this is only so if it comes from Thai Shan.' What he means is that from Mount Thai rain-clouds can spread all over the empire, but from small mountains only over a single province – the distance depends on the height. As to this coming of the rain from the mountains, some hold that the clouds carry the rain with them, dispersing as it is precipitated (and they are right). Clouds and rain really are the same thing. Water evaporating upwards becomes clouds, which condense into rain, or still further

> into dew. When the garments (of those travelling on high mountain passes) are moistened, it is not the effect of clouds and mist (through which they are passing), but of the suspended rain water.

There follows next a passage on seasonal lunar and stellar connections, but the main interest of the passage is both in its clear understanding of the water-cycle and in its appreciation of mountain ranges in the precipitation process. In later times the water circulation was well understood. Soon after Wang Chhung's time, clouds were defined as the 'moisture evaporated from marshes and lakes', and in the third century A.D. the cycle was again discussed, while many statements of it can be found in the Sung and Ming.

In Greece the recognition of the water-cycle goes back to the sixth century B.C., with Anaximander of Miletus. Aristotle built his *Meteorologica* round the idea of two emanations from earth, one watery and the other gaseous. The watery corresponds rather closely to the Chinese conception of the ascending watery *chhi* of earth; the gaseous may have originated from observations of such things as the sulphur-depositing gases of fumaroles, and was called upon to explain the formation of minerals and metals in the rocks. We shall meet with its close parallel in what the *Huai Nan Tzu* book has to say on the subject (page 307). One can only leave open the possibility of transmission in either direction, but the very early date at which these ideas were developing seems to make it most unlikely.

In different ages individuals naturally acquired particular fame in rain forecasting. One such was the magician Ching Fang of the first century B.C.; his lore was recorded to some extent in his *I Chang Chü* (Commentary on the Book of Changes). Another was Lou Yuan-Shan of the Sung. Then some books of the tenth century A.D. quote an otherwise mysterious lost book by one Huang Tzu-Fa of uncertain date, probably Han, on rain prediction. Lacking our cloud classification (cirrus, cumulus, etc.) the medieval Chinese devised many technical terms such as yellow clouds covered like chariots (*fu chhê*), shuttle-shaped (*shu chu*, lenticular) clouds, and so on. Another weather sign to which much attention was paid was a lunar halo (*yüeh yün*) caused by cirro-nebula at great heights, and regarded by the Chinese as a sure sign of wind. Lunar haloes showing colours were termed *chu mu*, 'typhoon mothers'. Solar haloes were also studied.

The first hygrometer in China was not introduced by the Jesuits as sometimes thought, for the Chinese had from quite early times taken advantage of the hygroscopic properties of such things as feathers and charcoal for indicating humidity. In the second-century B.C. *Huai Nan Tzu* book there are at least two mentions of a practice of weighing elm charcoal with a view to rain prediction by testing the moisture in the atmosphere. The test is met with in later books. It is interesting that in fifteenth-century

Europe, Nicholas of Cusa used exactly the same method, weighing wool against stones.

So much for the forecasting of rain. Since it had its inevitable sequel in the rising of rivers and canals, with the danger of floods, always so serious in China, it would not be surprising to find that the Chinese made use of rain-gauges from an early period. In Europe the very simple idea of catching rain in some kind of container so as to permit measurement dates only from A.D. 1639, but there were rain-gauges in Korea in the fifteenth century. What has not hitherto been realised, however, is that the rain-gauge is not a Korean invention, but goes back a good deal earlier in China. The chief evidence for this is that the *Shu Shu Chiu Chang* (Mathematical Treatise in Nine Sections) by Chhin Chiu-Shao of A.D. 1247 contains problems on the shape of rain-gauges. At that time there would seem to have been one in each provincial and district capital.

Still more remarkable, the same book shows us that snow-gauges were also in use. These were large cages made of bamboo, doubtless placed beside mountain passes and on uplands. If local magistrates in the Sung really transmitted to the capital readings on rain- and snow-fall, the high officials must have been greatly assisted in making their calculations concerning the maintenance and repairs likely to be required for dykes and other public works.

Needless to say, Chinese records have a very long and abundant list of floods and droughts, as Table 35 indicates.

RAINBOWS AND ASSOCIATED EVENTS

There are references to the rainbow in the Anyang oracle-bone inscriptions. The modern term for it, *hung* (虹), perpetuates, because of the 'insect' radical which it contains, the idea of the Shang people that it was a visible rain-dragon. In the Sung period, Shen Kua described a double rainbow which he saw when on a diplomatic mission about A.D. 1070 to the Chhi-tan Tartars in Kansu. Both he and Sun Yen-Hsien considered that the rainbow was due to the reflection of sunlight from suspended water-drops. Two centuries had to elapse before Quṭb al-Din al-Shīrazī (A.D. 1236 to 1311) gave in Persia the first satisfactory explanation of the rainbow, in which the light is said to be refracted twice and reflected once through a solid sphere (a water-drop).

But the rainbow in all its beauty is far outdone by the strange complex phenomenon which includes haloes and 'mock suns' or parhelia. Under certain atmospheric conditions, when the sun shines through high clouds of hexagonal or pyramidal ice-crystals, it is seen surrounded by haloes, and flanked by as many as four centres of bright light (the 'mock suns'), while there may more rarely be a fifth sun image opposite in the sky

Table 35. *Ratio of droughts to floods by centuries*

Century (A.D.)	Raininess ratio droughts per century / floods per century
2nd	1·98
3rd	1·60
4th	8·20
5th	2·06
6th	4·10
7th	3·30
8th	1·32
9th	1·80
10th	1·80
11th	1·70
12th	1·04
13th	1·80
14th	1·05
15th	2·25
16th	1·95

to the sun, and mock suns and haloes in this position too. A vertical column or pillar may also cross the sun. The first European description of such a display was given in A.D. 1630, but in China astrologically minded star-clerks had been devoting meticulous attention to halo phenomena a whole thousand years earlier. Every one of the sixteen or seventeen component arcs and haloes was actually named in the *Chin Shu* and other seventh-century A.D. writings. And so impressive did the Chinese find them that an emperor himself did not disdain to write an illustrated book which dealt, among other things, with this subject. This is the *Thien Yuan Yü Li Hsiang I Fu* (Essays on (Astronomical and Meteorological) Presages) by Chu Kao-Chih who reigned (for one year only) as Jen Tsung in A.D. 1425 (Fig. 119).

Of course, all kinds of presages were inferred from these appearances, but the precision of the observations is astonishing. Some 26 technical terms were in use by the seventh century A.D., and we cannot but conclude that the Europeans of the seventeenth century were long anticipated in the close study of halo phenomena.

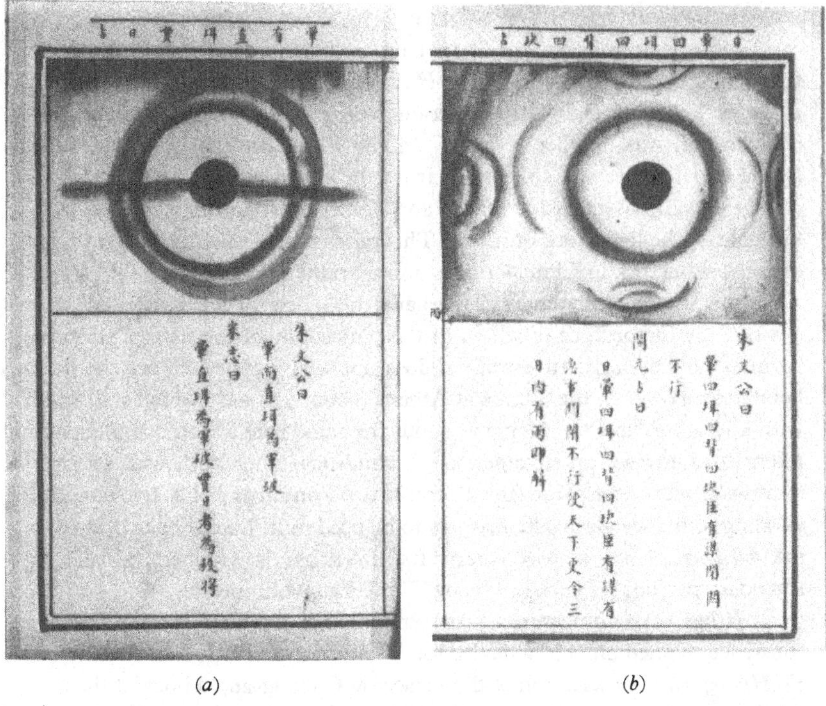

(*a*) (*b*)

Fig. 119. Two pages from the manuscript *Thien Yuan Yü Li Hsiang I Fu* by Chu Kao-Chih, emperor of the Ming, *c.* A.D. 1425, showing parhelia. (Photograph, Cambridge University Library).

Translation:

(*a*) '*Haloes* (*Yün*) *having straight Erh and threading the sun.* Chu Wên Kung says that when there is a *yün* with vertical *erh* an army will be defeated. The *Sung History* Memoirs also say that it means this, adding that if the sun is threaded through, a commander will be killed.'

(*b*) '*A solar halo* (*Yün*) *with four Erh, four Pei and four Chüeh.* Chu Wen Kung says that this signifies conspiracies on the part of ministers; let the gates be shut and (the emperor) not stir forth. The *Khai-Yuan Chan Ching* says exactly the same, further advising the issue of (emergency) orders throughout the empire. If within three days there is rain, the orders are to be cancelled.' Thus although the motive was often astrological, the observations of haloes were precise. In the above drawings one can recognise the parhelic circle, haloes at 22° and 46°, and a variety of tangent arcs.

WIND AND THE ATMOSPHERE

Observations on the wind, more or less fragmentary, were being made throughout Chinese history. The *Huai Nan Tzu* book (about 120 B.C.) contains a list of eight wind 'seasons' during the year, and later this classification was much elaborated. In the Ming Book *Li Hai Chi* (The Beetle and the Sea – a title taken from the proverb that the beetle's eye cannot encompass the wide sea) Wang Khuei mentions experiments with kites to test the behaviour of winds. This raises the question of the antiquity of the weathercock in China – not as unimportant a matter as it might seem, since this simple instrument is perhaps the oldest of all pointer-reading devices, the importance of which in the philosophy of the natural sciences requires no emphasis. In Europe it does not seem possible to trace it back before the Tower of the Winds at Athens (about 150 B.C.) which was fitted with a weather-vane. In China at about the same date a Han commentator refers to a thread or streamer as 'a wind-observing fan', and military treatises from the San Kuo (third century A.D.) onwards call it 'five ounces', alluding to the weight of the feathers to be used in it. Later books bring out the bird-like form it was given. Its invention is ascribed to various legendary personages such as Huang Ti the Yellow Emperor.

It has been conjectured that some attempt was made in the Han to construct an anemometer, a device for measuring wind velocity. In the *San Fu Huang Thu* (Description of the Palaces at Chhang-an), a book of the late third century A.D. attributed to Miao Chhang-Yen, there is the following:

> Kuo Yuan-Sêng in his *Shu Chêng Chi* (Records of Military Expeditions), says that south of the palaces there was a Ling Thai, fifteen *jen* (37 m) high, upon the top of which was the armillary sphere made by Chang Hêng. Also there was a wind-indicating bronze bird which was moved by the wind; and it was said that the bird moved (only, or faster?) when a 1000 *li* (very strong?) wind was blowing.

The evidence of this passage is not very clear, though unless the movements of the bird had something to do with the strength of the wind there would hardly be any point in referring to them. Moreover, elsewhere in the same book there is a mention of a bronze phoenix set on a roof tower 'which faced the wind on a turning axle above and below, as if flying'; this suggests the continuation of the axis to a lower floor where it could have been fitted with a device to indicate, if not record, the speed of rotation by the wind. It is not out of place to reflect that our present anemometer with hemispherical cups on the ends of its arms is a version of the paddle-wheel theme, and that it was in the Han (as we shall in a later volume) that the first water-wheels appeared. Besides, the bronze phoenix is said to have been 1·5 m high, which seems rather excessive for a weather-cock, but not for a device where wind resistance was sought. If this interpretation is justified, the Han anemometer may have been an anticipation of the modern four-cup type, for the early Renaissance anemometers were of a pendulum pattern not used today.

A good description of a whirlwind with waterspouts is given by Yang Yü in his *Shan Chü Hsin Hua* (New Discourses from the Mountain Cabin) of A.D. 1360.

> On the fifteenth day of the twelfth month, in the eighth year of the Chih-Chêng reign-period (A.D. 1348), at about 3 o'clock in the afternoon, there appeared in the south four black 'dragons' coming down from the clouds, and taking up water. Shortly afterwards another one appeared in the south-east, and lasted a considerable time before it disappeared. This was seen at Chia-hsing city.

As to effects of the earth's atmosphere, we have already seen something of this in the question of the apparent size of the sun when at different altitudes (page 89). About A.D. 400 the astronomer Chiang Chi undertook to explain why the sun rises and sets as a large red globe but appears small and white at noon.

> The terrestrial vapours do not go up very high into the sky. This is the reason why the sun appears red in the morning and evening, while it looks white at midday. If the terrestrial vapours rose high into the sky, it would still look red then.

He thus understood that when at low altitude, the sun was seen through a thicker layer of the earth's atmosphere than when it was high in the heavens. A century earlier Kuo Pho had spoken of the dawn and sunset mists in a way which suggests he thought of them as composed of minute particles. As for mirages, an explanation substantially correct was given by Chhen Thing in the Ming period.

THUNDER AND LIGHTNING

It was natural that the ancient Chinese should have conceived of thunder and lightning as the result of a clash between the two deepest physical forces which they could imagine, the Yin and the Yang. In the broadest sense, when one bears in mind what this theory contributed to the conception of positive and negative in Nature, including ultimately the effects of electricity and chemical combination, there was a great element of truth in the old Chinese ideas. The Huai Nan Tzu book of about 120 B.C. says:

> The Yin and Yang hurl themselves upon one another and this is the cause of thunder. Their forcing their way through each other produces lightning.

But such naturalistic theories did not still the superstitious fears of the Han people, who viewed these electrical discharges in the heavens as 'heavenly reprimands' for improper governmental or private behaviour. Wang Chhung, so against this phenomenalist outlook (volume I of this abridgement, page 203), took a firmly naturalistic approach in his *Lun Hêng* (Discourses weighed in the Balance) of A.D. 83. We see him at his best in this passage:

> At the height of summer, thunder and lightning come with tremendous force, splitting trees, demolishing houses, and from time to time killing people. The common idea is that this ... is due to Heaven setting a dragon to work. And when lightning strikes a person and kills him, this is attributed to some secret faults which he must have committed, such as eating unclean things. The roar and roll of thunder, they say, is the voice of heaven's anger, like men gasping with rage. Ignorant and learned alike talk thus,...
>
> But all this is nonsense. The genesis of thunder is one particular kind of energy (*chhi*), and one particular kind of sound...
>
> To speak truly, thunder is the explosion (*chi*) of the *chhi* of the solar Yang principle. This may be understood by the fact that in January, when the Yang begins to grow, we hear the first thunders, ... In summer when the Yang is reigning, the Yin disputes its supremacy, so that there is collision, friction, explosions and shootings...
>
> How can we test this? Throw a ladle of water into a smelting-furnace. The *chhi* will be stirred and will explode like thunder ... You may consider heaven and earth as like a furnace, the Yang *chhi* as the fire, and clouds and rain as abundant water...
>
> Lightning is essentially fire. Such *chhi*, burning a man, leaves a mark. If the mark looks like some writing, people, seeing it, are tempted to regard it as a statement concerning his guilt written by Heaven. Again this is empty nonsense ... Marks made by lightning are certainly not characters written by heaven...

There is a parallel here with a section of the poem *De Rerum Natura* by the Latin poet Lucretius who in the first century B.C. held similar opinions. Other Han scholars, such as Huan Than (before A.D. 30), had urged the same point of view.

In later centuries the denial of any powers of foretelling the future by thunder and lightning became a commonplace of the sceptical rationalism of the Confucians. In the eleventh century Su Hsun, alluding to a popular belief that death by lightning was a punishment for lack of filial piety, remarked that it would have its work cut out to punish all those who deserved it, and evidently it did not do so. In the same century, about 1078, Shen Kua recorded a description, so carefully written that it might have been destined for the columns of a scientific journal:

> A house belonging to Li Shun-Chu was struck by lightning. Brilliant sparking light was seen under the eaves ... After the thunder had abated, the house was found to be all right, though its walls and the paper on the windows were blackened. On certain wooden shelves, certain lacquered vessels with silver mouths had been struck by the lightning, so that the silver had melted and dropped to the ground, but the lacquer was not even scorched. Also a valuable sword made of strong steel had been melted to liquid, without parts of the house nearby being affected...

Similar precise and objective descriptions were given in later centuries, but the explanation of the true nature of thunder and lightning had to await the full flood of post-Renaissance science.

THE AURORA BOREALIS

In the Chinese historical records there are a number of 'strange lights', 'coloured emanations', etc. which can only refer to auroral phenomena. The earliest dates from 208 B.C. and the latest from A.D. 1639. The *Khai-Yuan Chan Ching* (Khai-Yuan reign-period Treatise on Astrology and Astronomy) of A.D. 718 quotes a host of books such as the *Yao Chan* (Divination by Weird Wonders) of the Han fortune-teller Ching Fang, mentioning great displays in 193 B.C. and 154 B.C., which were attributed, like earthquakes, to excessive Yin. No complete listing of the Chinese observations has yet been made, since the phenomenon was not recognised very clearly as an entity, and went by many names – 'red vapour', 'north polar light' etc. – and a vast literature would thus have to be searched. One technical term often encountered is *thien lieh* (cracks in the heavens). No explanation of the aurora was, however, possible until modern times.

SEA TIDES

Until modern times there was, on the whole, more knowledge of, and interest in, the phenomena of tides in China than in Europe. As has often been pointed out, this was probably due to the fact that the tides in the Mediterranean are slight and did not attract much notice by the ancient naturalists. The coasts of China, however, have tides of considerable range, for example, about $3\frac{1}{2}$ m off the mouth of the Yangtze in the Spring. Moreover, China has one of the only two great tidal bores in the world, that on the Chhien-Thang River near Hangchow (Fig. 120), the other being in the northern mouth of the Amazon, far away from any ancient civilisation.

Fig. 120. Lin Chhing's picture of the bore on the Chhien-Thang River (from *Hsung Hsüeh Yin Yuan Thu Chi*, 1849).

The bore on the Severn is much smaller. From very early times, therefore, a natural event of great impressiveness, obviously closely allied to coastal tides in general, invited explanation by Chinese thinkers.

The normal behaviour of tides, is, of course, a rise, slow at first, fastest at mid sea level, and diminishing to high water; then a fall returning to the original low level. The total cycle occupies about 12 hours 25 minutes, this being the interval between successive transits of the moon across the meridian. The rise is greatest at the beginning of each lunation, and least about two weeks after it. The range of the tides, the intervals between high waters, and the difference between spring and neap tides, greatly depend upon local conditions such as coastline and sea bottom near the shore. Places in north China may have but one tide each 25 hours. As one proceeds up a tidal estuary such as the Yangtze the period of rise shortens and the period of fall increases. A bore is simply an extreme example of this in which the first half of the rise is almost instantaneous. That on the Yangtze has a total range of some 6 m. Out at sea where the bore commences, the speed of the waves is as much as seven knots; in the river, after the passage of the bore, a flood current of ten knots occurs in the river. A standing wave of perhaps 9 m high is generated at the point where the two wave fronts intersect, and if this happens near the shore, it may flood over the massive sea walls. A thunderous noise is heard long before the bore arrives.

Already in the early second century B.C. the Chinese expected a good bore at full moon, as appears from the poetical composition *Chhi Fa* (The Seven Beguiling Tales) by Mei Shêng (died 140 B.C.):

> The guest said: 'At full moon in the eighth month I hope to go to Kuang-Ling with the feudal lords and with companions from far away, and with my brothers, to see the bore on the River Chhu. When we reach the spot, before we see the bore itself, but only the places where the strength of the water has come, it will be alarming enough...'

But a causal connection with the moon is not stated, and when the sick prince asks the guest what the force is which drives the wave, he replies that, though not supernatural, it is not recorded. In the first century A.D., however, the dependence of the tides on the moon was clearly indicated, and by none other than Wang Chhung in his *Lun Heng*. He takes the popular belief that the vengeful spirit of Wu Tzu-Hsü, an unjustly killed minister of State, whose body had been thrown into the river, regularly roused the waves to their periodical wrath and havoc, and proceeds to tear it limb from limb. He also gives a scientific view:

> Now the rivers in the earth are like the pulsating blood-vessels of a man. As the blood flows through them they throb or are still in accordance with their own times and measures. So it is with the rivers. Their rise and fall, their going and coming are like human respiration, like breath coming in and out.
>
> Finally the rise of the wave follows the waxing and waning moon, smaller and larger, fuller or lesser, never the same. If Wu Tzu-Hsü makes the waves, his wrath must be governed by the phases of the moon!

In Wang Chhung's mind, the influence of the moon on the tides must have been combined with the microcosm–macrocosm respiration theory. But these naturalistic views had to contend in his time with others far more primitive.

The next advance was made in the Thang by Tou Shu-Mêng, about A.D. 770, who seems to have been the first to deal with the lunar theory of the tides with any scientific detail, although he may have thought that the moon caused the water to swell and contract. Then, in the eleventh century Yü Ching made a notable advance with his *Hai Chhao Thu Hsü* (Preface to Diagrams of the Tides) of 1025. He believed both sun and moon affected the tides, although the moon was the more important '... since the moon is the essence of the Yin, and water is Yin,...', and his text gives evidence of much careful observation. Later in the same century Shen Kua wrote that he himself had given much study to the periodical motion of the moon and the behaviour of the tides, after which he points out in his *Mêng Chhi Pi Than* (Dream Pool Essays) 'Here I am referring to the tide on the coast itself; the farther you are away from the sea (i.e. up an estuary or the like) there will be a delay varying according to the place...'. Thus in 1086 Shen Kua clearly defined what we now call the 'establishment of the port', i.e. the constant interval between the theoretical time of high tide and the time when it occurs at the place in question.

In the West it seems that Antigonus of Carystos (about 200 B.C.) was the first of the Greeks to suggest that the moon was the primary influence on the tides. His position was thus similar to that of Mei Shêng, but advance was more rapid than in China, since Seleucos the Chaldean of Seleuceia on the Persian Gulf tied in the tides with the moon's motions in about 140 B.C., and half a century later Poseidonius of Apameia stated the meridian rule and the rule of spring and neap tides. This level was not reached in China until the Thang and Sung. But later, while the Chinese were advancing, Europe forgot the progress which had been made. With the exception of the Venerable Bede, who lived about A.D. 700, nothing was done between 100 B.C. and A.D. 1000. Bede recognised differences in local tides, but the 'establishment of the port' only came to full expression in 1188 in Giraldus Cambrensis' *Topographica Hibernica*. We met with it in China in 1086.

Certainly the Chinese had clear priority as to the systematic preparation of tide-tables, which go back to the ninth century A.D. at least, and in the eleventh century they were much more enlightened on the theory of the subject than the Europeans until the Renaissance. The crowning irony was that Galileo rejected Kepler's lunar theory of the tides on the ground that it was astrological. It was not until the time of Newton that the true gravitational explanations of the tides were worked out and accepted. Yet it would never have occurred to any Chinese observer that the moon could not have an effect on terrestrial events – such a separation would have been contrary to their whole world-view of organic naturalism.

4

The sciences of the earth: (i) Geography and map-making

After the expanses of the stars and the deeps of sky and sea, it is quite natural to turn to the homelier features of the mundane world, to the realm of the explorers and geographers. But geography is a subject which lies on the borderline between the natural sciences and the humanities. Any systematic treatment of the cumulative growth of geographical knowledge in China would far exceed the limits of the space we have. There is a vast literature, both in Chinese and Western languages, but it belongs rather to history itself than to the history of science. What we shall do, therefore, is to concentrate mostly on the development of scientific map-making in China.

Whoever sets out to write on any aspect of the history of geography in China faces a quandary, however, for while it is indispensable to give the reader some appreciation of the immense amount of literature which Chinese scholars have produced on the subject, it is necessary to avoid the tedium of listing names of authors and books, some of which have indeed been lost. Only a few examples will be given here, and they must stand as representatives of a whole class of works.

Many geographical symbols of great antiquity are embedded in the Chinese language. The character for river (*chhuan* 川) is a graph of flowing water, the character for mountain (*shan* 山) was once an actual drawing of a mountain with three peaks (ᗰ), and that for fields (*thien* 田) shows enclosed and divided spaces. Political boundaries are seen in the character for country (*kuo* 國) where the frontier encloses the symbols for 'mouths' and 'dagger-axes', the eaters and defenders. Bone and bronze forms of the character which came to mean 'map' (*thu* 圖) actually show a map. Unfortunately this word acquired a general meaning to refer to any kind of diagram or drawing, so in cases where a book disappeared at an early time it is not possible to be sure whether the *thu* which it was said to have had really were maps. But it would not be far off the mark to guess that the pictographic character of Chinese encouraged the idea of mapping.

As for ideas about the shape of the earth current in ancient Chinese thought, mention has often been made of the prevailing belief that the heavens were round and the earth square. But there was always much scepticism about this. For instance, in the first and second centuries A.D. it was often said that the universe was like a hen's egg, and the earth was like the yolk in the midst of it. And Chinese thinkers of all ages joined Yü Hsi (about A.D. 330) when he pointed out that a square earth would hinder the motions of the heavens: in his view it was spherical like the heavens, but smaller, and all supporters of the Hun Thien theory (page 85) must have tended to believe this. The influence of these views on Chinese map-making, however, remained slight, for, as we shall see, it revolved round the basic plan of precise rectangular grid, taking no account of the curvature of the earth's surface. At the same time Chinese geography was always thoroughly naturalistic.

GEOGRAPHICAL CLASSICS AND TREATISES

Ancient writings and official histories

The oldest Chinese geographical document that has come down to us is presumably the Yü Kung (Tribute of Yü) chapter of the *Shu Ching* (Historical Classic) which is probably of the fifth century B.C. Yü the Great was the legendary hero-emperor who 'mastered the waters' and became the patron of hydraulic engineers, irrigation experts and water-conservancy workers in after ages. This chapter lists the traditional nine provinces, their kinds of soils, their characteristic products and the waterways running through them. The accepted view is that this primitive economic geography covers hardly half the region which Chinese civilisation was later to occupy. However, this ancient inventory of the Chou 'empire' is essentially a physical geography; the boundaries of the 'provinces' are natural, not political. It is also completely devoid of magic, and even of fantasy or legend, apart from the appearance of Yü himself.

It is usual to hold that the Yü Kung chapter contains, by implication, a naive map of concentric squares (Fig. 121). This is based on the concluding sentences of the chapter, where it is said that throughout a zone of 500 *li* (presumably in all directions) from the capital there are the 'royal domains', within the next concentric zone of 500 *li* are the 'princes' domains', then come the 'pacification zone', the zone of 'allied barbarians', and lastly the zone of 'cultureless savagery'. Nevertheless, there is nothing in the text to justify the traditional view that these zones were concentric squares; this was probably just assumed on the basis of the idea of a square earth. The point is more important than it may seem, for if the cores were thought of as concentric circles, this ancient gradient system might have been one of the sources of the East Asian tradition of a disc-shaped world in the 'religious cosmography' we shall come to later (page 274). On the other

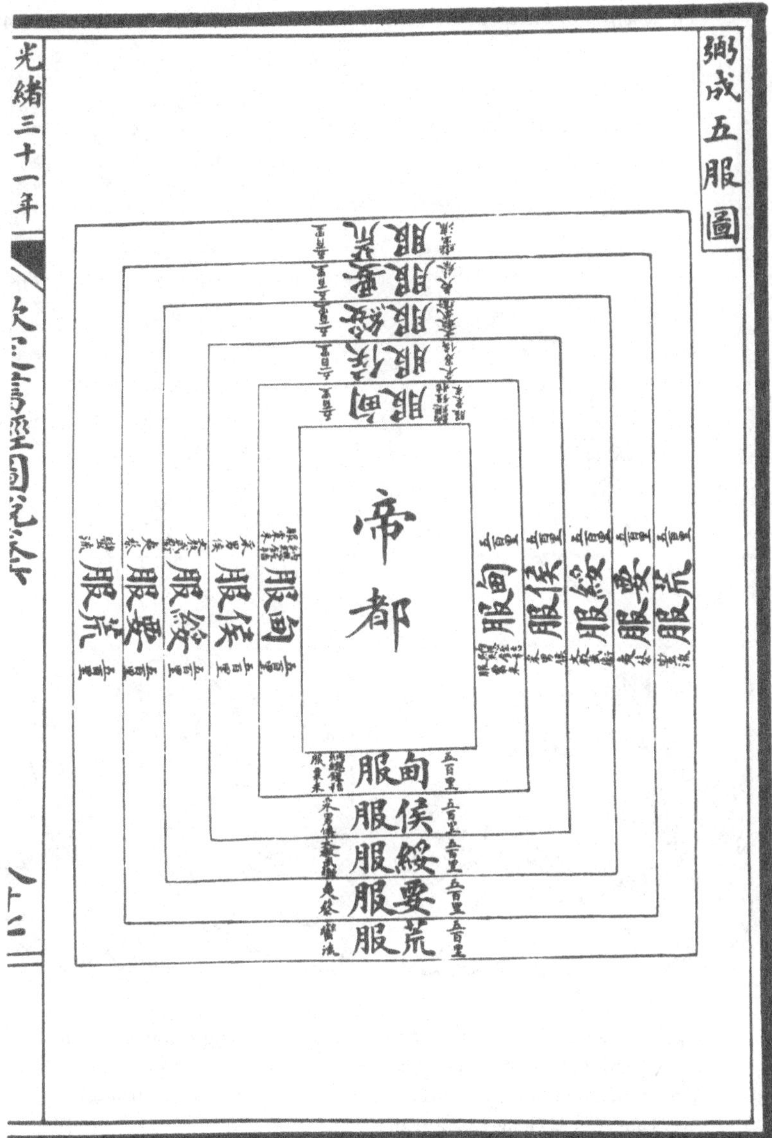

Fig. 121. The traditional conception of the radiation of ancient Chinese culture from its imperial centre (from the *Shu Ching Thu Shuo*). Proceeding outward from the metropolitan area, we have in concentric rectangles, (*a*) the royal domains, (*b*) the lands of the tributary feudal princes and lords, (*c*) the 'zone of pacification', i.e. the marches, where Chinese civilisation was in course of adoption, (*d*) the zone of allied barbarians, (*e*) the zone of cultureless savagery. The systematisation can never have been more than schematic but Egypt and Rome might have used a similar image, all unconscious of the equally civilised empire at the eastern end of the Old World.

Fig. 122. Comparative representations of fabulous beings from the *Shan Hai Ching* (sixth century B.C. to first century A.D.) and from the *Collectanea Rerum Memorabilium* of Solinus (third century A.D.). (*a*), (*b*) headless beings (acephali); (*c*), (*d*) long-eared men.

hand, concentric squares would foreshadow the rectangular grid which, as we shall see, was so characteristic of Chinese map-making.

In general, it may be said that the Yü Kung, the first naturalistic geographical survey in Chinese history, is approximately contemporary with the first map-making in Europe, which is associated with Anaximander (sixth century B.C.). But the Chinese document is much more detailed and elaborate than anything which has come down to us from Anaximander's time. Throughout Chinese history the influence of the Yü Kung was enormous; all Chinese geographers worked under its aegis, drew the titles of their books from it, and tried unceasingly to reconstruct the topography which it contained.

Brief consideration must now be given to the curious subject of the Nine Cauldrons of the Hsia and the *Shan Hai Ching* (Classic of the Mountains and the Rivers). The cauldrons were supposed to have been cast with pictures or maps illustrating various places or regions and the strange things found in them. The classical reference is the *Tao Chuan* (Master Tsochhiu's Enlargement of the Spring and Autumn Annals) and purports to relate to the year 605 B.C. It claims that the cauldrons were made when the legendary Hsia kingdom had attained its full greatness, and that they acted as a form of instruction to the people 'so that they could recognise all things and spirits both good and evil. And thus when they travelled over the rivers and marshes, and through mountains and forests, they did not meet with any adversities'. We are certainly in the presence here of an ancient tradition, perhaps magical and ritualistic rather than geographical. A similar atmosphere pervades the *Shan Hai Ching* which, however, also bears a resemblance to the Yü Kung in that it often mentions quite reasonable minerals, plants and animals. The date of the book is uncertain, but it was certainly current in some form in the Former Han period, and on internal evidence a good deal of the material goes back to the fourth century B.C.; indeed, some of the contents are likely to be much older than this. Although it is a veritable mine of information concerning ancient beliefs about natural things like minerals and drugs, the chief discussion about it has centred on the fabulous beings and peoples described.

Taking the view that the *Shan Hai Ching* is the oldest 'travellers' guide' in the world, an attempt has been made to identify some of the naturalistic identifications – thus the *pai min kuo* and *mao jen* (hairy white people) were probably the Ainu, and the *yü i kuo* the 'malodorous barbarians' of the Siberian coast from whom the Chinese imported fish-glue for bows in very early times. But a large proportion of the peoples mentioned are clearly fabulous – heads that fly about alone, winged men, dog-faced men, bodies with no heads and the like. Since a great many of these appear in Greek mythology (Fig. 122), the problem of transmission at once presents itself. Herodotus (fifth century B.C.) was one of the earliest sources, but there is much in Strabo (first century B.C.) and Pliny (first century A.D.), with Gaius Julius Solinus in the third century A.D. collecting the fabulous material together in his *Collectanea Rerum Memorabilium* (Anthology of Remarkable Things) which, in a sixth-century revision, supplied abundant 'marvels' for geographers throughout the European Middle Ages. Western scholars have been strongly inclined to consider it all as Greek in origin. For certain tales they may be right, but some may well go back in China beyond the time of Herodotus. It could be, too, that a few are Indian or

Iranian in origin, while it should be remembered that the Babylonians also had an interest in such accounts.

This material would seem to have some interest from the point of view of the history of biology. To an embryologist it appears obvious that most of the different abnormalities which form the basis of this body of legend could have been derived from human and animal monstrosities occurring naturally. For example, research has shown that Eastern and Western stories of dog-faced or hairy-faced human beings could well have been derived from living abnormal examples, and from species of dog-faced monkeys. If this point of view should prevail there would be no reason to assume any transmission at all, at any rate to account for origins. Nor would it run counter to the suggestion that the existence of these myths is related to the kind of hatred and fear of strangers present among all ancient peoples.

Anthropological geographies

The human geographies developed rather naturally from the descriptions of fabulous peoples. The generic name for this literature, Chih Kung Thu (Illustrations of the Tribute-Bearing Peoples), does not, however, appear before the Liang (about A.D. 550), when it occurs as the title of a book prepared by Chiang Sêng-Pao or Hsiao I, but Chuko Liang is said to have written a *Thu Pu* on the southern tribes already in the third century A.D. In the Thang and other dynasties it was customary for foreign tribute-bearers to attend at an office known as the Hung Lu, where the officials took notes of the geography and customs of their countries, and committed these to writing as part of an ever-growing body of government intelligence (Fig. 123). In due course this gave rise to numerous books.

Descriptions of southern regions and foreign countries

There were no very sharp lines between ethnological geography and what might be described as ethnology and folklore proper, since there were in all centuries large enclaves of territory occupied by tribal peoples only partially affected by Chinese culture. Moreover, as Chinese civilisation expanded to the south, more and more descriptions were written of the strange things, as well as the topography, of the southern regions. Works on folk customs and their geographical distribution were termed Fêng Thu Chi, and descriptions of unfamiliar regions I Wu Chih. Of the former, the oldest is the *Chi-chou Fêng Thu Chi* (Customs of the Province of Chi) by Lu Chih about A.D. 150. An early description of the southern regions was Yang Fu's second-century A.D. *Nan I I Wu Chih* (Strange Things from the Southern Borders) and Wan Chhen's *Nan Chou I Wu Chih* (Strange Things of the South), two centuries later.

Fig. 123. A late Chhing representation of the attendance of barbarian envoys presenting tribute at the Hung Lu Department where details of their countries and products were recorded. From the *Shu Ching Thu Shuo*.

Fig. 124. Two pages from the *I Yü Thu Chih* (Illustrated Record of Strange Countries) of *c.* A.D. 1430, perhaps written by the alchemist, mineralogist and botanist Ning Hsien Wang (Chu Chhüan), a prince of the Ming, and almost certainly profiting from the zoological and anthropological knowledge gained from the expeditions of Chêng Ho.

(*a*) A zebra. (*b*) An inhabitant of the country of Black-clothed people, no doubt some outpost of Arab culture. The description says that they hide their faces from Chinese visitors, and anyone who sees them is killed. Bargainers in trade are separated by a curtain, but one has to be careful for if the native merchants are dissatisfied with the deal one is likely to be pursued and slain. (From the unique copy in the Cambridge University Library.)

In spite of all losses, an enormous literature exists on the geography of foreign countries. Among its earliest representatives was the *thu shu* (illustrated account) of the Huns presented in 35 B.C. (of which we shall have more to say later, page 260); though it is unlikely that Chang Chhien in the previous century had failed to provide some charts and illustrations of the peoples whom he had visited in the West. In the second century A.D. Tsang Miu produced an elaborate memorandum dealing with 55 countries, but the climax of this literature was reached in the Ming, with the fifteenth-century voyages of Chêng Ho, the famous admiral. Military and other expeditions as well as trade gave rise to many books, of which the *I Yü Thu Chih* (Illustrated Record of Strange Countries) of about 1430 is perhaps the most celebrated (Fig. 124). There is some doubt about the author, but it may have been written by the alchemist, mineralogist and botanist Chu Chhüan, a prince of the Ming, and almost certainly profited from the zoological and anthropological knowledge gained in the expeditions of Chêng Ho.

How does this general picture compare with the development of descriptive geography in the West? The Chinese had nothing of the quality of Herodotus or even of Strabo at times contemporary with them, but during the gap between the third and thirteenth centuries A.D., when European learning sank so low, the Chinese were far more advanced and steadily progressing. The floor was held in Europe by Solinus and his myths, almost as if the *Shan Hai Ching* had continued to dominate in China without competition. In the Thang period (A.D. 618 to 906) almost the only reasonable representative that the West could produce was the Syrian bishop Jacob of Edessa (633 to 708). The Arabs, however, match up better and by the Sung, about 950, they were laying the foundations of later Western geography. Arab geography reached its climax with al-Idrīsī in the twelfth century but still yields many good names in the thirteenth. Of course, the West had its pilgrim literature, analogous to that of the Chinese Buddhists, beginning with 'the first of the Christian guide-books', the *Itinerary from Bordeaux to Jerusalem* of A.D. 333 and its records of trading voyages a couple of centuries later; but when one reads the careful chronicles of the Renaissance, one feels that the West was only then beginning to follow the path of objective criticism which the Chinese had been treading for the previous millennium and a half. We have already met with this pattern, when examining the growth of knowledge of the tides (pages 233–6), and it will manifest itself again throughout the present section with especial clarity when we discuss map-making.

Hydrographic books and descriptions of the coast

In view of the great importance of waterways for the Chinese social and economic system at all times, it was natural that close attention should be paid to them. The first treatise of the kind was that of Sang Chhin of the first century B.C., the *Shui Ching* (Waterways Classic), but the text as we now have it is thought to be from the hand of some geographer of the San Kuo period, at any rate before A.D. 265. It gives a brief description of no less than 137 rivers. About the beginning of the sixth century it was enlarged to nearly forty times its original size by the great geographer Li Tao-Yuan, and given the title *Shui Ching Chu* (The Waterways Classic Commented). This constitutes a work of the first importance. From the titles of several books it would seem that rivers were being mapped from the Chin onwards. Such was a book on water-conservancy by Shan O (1059) who spent more than thirty years exploring the lakes, rivers and canals in the region of Suchow, Chhangchow and Huchow, while a century later Fu Yin wrote the *Yü Kung Shuo Tuan* (Discussions and Conclusions regarding the Geography of the *Tribute of Yü*) in which he dealt mainly with the Yellow River valley. Some diagrammatic charts, presumably of the twelfth century, are included in this book (Fig. 125). In the Chhing dynasty much larger books were produced, but here we are in the eighteenth century. In Europe, though, there seems to be no class of literature which quite corresponds to it (Fig. 126).

Allied to the group of books on the greater and lesser waterways is another on the Chinese coast. But most books on this subject are late. In 1562 Chêng Jo-Tsêng published a large work on it, *Chhou Hai Thu Pien* (Illustrated Seaboard Strategy), stimulated by the marauding raids of Japanese pirates which were a great affliction to all the maritime provinces during that century. Coastal protection against encroachment of the sea also gave rise to a special literature.

Local topographies

The series of topographical writings produced in China is probably unrivalled by any nation for extent and systematic comprehensiveness. Anyone at all acquainted with Chinese literature is familiar with the host of 'gazetteers', as they came to be called, which are really local geographies and histories. In other literature there is little comparable to this forest of monuments which the industry of local scholars erected over the centuries. This work is generally regarded as beginning in A.D. 347 with the *Hua Yang Kuo Chih* (Historical Geography of Szechuan) which contains much about rivers, trade-routes and the various tribes. Szechuan had also been subjected to mapping, for in the book we read of a map of the province apparently made about A.D. 150 in the Later Han.

Such local map-making was slow to spread, for this kind of title does not appear much until the book list of the *Sui Shu* (late sixth century) when suddenly many are mentioned. With the growth of a stable bureaucracy, in which men were generally sent to serve far from home, local topographies acquired social importance, and in about 610 the emperor ordered officials all over the country to compose records of customs and products, illustrated by maps or diagrams, and to present these to the imperial secretariat. The Sung dynasty energetically continued this compilation. Not long after it came to power, the emperor ordered Lu To-Sun in 971 to 're-write all the Thu Ching in the world', and this official, pursuing his colossal task, travelled through the provinces collecting all the relevant available texts.

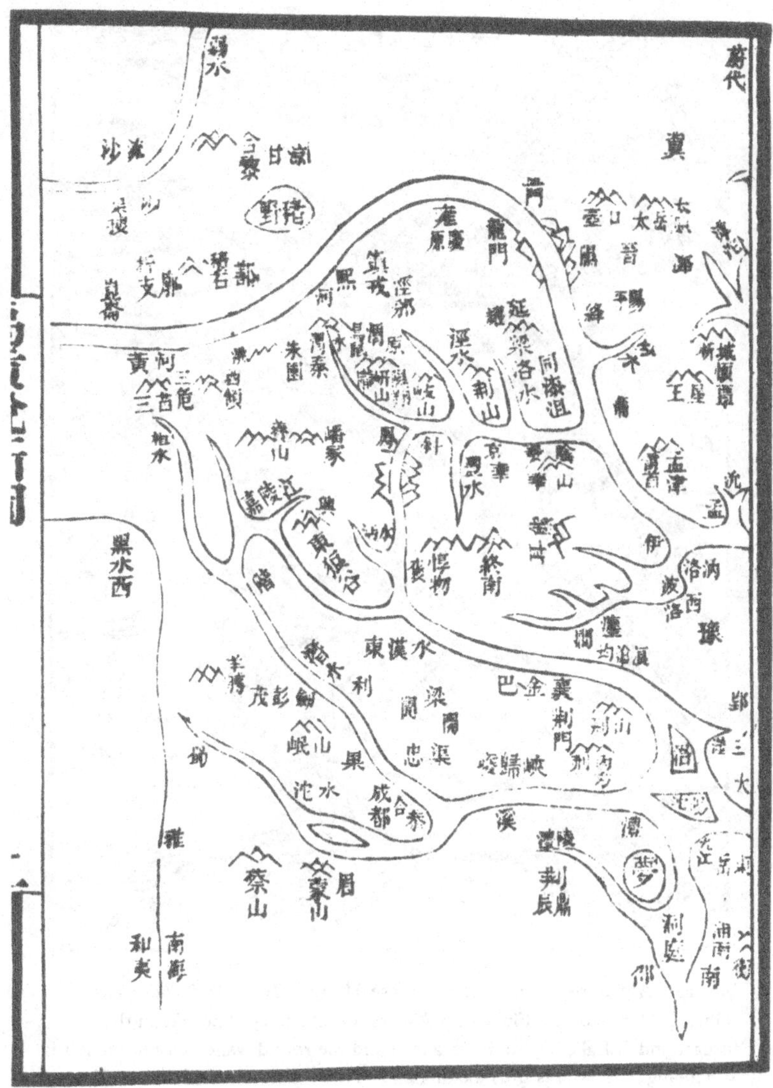

Fig. 125. A diagrammatic chart of the river systems of West China, from Fu Yin's *Yü Kung Shuo Tuan* (Discussions and Conclusions regarding the Geography of the *Tribute of Yu*) of *c.* A.D. 1160. The great bend of the Yellow River round the Ordos Desert and its passage through the Lung Mên gorges will be seen at the top, while lower down the old road and canal connecting the Wei River with the Han River (i.e. the Yellow and Yangtze River systems) through the Chhinling mountains can be made out. In Szechuan, Chhêng-tu and Mt Omei are marked.

Fig. 126. A panoramic map from Fu Tsê-Hung's *Hsing Shui Chin Chien* (Golden Mirror of the Flowering Waters) of A.D. 1725. The lake in the foreground is Lake Taihu in Chiangsu and the round walled city to the left is Wuchiang; the vista is thus looking east. The Grand Canal is shown crossing the panorama towards Hangchow past the village of Wang-ching. Beside the city of Wuchiang the famous 'Drooping Rainbow Bridge' (Chhui Hung Chhiao) is marked, and two other bridges are shown. A reproduction such as this cannot do justice to the delicacy of the original.

They were then worked together by Sung Chun and by 1010 no less than 1,566 chapters were finished. Some kind of map-making survey was included in these activities.

Even before the end of the Sung the number of titles of topographies amounted to 220, and the quantity then increased from the Ming onwards, till today there is hardly a town, however small, which does not have its own historical geography. Nor were these local books confined only to towns, cities and their administrative districts; similar books were devoted to famous mountains. There were also books dealing with cities alone, or aspects of cities, as in the *Lo-Yang Chhieh Lan Chi* (Description of the Buddhist Temples of Loyang) of A.D. 500.

What is there in the West to compare with this vast mass of literature? Greek or Hellenistic antiquity supplies no real parallel, and there seems little to adduce from the early Middle Ages. By the thirteenth century A.D. there is a French description of the state and cities of Jerusalem which might offer a comparison with the description of Loyang seven centuries before, and there are the works of Giraldus Cambrensis, but one has really to wait until Renaissance Europe to find good parallels. On the whole European topography shows the same gap of a thousand years as we noted before and shall shortly see again, with the difference that the ancient Western world had not made much progress in this field beforehand.

Geographical encyclopaedias

It only remains to mention some of the great compendiums of geographical information which the Chinese produced from the Chin dynasty (third and fourth century A.D.) onwards. They were mainly descriptive, though during the Thang and Sung they certainly contained some maps, now long lost. Perhaps one might name as the earliest of the kind the *Shih San Chou Chi* (Record of the Thirteen Provinces) by Khan Yin, written between A.D. 300 and 350. The oldest extant general work, however, is the *Yuan-Ho Chün Hsien Thu Chih* (Yuan-Ho reign-period Illustrated Geography), by Li Chi-Fu, written about A.D. 814. This was towards the end of the Thang. The period between Thang and Sung produced one important geographer, Hsü Chiai, but shortly after the beginning of the Sung there appeared (between A.D. 976 and 983) a great compendium which is still consulted, the *Thai-Phing Huan Yü Chi* (General Description of the World in the Thai-Phing reign-period) by Yüeh Shih, in 200 chapters. During the eleventh and twelfth centuries more encyclopaedias appeared, but the treatment became increasingly literary and historical-biographical and less geographical in the scientific sense. The thirteenth century, however, saw two useful geographies.

Later centuries produced less material of this kind. Only three comprehensive geographies of the empire need mention; the first was started in 1310 but never printed, the second came out in 1450 and the third in the eighteenth century, having been initiated in 1687. This last includes every kind of geographical detail for all the provinces, and also a mass of information about dependencies and tributary states beyond the frontier of China proper. Also early in the eighteenth century appeared the *Thu Shu Chi Chhêng* encyclopaedia, in which a great quantity of geographical material is included. By this time, of course, the Chinese work had to face what is in many ways a damaging comparison with post-Renaissance Europe. Again we see that it was during the earlier medieval time that China was more advanced. Between the Han and the Thang the West had nothing to set beside the Chinese map-makers, and during the Sung there was no competition apart from the Arabs. Only with the Ming decadence and the rise of modern science in Europe did the West draw considerably ahead.

A NOTE ON CHINESE EXPLORERS

The immense mass of geographical knowledge represented by the literary works just described, as well as the achievements of scientific map-making which will occupy the remainder of this section, was certainly not gained except by the accumulated observations of countless travellers and explorers. Some were engaged on official or diplomatic missions, others were travelling in the cause of religion, but they all added their store of experience and their observations, more or less accurate and complete, to the growth of knowledge about the terrestrial world. On account of the great importance for contacts between China and Europe, we have already described (in volume 1) the exploits of Han officials such as Chang Chhien and Kan Ying and the eunuch admiral Chêng Ho.

So great has been the fame of Marco Polo and the other European travellers of the thirteenth century A.D. that their Chinese counterparts who also made important journeys have generally been overlooked. Yet there is the *Hsi Yu Lu* (Record of a Journey to the West) by the statesman and patron of astronomers, Yehlü Chhu-Tshai, who accompanied the Mongol emperor, Chinghiz Khan, on an expedition to Persia (A.D. 1219 to 1224), and the *Pei Shih Chi* (Notes on an Embassy to the North) by Wukusun Chung-Tuan, a Chin Tartar who returned from a mission to Chinghiz Khan in 1222. Another famous journey was that of the Taoist adept and alchemist Chhiu Chhang-Chhun who was summoned to the court of Chinghiz Khan, then in Afghanistan, and made the journey there and back between 1219 and 1224. The record of it, *Chhang-Chhun Chen-Jen Hsi Yu Chi* (Western Journey of the Taoist Chhang-Chhun) was prepared by his secretary Li Chih-Chhang, who 'kept a record of their experiences throughout the journey, noting with the greatest care the nature and degree of the difficulties – such as mountain passes, river-crossings, bad roads and the like – with which they had to contend; also such differences and peculiarities of climate, clothing, diet, vegetation, bird-life and insect-life as they were able to observe'. Indeed during the journey through Mongolia and Central Asia, Chhiu Chhang-Chhun's party made observations of a eclipse and took gnomon solstice shadow measurements at a more northerly point than any Chinese astronomers had reached before them.

There are other such valuable journeys, but of greater scientific interest were journeys where the main motives were geographical. For example, there was the question of the source of the Yellow River. In the time of Chang Chhien (second century B.C.) it was supposed that the Khotan River, issuing from the Khun-Lun Mountains (which form the northern escarpment of the Tibetan plateau) and making its way round the north of the Tarim basin to fall into Lop Nor, was the source of the Yellow River. Between Lop Nor and a pass near Lanchow the river was supposed to flow underground. But by the time of the Thang the true situation was clearly understood. In A.D. 635 general Hou Chün-Chi, in charge of a punitive expedition against a Tibetan tribe, pushed as far as Djaring Nor and 'contemplated the sources of the Yellow River'. The river does in fact take its origin near this lake, and the identification was confirmed not long afterwards by Liu Yuan-Ting, who was sent as ambassador to the Tibetans in 822. Hou and Kiu probably did not realise that the Yellow River makes an enormous detour around the Amne Machin range before doubling back again, but otherwise they were perfectly correct. In 1280 Khubilai Khan sent out a scientific expedition under Tu Shih to clear up this question, and the results were embodied in a book *Ho Yuan Chi* (Records of the Source of the (Yellow) River) of 1300 by the geographer Phan Ang-Hsiao.

Later we shall touch on the road guides which exist in Chinese literature and were prepared primarily for practical purposes. It is in the Ming, however, that we find the prince of all Chinese itinerary-makers, the traveller Hsü Hsia-Kho (1586 to 1641), a man whose interests were neither official nor religious, but scientific and artistic. For more than thirty years he walked across the most obscure and wildest parts of the empire, exposed to all kinds of difficulties and sufferings, often dependent on the patronage of local scholars who helped him after he had been robbed of all his belongings, or local abbots who were willing to pay him for composing a history of their monastery. On sacred mountain and snowy pass, beside the rice terraces of Szechuan and in the semi-tropical jungles of Kuangsi, there inevitably to be seen was Hsü Hsia-Kho with his notebook. His biographers agree in saying that he thoroughly disbelieved in the theories of the geomancers, and wished to go and see for himself the dispositions of the great mountain regions radiating from the Tibetan massif. His notes read more like those of a twentieth-century field surveyor than of a seventeenth-century scholar. He had a wonderful power of analysing topographical detail, and made systematic use of special terms which enlarged the ordinary usage, such as 'staircase' (*thi*) and 'basin' (*phing*). Everything was noted carefully in feet or *li*, without vague stock phrases. Hsü Hsia-Kho's chief scientific achievements were, first, the discovery of the true source in Kweichow of the West River (Hsi Chiang) of Kuangtung. Secondly, he established that the Mekong and the Salween were separate rivers. Thirdly, he showed that the Chin Sha Chiang river was none other than the upper waters of the Yangtze; this had long been very confusing on account of the enormous detour which it makes around the Lu Nan Shan mountains south of Ningyuan.

A curious factor which led to the increase of geographical knowledge in the Chhing dynasty was the custom (of course not new) of banishing scholars to remote parts of the empire for real or alleged misdemeanours. Thus in 1810, Hsü Sung, a good scholar, then Commissioner of Education for Hunan, was accused of using his office to promote the sales of his own books, and of failing to assign topics for essays on the classics by degree aspirants. Two years later he was banished to Sinkiang, where he remained for seven years. But the result was that Chinese geographical literature was enriched by four excellent books on the remote parts of that dominion.

PRECISION MAP-MAKING IN EAST AND WEST

Introduction

The history of scientific geography and map-making is usually presented as containing an unaccountable gap between the time of Ptolemy (second century A.D.) and about A.D. 1400. Books on the subject contain what seem to be certain conventions about the participation of China – there are discussions of medieval European knowledge of China, what the Arabs said about it, and the stimulus of visits made by the merchants and diplomatic envoys in the thirteenth century A.D. – but never by any chance the story of Chinese map-making itself. Yet during the whole of the millennium when scientific cartography was unknown to Europeans, the Chinese were steadily developing a map-making tradition of their own, not strictly astronomical, but as quantitative and exact as they could make it.

Scientific cartography; the interrupted European tradition

The development of Greek map-making began with Eratosthenes (276 to 196 B.C.), the contemporary of Lü Pu-Wei. The coordinate system he applied to the earth's surface originated from his determination of the size of the earth and so of the earth's curvature. By measuring gnomon shadows at the summer solstice at Alexandria and Syene, Eratosthenes obtained a value of 40,225 km for the earth's circumference. It should be noted here that a spherical earth was at the basis of Greek cartography just as the flat earth was at the basis of the Chinese. But in practice it made less difference than would seem at first sight, for the Greeks never developed a satisfactory 'projection' for describing the spherical surface on a flat sheet.

The *oikoumene*, or inhabited world, of Eratosthenes was oblong, some 12,500 km in length and 6,000 km from north to south. This was crossed by a series of parallels of latitude, chosen according to gnomon shadow lengths at solstice, and a series of meridians, chosen arbitrarily. The fundamental parallel ran through Rhodes and the fundamental meridian through Syene, Alexandria, Rhodes and Byzantium but neither was precise, both involving considerable distortion of the true position of the various places through which each was supposed to pass. There was no way of determining the distance between meridians except by dead reckoning (i.e. calculating distance by noting speed and time taken in travelling), mostly by sea voyages. But such a method is not exact, and in consequence the length of the Mediterranean came out about one-fifth too great.

Hipparchus, who worked sometime between 162 and 125 B.C. and was a contemporary of Liu An and his school, criticised the work of Eratosthenes and introduced various improvements, including the term *climata* for the areas between parallels. The parallels of Eratosthenes had been arbitrary, but Hipparchus made them equal and astronomically fixed. With Ptolemy, who was active between A.D. 120 and 170 and therefore worked at the same time as Tshai Yung, accurate or scientific cartography reached its greatest height. No less than six out of the eight books of his *Geography* are occupied with tables of latitude and longitude of specific places, given to a precision of one-twelfth of a degree. But the longitudes were really only guesswork. Hipparchus, indeed, had suggested a way of measuring by observations, at different places, of the onset of lunar eclipses, but only one or two experiments of this kind were available to Ptolemy. The ancient world was not able to organise scientific observations on the scale required. However, Ptolemy greatly reduced the estimate of the length of Asia which had been given by Marinus of Tyre, and in this he was fully justified. On his largest map, which covered 180° of longitude and 80° of latitude, he made an attempt to show the meridians and parallels as curved lines (Fig. 127).

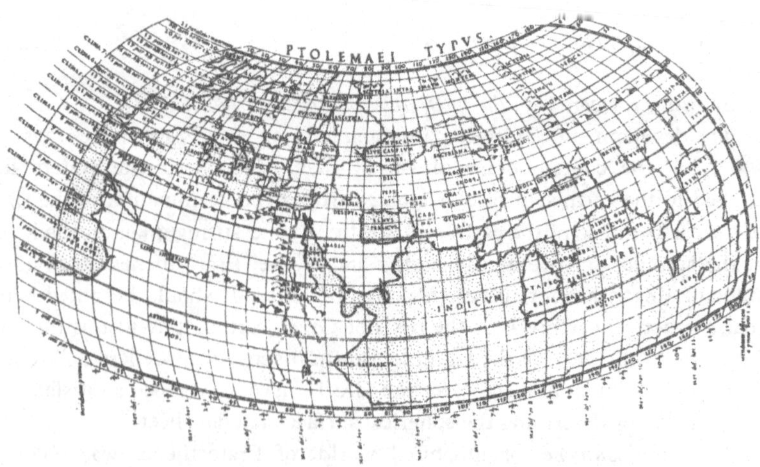

Fig. 127. Ptolemy's world-map as reconstructed by the Venetian Ruscelli (A.D. 1561). Longitude is expressed in fractions of hours east of the Fortunate Islands, latitude designated by the number of hours in the longest day of the year (from A. Lloyd Brown).

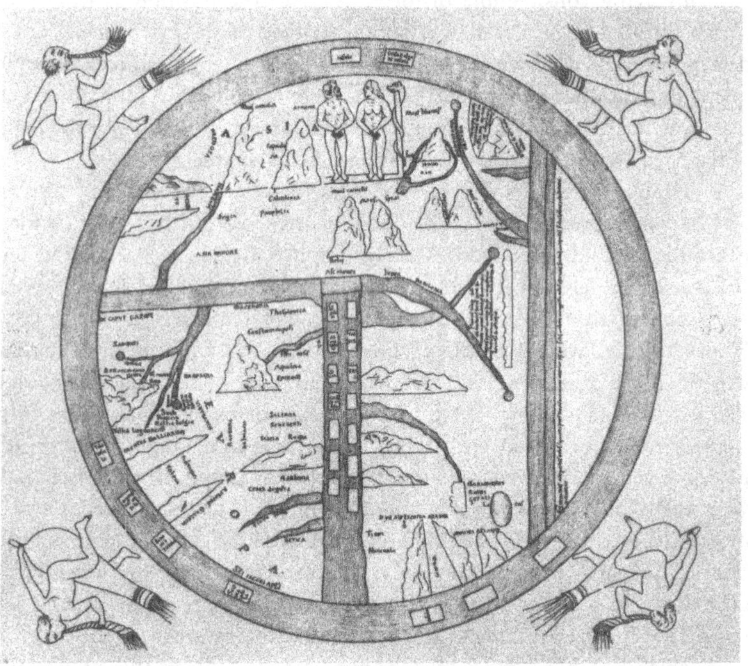

Fig. 128. A world-map of the type due to Beatus Libaniensis (*d.* A.D. 798), from a Turin manuscript of A.D. 1150 (from Lloyd Brown). Here the T-O shape is seen. Paradise and Eden are placed in the Far East at the top; below the Mediterranean with its many islands separates Europe from Africa.

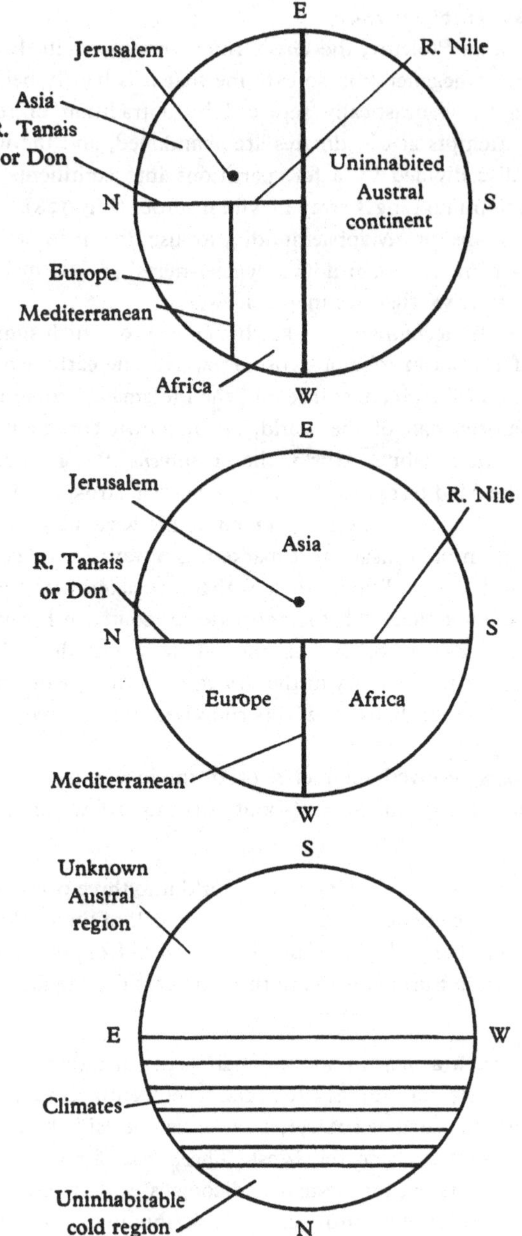

Fig. 129. Diagrams to illustrate the chief forms of the European tradition of religious cosmography.

Religious cosmography in Europe

After the time of Ptolemy, the Great Interruption sets in. European map-making suffers a degeneration so extreme that it is hardly believable. Scientific cartography is drastically replaced by a tradition of religious cosmography, all attempts at coordinates are abandoned, and the world is represented as a disc divided by a few partitions into continents, across which rivers and mountain ranges stray in wild disorder (Fig. 128). There is no lack of these maps or 'Mappaemundi', to use the medieval term, although the usual name for them now is 'wheel-maps' or T-O maps – the significance of the latter we shall see in a moment.

Let us look at the accompanying sketches (Fig. 129) which summarise the main forms of European religious cosmography. The earliest of them, which has been called Macrobian, still retains the Ptolemaic recognition of a southern sub-equatorial half of the world, an unknown continent in the antipodes. But for the inhabited world, the *oikoumene*, the geographer is content with a T, the Mediterranean forming its verfical stroke, and the two parts of the horizontal stroke being represented by the Rivers Tanais (Don) and Nile. The position of Jerusalem, if marked, is always in the centre of the inhabited globe. Later traditions, such as that associated with the name of Beatus, dispense altogether with the antipodes or southern hemisphere, and draw the *oikoumene* as occupying the whole world disc. A third tradition preserved a vague memory of the *climata*, showing them simply as parallel lines without meridians and superimposed on no geographical features.

The T-O maps received their classical form in a fifteenth-century poem by Leonardo Dati, *La Sfera* (about A.D. 1420), which may be translated as:

> A T in an O gives us the division of the world into three parts. The upper part and the greatest empire take nearly half of the world. It is Asia; the vertical bar is the limit dividing the third from the second, Africa, I say, from Europe; between them appears the Mediterranean Sea.

Fig. 130 illustrates such a map in an extremely schematic form. The T-O map tradition did not die out until as late as the seventeenth century.

Before leaving this curious subject, a word may be said about certain other aberrant representations in the West. There was, for instance, the *Christian Topography* written by Cosmas Indicopleustes between A.D. 535 and 547. This was essentially a controversial work, intended to expose 'the wicked folly of the Greeks in Geography'. It therefore has importance in connection with the exact reasons why Christian Europe, under the in-

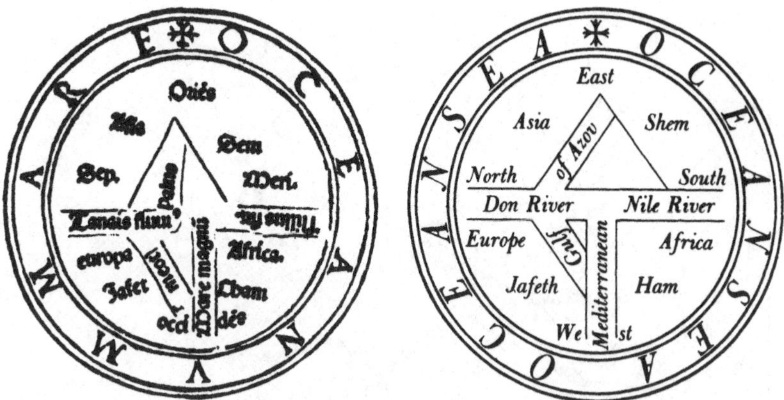

Fig. 130. The extremely schematised world-map of Isidore of Seville (A.D. 570 to 636) in a Venetian edition of his *Etymologiae* of 1500. Interpretation on the right from Lloyd Brown.

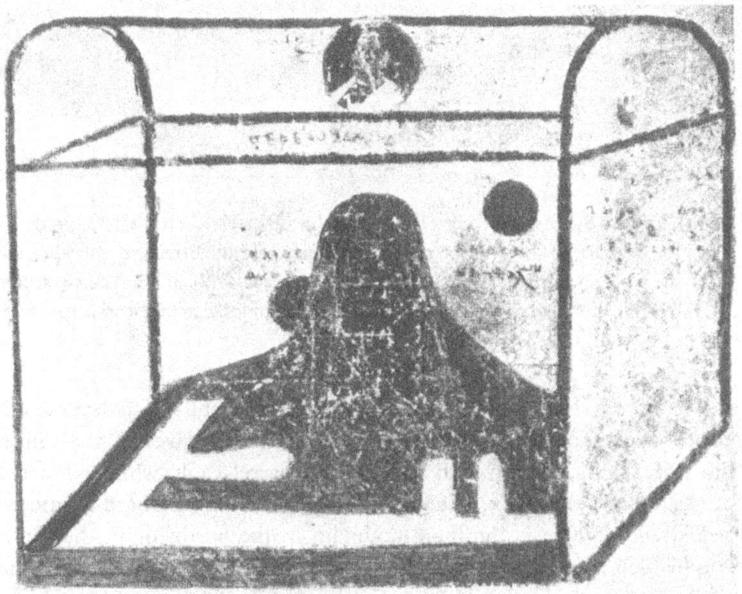

Fig. 131. The universe according to Cosmas Indicopleustes' *Christian Topography*, *c.* 540. The rising and setting suns move round the great mountain in the north; the inlets of the Mediterranean, Red Sea and Persian Gulf are shown below, the heavens have the shape of a barrel vault, and within them the Creator surveys his works (from C.R. Beazley).

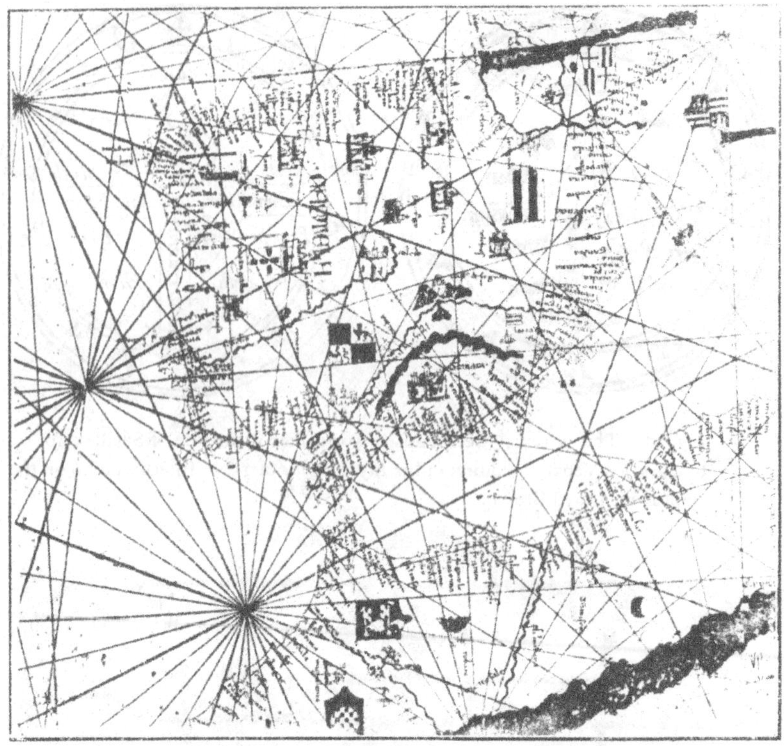

Fig. 132. Spain on the portolan of Angelino Dulcerto, A.D. 1339 (from de Reparaz-Ruiz). Behind the rhumb-lines and grid the Straits of Gibraltar and the whole outline of the peninsula clearly appear, with an abundance of names of ports and coastal features, marked by flags to indicate territorial authorities.

fluence of the Fathers, threw away all the Hellenistic achievements in precise cartography. The principal picture Cosmas gives of the universe (Fig. 131) shows Heaven with its walls and barrel-vault, while below is the flat surface of the earth, with some schematic sea and land hard to interpret. Particularly notable is the high mountain in the north (or in the centre?) round which the sun is pictured both rising and setting, for this was an idea well known and widespread in Asian thought, and Cosmas himself, as his second name Indicopleustes indicates, had made the voyage to India, and perhaps to Ceylon. We shall come back later to this mountain of Cosmas.

Among other European maps which deviated from the general sort was the eighth-century map – perhaps the most degenerate of all – which shows merely a horseshoe-shaped continent surrounding the Mediterranean, and, conversely, the late tenth-century 'Cottoniana' map which shows the contours of the European land-masses on a square plan better than any Western map before the 'portolans' or handy sea-charts.

The role of the navigators

About A.D. 1300 there began to appear in the Mediterranean region, first slowly and then in increasing numbers, sea-charts for practical use. These showed mainly the contours of the land-masses, along the edges of which the names of all havens and coast towns were marked, often with flags to indicate their political adherence (a point which it might be very useful for a pilot to know; Fig. 132). The first of these which is dated is the Vesconte portolan of 1311, but some extant specimens are thought to go back about thirty years previously.

The characteristic graduation of the portolans was not meridians or parallels but an interlocking network of rhumb-lines or loxodromes, i.e. lines on which a ship sails when keeping to a particular compass-bearing. The connection of more precise cartography with the use of the mariner's compass is thus obvious, magnetic polarity having become known in Europe just before 1200. The stimulus for the portolans has long been suspected to lie with the Arabs, but there is now a tendency to look further east. Developments following the portolan period were due to the effects of the Renaissance: the work of Prince Henry the Navigator, of Portugal (1394 to 1460), who organised many celebrated expeditions, and the revival in knowledge of Ptolemy's geographical ideas. Then came the time of Gerard Mercator (1512 to 1594), and with the maps of Waghenaer and Ortelius towards the close of the sixteenth century we are fully in the modern period.

Scientific cartography: the continuous Chinese grid tradition

The general point now to be made is that just as the scientific cartography of the Greeks was disappearing from the European scene, the same science in different form began to be cultivated among the Chinese – a tradition which continued without interruption down to the coming of the Jesuits. It was born in the work of Chang Hêng (A.D. 78 to 139), the contemporary of Marinus of Tyre, who was one of Ptolemy's principal sources. But before looking at his work, it will be more logical to glance at the evidence for Chinese map-making before his time.

Origins in Chhin and Han. The *Chou Li* (Record of the Rites of Chou) can be taken as telling us something about the ideas of the Former Han period, if not about the Chou. Its references to maps are remarkably numerous. The Director-General of the Masses has the function of preparing maps of the feudal principalities and of registering their populations. The Directors of Regions are placed in charge of the maps of the empire, in accordance with which they supervise the lands in its different districts. The Geographer-Royal takes care of the maps of the provincial circuits, and when the emperor is making a tour of inspection he rides close to the imperial vehicle to explain the characteristics of the country and its products. The Antiquary-Royal is likewise in close attendance in order to explain points of archaeological interest. Then there are the Frontier Agents who register the boundaries of principalities and domains. Maps for special purposes are also mentioned. Thus the Superintendents of Mining observe the occurrences of the ores of metals, and make maps of the locations, which they give to the miners. Military maps are also discussed.

That the Han people should have said so much about maps in their idealised picture of the perfect imperial bureaucracy is not surprising, since the first historical reference to a map in China goes back to the third century B.C. In 227 B.C. the crown prince of the State of Yen induced a certain Ching Kho to attempt to assassinate the Prince of Chhin, who was afterwards to found the Chhin Dynasty. This man came into the Prince's presence with a map of a district on the pretext that this domain was to be offered to the king. The map was painted on silk and contained in a case, but when it was taken out a dagger appeared as well; this Ching tried to use but failed. The scene was a favourite with the decorators of Han tombs.

When Chhin Shih Huang Ti became emperor, he assembled all available maps of the empire. These were of inestimable advantage to the Han dynasty, and there is evidence that they lasted until the end of the first century A.D. But by the third century they had disappeared. All through the Han there are references to maps. By a wonderful chance, several elaborate and detailed maps of Chhangsha and the region afterwards called Hunan province, made in the second century, have recently been recovered from the tomb of the bibliophile son of the Lady of Thai, a young man who died in 168 B.C. When Chang Chhien returned from the West (126 B.C.), the emperor we are told in the *Chhien Han Shu* consulted maps and books; in 117 B.C. maps of the whole empire were submitted to Han Wu Ti in connection with the investiture of his sons as feudal princes. There was military map-making in 99 B.C. when the general Li Ling was campaigning against the Huns, and some fifty and more years later when an unauthorised expedition stormed the city of the Hun leader Chih-Chih and killed him in retaliation for his murder of a Chinese diplomatic envoy in 43 B.C. There has been argument about precisely what the Chinese brought back with them to show at court, for the words *thu shu* were used to describe the documents, and could refer to pictures rather than maps. However, a careful analysis of this notable campaign, when Chinese soldiers fought with Roman legionaries turned Hunnish mercenaries, indicates that maps embellished with drawings is the best interpretation. Certainly in the next century the reforming emperor Wang Mang (A.D. 9 to 23) was aware of the importance of maps, and delegated a special official to assemble and study them in connection with the problems of fiefs. Again, in 26, when Kuang Wu Ti was fighting to establish a new dynasty, he used a large map painted on silk, and when securely established on the throne, held a special annual ceremony at which he was presented with a map of the empire. In 69, when breaches in the Yellow River dykes at Khaifêng were repaired, the engineer in charge was given a set of maps to assist him.

Establishment in Han and Chin. This brings us to the time of Chang Hêng, whose work in astronomy has already been described and whose achievements in seismology we shall discover later. None of the surviving fragments of his writing deals with map-making, but that it was he who originated the rectangular grid system seems very probable from the pregnant phrase used about him by the astronomer Tshai Yung. He is said to have 'cast a network (of coordinates) about heaven and earth, and reckoned on the basis of it'. The celestial coordinates must have been the *hsiu*; unfortunately we cannot tell exactly what the terrestrial ones were. That Chang Hêng occupied himself with map-making is sure, for a map of physical geography was presented by him in A.D. 116.

But the San Kuo and early Chin periods were more important than the Han for the attainment of the definitive style of Chinese cartography. The father of scientific cartography in China was Phei Hsiu (A.D. 224 to 271). In 267 he was appointed Minister of Works, and the *Chin Shu* (History of the Chin Dynasty) preserves particulars of the map-making in which he was then engaged, together with his preface to the maps. Parts of the preface are of great interest:

> The origin of maps and geographical treatises goes far back into former ages. Under the three dynasties (Hsia, Shang and Chou) there were special officials for this. Then, when the Han people sacked Hsien-yang, Hsiao Ho collected all the maps and documents of the Chinn. Now ... we have only maps, both general and local, from the (Later) Han time. None of these employs a graduated scale and none of them is arranged on a rectangular grid ...
>
> In making a map there are six principles observable:
>
> (1) The graduated divisions which are the means of determining the scale to which the map is to be drawn.
>
> (2) The rectangular grid (*chun-wang*) (of parallel lines in two dimensions), which is the way of depicting the correct relations between the various parts of the map.

(3) Pacing out the sides of right-angled triangles, which is the way of fixing the lengths of derived distances (i.e. the third side of the triangle which cannot be walked over).
(4) (Measuring) the high and the low.
(5) (Measuring) right angles and acute angles.
(6) (Measuring) curves and straight lines. These three principles are used according to the nature of the terrain, and are the means by which one reduces what are really plains and hills (literally cliffs) to distances on a plane surface . . .

Although none of Phei Hsiu's maps survives in any form, attempts to reconstruct them have been made, and he has been considered quite worthy to be compared with Ptolemy.

The rectangular grid is certainly as old as Phei Hsiu, but there is room for some speculation as to the source which may have suggested the coordinate system to him and to Chang Hêng. The characters combined in the ancient phrase *ching thien* (井田) (well-field system), which had been the subject of social and economic debates since feudal times, both plainly betray an ancient manner of thinking of land-allotment in coordinate terms. Then there was the system of concentric squares (Fig. 121) described in the *Shu Ching*. Nor should the origins of the abacus be forgotten; it may be significant that the first name in Chinese history to be associated with this device was that of the Taoist mathematician Hsü Yo, whose work was just between the times of Chang Hêng and Phei Hsiu. At what time the terms *ching* and *wei* were used for the *chun-wang* coordinates (as Phei Hsiu called them) is hard to say, but they certainly meant textile warp and weft before they were employed by map-makers. The very fact of drawing maps on silk, from the Chhin onwards, would invite the suggestive idea that the position of a place could be fixed by following a warp and a weft thread to their meeting place. This was perhaps the significance of the report by Wang Chia that the emperor Sun Chhüan had a girl cartographer, and we know that both Phei Hsiu and Chia Tan (the great Thang map-maker) used silk for their maps.

Another set of related ideas and techniques was that of the diviner's board, the magnetic compass and chess. Evidence on these will be discussed in the next volume of the abridgement; here it is enough to note that compass indications given on the diviner's board (Fig. 133) and on Han 'cosmic mirrors' (where names of divinities, trigrams, the duodenary cyclic characters, etc.) were so arranged that if coordinate lines were drawn across the space occupied by the azimuth points, nine small squares (*yeh* or *chiu kung*) would be formed. *Yeh*, as we shall see shortly, is met within some astronomical-geographical work.

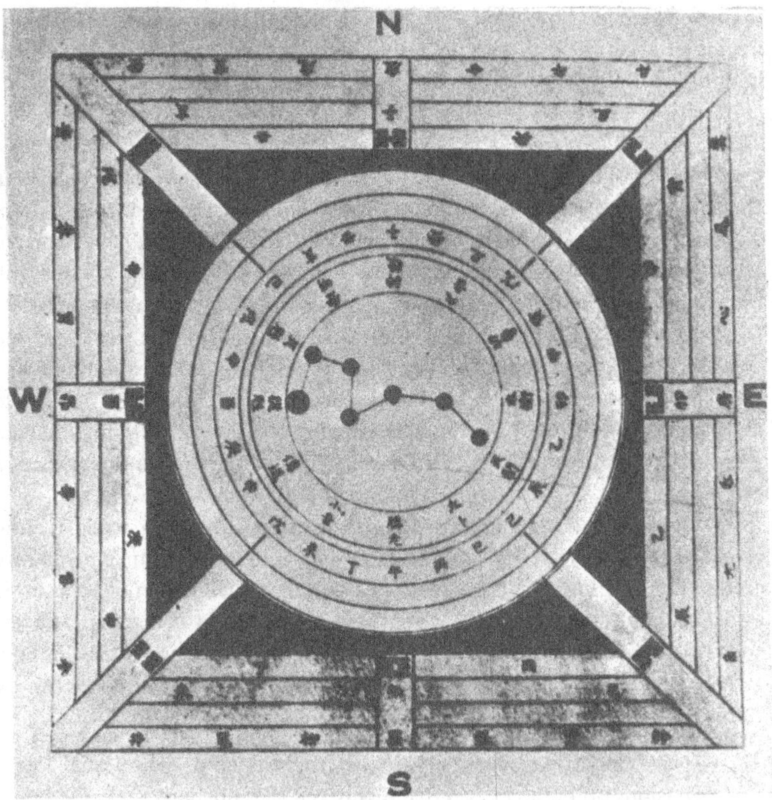

Fig. 133. A reconstruction of the *shih* or diviner's board of Han times (Rufus after Harada & Tazawa). This is a symbolic representation of heavens and earth which is connected with the origins of the magnetic compass. The square plate of the earth was surmounted by a rotating round plate signifying the heavens. The former was marked with cyclical characters and *hsiu*, the latter carried the figure of the Great Bear, with cyclical characters and prognosticatory signs. Fragments of these boards were first found in two Han tombs in Korea, the tomb of Wang Hsü (*d.* A.D. 69) and the Painted Basket tomb.

Mention of astronomical aspects of map grids raises at once the question of to what extent the Chinese cartography of Phei Hsiu and Chang Hêng was keyed to astronomical phenomena. In this respect there would seem to have been little difference between the Chinese and the Greeks. The Greeks used the gnomon shadow and the length of the day at solstice to determine latitude; the Chinese were also perfectly aware that the shadow length varied continuously in the north–south line. As for longitude, the

Chinese were no worse off than the Greeks. Its measurement with any degree of accuracy did not come until the eighteenth century; throughout antiquity and the Middle Ages dead reckoning was the only way.

Development in Thang and Sung. The Thang saw a great development in map-making. The wide extensions of the empire which were added early in the dynasty stimulated the mapping of Central Asia. Some books and maps were produced in A.D. 658, 661 and 747, but all are lost. Even the work of the greatest Thang cartographer, Chia Tan (730 to 805), did not survive, but we have a good deal of information about it. In 785 he was entrusted by the emperor to prepare a map of the whole empire, but the work was not completed until 801, smaller sections being presented in the meantime. The map was 9 m long and 10 m high, constructed on a grid scale of 1 cm to 40 *li*. It must therefore have covered an area 30,000 *li* from east to west and 33,000 *li* from north to south, and have been, to all intents and purposes, a map of Asia.

It is possible that in the Thang a further effort was made to link geographical with celestial coordinates. The hint arises from an obscure group of writings in which we hear of certain maps which do not seem to be associated with local topographics as their names would at first sight imply, and which are associated with Taoist and Buddhist scholars. Then, in 800, an official, Lü Wên, wrote a preface to a geographical work *Ti Chih Thu* by Li Kai in which he said:

> The boundaries of every square inch correspond to the celestial divisions on high. The (relation between the) phenomena in the heavens and the aspect (i.e. the physical features) of the earth can thus be brilliantly seen.

There seem to be three possible explanations of this cartographic movement in the Thang. First, there may have been an astrological reason. The association of particular earthly regions with sections of the sky was a very old idea in China, found in early Han and pre-Han texts, and in this context it may have emphasised anew the importance of showing physical features as opposed to political territorial divisions. To the fixed stars would correspond the fixed mountains and rivers, not the changeable city names. As we shall see in a moment this may have been the background of the greatest extant achievement of twelfth-century map-making. Second, the efforts of the Taoists and Buddhists may have been directed towards a really astronomical coordinate system. It would have been just as easy to draw meridians parallel to the hour-circles separating the different *hsiu*, as it would have been for the Greeks to draw them on the model of celestial longitudes. This does not necessarily imply accepting a spherical earth,

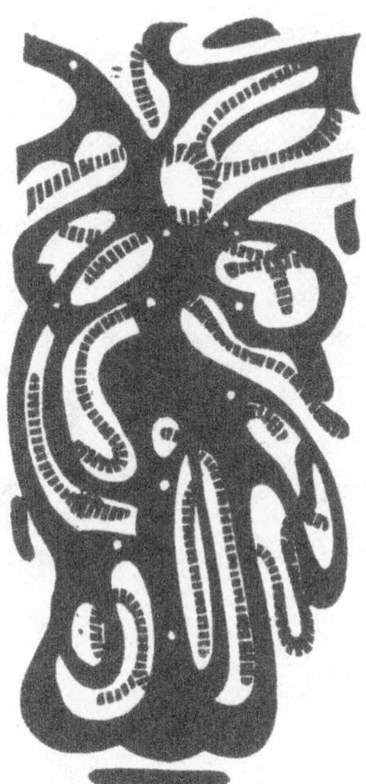

Fig. 134. Early attempt at contour mapping; on the right a representation of the Thai Shan mountain range from a seventeenth-century edition of the *Wu Yo Chen Hsing Thu*, the text of which is of much earlier but uncertain date. On the left, for comparison, a contour map of modern type. (From Ogawa, *Chigaku Zasshi* (Journal of the Tokyo Geographical Society), 1910, vol. 22, p. 207.)

though by the time of the Thang there had been so many foreign contacts that Chinese philosophers can hardly have been unaware of the idea.

The third alternative may serve to alert us to the possibility of a primitive kind of contour map originating in the Thang. In the seventeenth-century editions of an anonymous work *Wu Yo Chen Hsing Thu* (Map of the True Topography of the Five Sacred Mountains), the method of delineating the mountain ranges does not compare unfavourably with the modern style placed beside it (Fig. 134). Quotations from this book in the *Thai Phing Yü Lan* (Thai-Phing reign-period Imperial Encyclopaedia) of A.D. 983 show that at some time or other quite precise measurements were made of the breadth and depth of gorges. The date of the anonymous text is uncertain but its title is mentioned in a Taoist romance of the fourth century A.D.

A great deal of geographical work was done in the Sung, and it is from this period that the oldest examples of Chinese cartography still extant

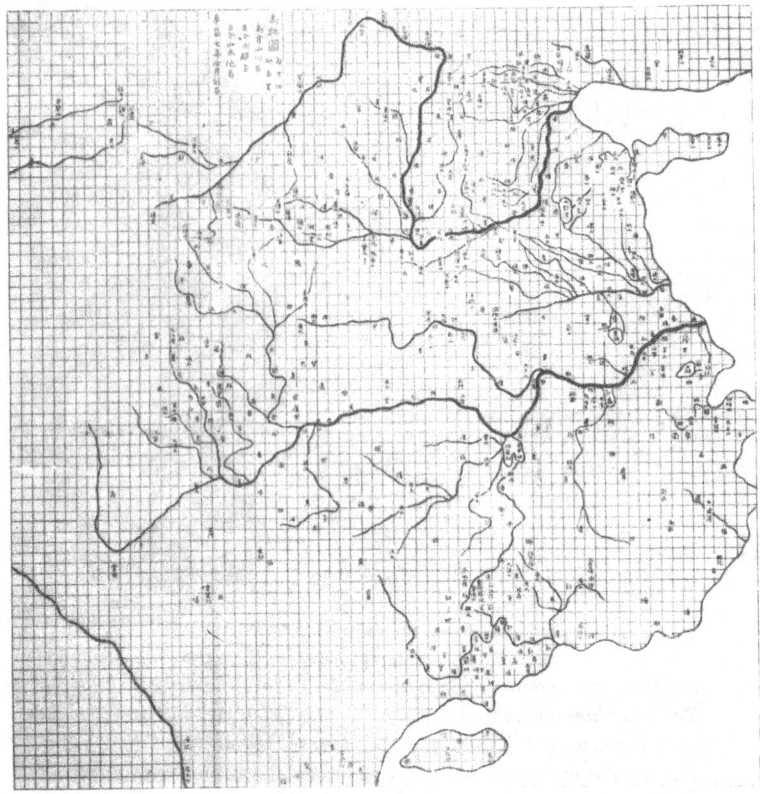

Fig. 135. The *Yü Chi Thu* (Map of the Tracks of Yü the Great), the most remarkable cartographic work of its age in any culture, carved in stone in A.D. 1137, but probably dating from before A.D. 1100. The scale of the grid is 100 *li* to the division. The coastal outline is relatively firm and the precision of the network of river systems extraordinary. The size of the original, which is now in the Pei Lin Museum at Sian, is about 0·9 m square. The name of the geographer is not known.

derive. Early in the dynasty (before A.D. 1000) there is an account of a map of the Western countries, and strategic maps of Kansu, prepared by a geographer in the imperial service, Shêng Tu. But all other records of the eleventh century are overshadowed by two magnificent maps which still exist. They were carved in stone in 1137, and one (Fig. 135) does the greatest credit to the Sung map-makers. Its inscription says that its grid scale is 100 *li* to each square, and comparison of its network of river systems with a modern chart shows at once the extraordinary correctness of the pattern. Anyone who compares this map with contemporary European religious cosmography (e.g. Fig. 129) cannot but be amazed at the extent to which Chinese geography was at that time ahead of the West. The other carved map appears to belong to an earlier tradition.

It is of interest that in both these maps north is at the top, and this is true for all Sung maps that have survived. The practice of placing south at the top seems to have originated among the Arabs rather than the Chinese, and to have become known only later in China. The first Chinese printed map also appeared in the Sung, the first printed European map not appearing until about two centuries later.

Climax in the Yuan and Ming. With the Yuan period we reach a man who might be regarded as the focal figure in the history of Chinese cartography, Chu Ssu-Pên (1273 to 1337). Inheriting the tradition of Chang Hêng and Phei Hsiu, he was able to summarise by its aid the large mass of new geographical information which the Mongolian unification of Asia had added to earlier knowledge possessed by the Thang and Sung. Between 1311 and 1320 he prepared his map of China, using older maps, literary sources and the results of personal travel. His map was a large one and bore the simple title *Yü Thu* (Earth-Vehicle (i.e. Terrestrial) Map). He was rather chary in mapping the more distant regions, saying, in sceptical words which deserve to be remembered:

> Regarding the foreign countries of the barbarians south-east of the South Sea, and north-west of Mongolia, there is no means of investigating them because of their great distance, although they are continually sending tribute to the court. Those who speak of them are unable to say anything definite, while those who say something definite cannot be trusted; hence I am compelled to omit them here.

Chu's successors had access to better information, but their cartography did not equal his.

For about two centuries the map, which was over 2 m long, existed only in manuscript or copies, but in 1541 it was revised and enlarged by Lo Hung-Hsien and printed in about 1555 under the title *Kuang Yü Thu* (Enlarged Terrestrial Atlas). Apart from the general map (Figs. 136, 137) there were sixteen sheets of the various provinces, sixteen of the border regions, three of the Yellow River, three of the Grand Canal, two of sea routes, and four devoted to Korea, Annam, Mongolia and Central Asia. The scale was usually 100 *li* to the division.

In spite of Chu Ssu-Pên's caution about far-distant regions, it is a remarkable fact that he and his contemporaries already recognised the triangular shape of Africa. In European and Arabic maps of the fourteenth century the tip of Africa is shown pointing eastwards, and this is not

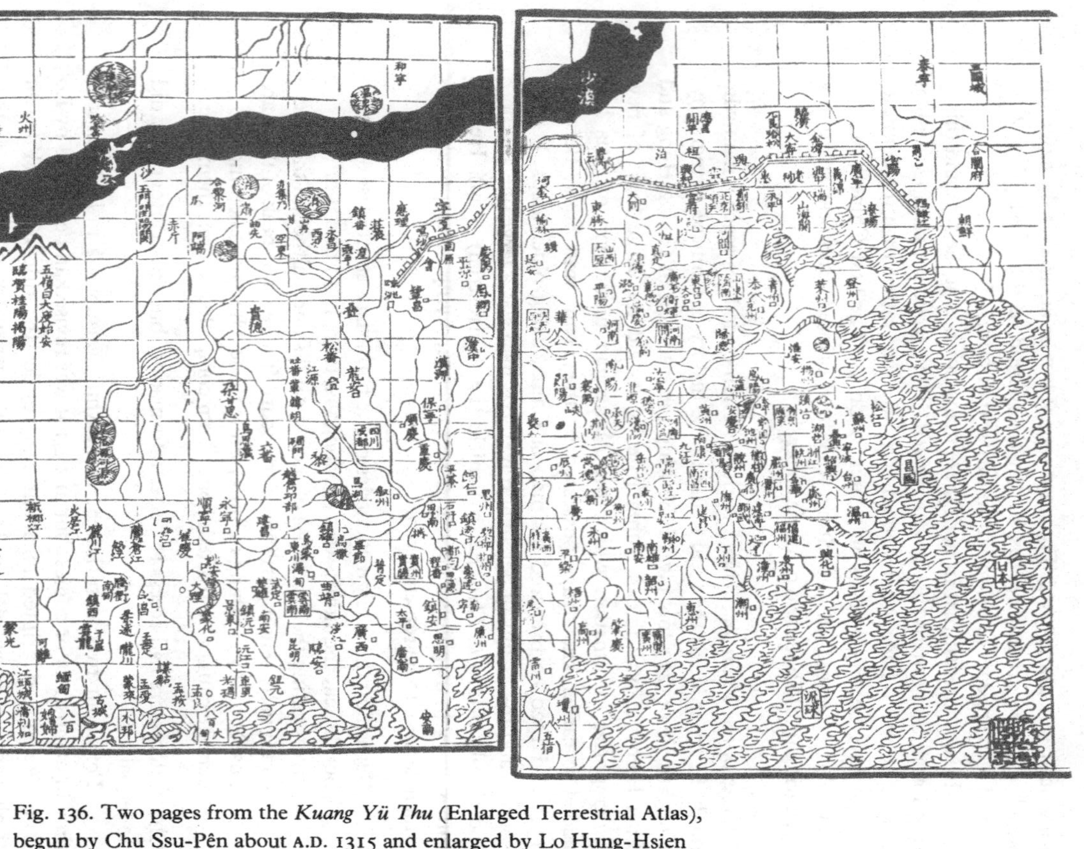

Fig. 136. Two pages from the *Kuang Yü Thu* (Enlarged Terrestrial Atlas), begun by Chu Ssu-Pên about A.D. 1315 and enlarged by Lo Hung-Hsien about 1555; the general map of China on a grid scale of 400 *li* to the division. The black band in the north-west represents the Gobi Desert. A map deriving from this was printed in *Purchas his Pilgrimes* (1625).

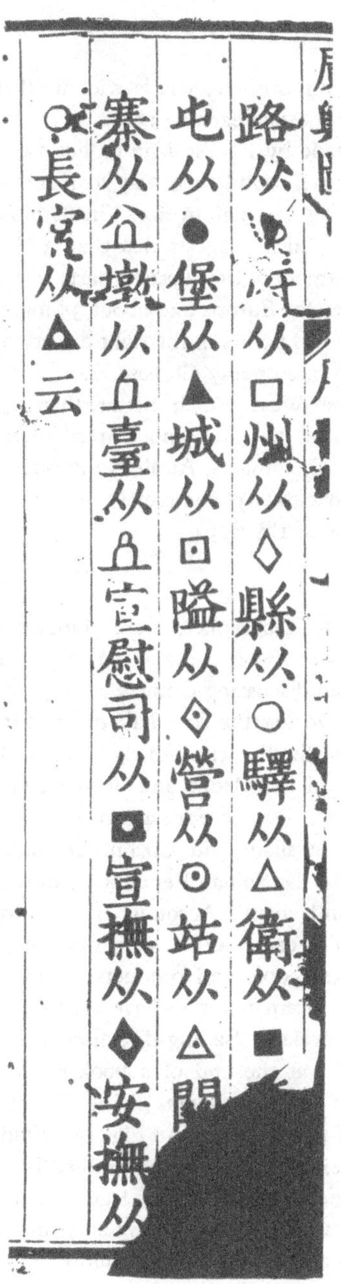

Fig. 137. The key to the symbols in the *Kuang Yü Thu*. Cities of the first order were indicated by a white square, those of the second order by a white lozenge, and those of the third by a white circle; post-stages by a white triangle and forts by a black square, etc. This must be one of the earliest occasions on which systematic symbolism of this kind was employed.

corrected until the middle of the fifteenth century; here is evidence that Chu Ssu-Pên must have drawn it pointing south as early as 1315.

Other maps of this century, made by Li Tsê-Min (about 1330) and the monk Chhing-Chün (1328 to 1392), found their way to Korea where they were combined into a large map 1·5 × 1·2 m in size. The names of the cities are the same as those in 1320, which suggests that they go back to around Chu Ssu-Pên's time. The treatment of the Western parts is interesting; it includes about 100 place-names for Europe and about 35 for Africa, which has its correct triangular shape and points in the right direction. The Sahara is shown black, like the Gobi in so many Chinese maps. The map displays extensive knowledge of the West, much better than that of Europeans about Chinese geography at the time, and it must have been gained from Arab, Persian and Turkish contacts. At all events such maps are magnificent examples of Yuan cartography, completely overshadowing all contemporary European and Arabic world-maps.

Chinese sailing charts

Just about the time the Korean geographers were combining the Yuan world-maps into one, there began the remarkable series of Chinese maritime explorations mentioned when discussing Chinese history (volume 1 of this abridgement, Chapter 5). During the first half of the fifteenth century there were expeditions to the South Seas and the Indian Ocean, greatly adding to Chinese geographical knowledge and bringing back all kinds of rarities to the imperial court. The expeditions visited Champa, Java, Palembang, Thailand, Ceylon, Calicut and Cochin in India, the Persian Gulf, Aden and Mogadiscio in the Somali Republic – 'all together more than thirty countries, large and small'. Accounts of their results appeared in four important books which came out between 1434 and 1520. All are full of information about the people and products of the places visited, but they do not contain (and seem never to have contained) any maps. However, one of the authors speaks of 'sailing-directions' which he used as a source of information, and at the end of a book on military technology written before 1651, there are certain maps giving the routes of the admiral Chêng Ho. Part of one of these is shown in Fig. 138. Evidently it does not derive from any of Phei Hsiu's grids, but rather is a true mariner's chart giving elaborate compass-bearings in the legends along the lines of travel, marking distances in 'watches', and noting all points along the shore such as rocks, shoals, tides and havens which could be of use to mariners. Neither south nor north is at the top, but east, and part of the distortion is in order to squeeze in the Indian Ocean.

Such charts correspond not only in nature but also in date to the portolan charts of Europe; the only difference is that they give their

Fig. 138. One of the sea-charts of the *Wu Pei Chih*. Though not printed until A.D. 1621, these charts date from the expeditions of Chêng Ho (A.D. 1405 to 1433). In the above map the coast at the top is that of western India, that at the bottom representing Arabia. To the left is the opening of the Persian Gulf, to the right is the entrance to the Red Sea. The Indian Ocean is thus compressed to a schematic corridor in which sailing tracks are marked with precise compass-bearings and other instructions. Places can be identified by the numbers, for example Muscat (81), Aden (62), Bombay (67). Alongside each one is given the altitude of the Pole Star in *chih* (finger-breadths). Such charts as these may be considered Chinese portolans but they are quite different in type from the portolans of the West (from Phillips).

compass-bearings in words instead of drawing rhumb-lines. Careful examination shows that they reach high standards of precision, an accuracy of 5° being reached in general, which is to be considered excellent for a pilot of 1425.

The role of the Arabs

The extensive Arab–Chinese contacts which there were around this time make it convenient to mention briefly the role of the Arabs in the history of map-making. Their close contact with the Byzantine world gave them an early contact with the remains of the Hellenistic geographers, and Ptolemy was available to them from the middle of the ninth century A.D. In the time of the Caliphate of al-Ma'mūn (the son of the famous Caliph Harūn al-Rashīd, immortalised in the *Arabian Nights*, i.e. 813 to 833), new lists of latitudes and longitudes were prepared, so the Arabs never quite lost the tradition of precise cartography.

Nevertheless, from the eighth to the eleventh centuries, the other tradition, that of religious cosmography, was strong, if not entirely dominant. Many examples of Arabic T-O maps and wheel-maps with 'climates' are known. There was also a tendency to draw more geometrical maps, so

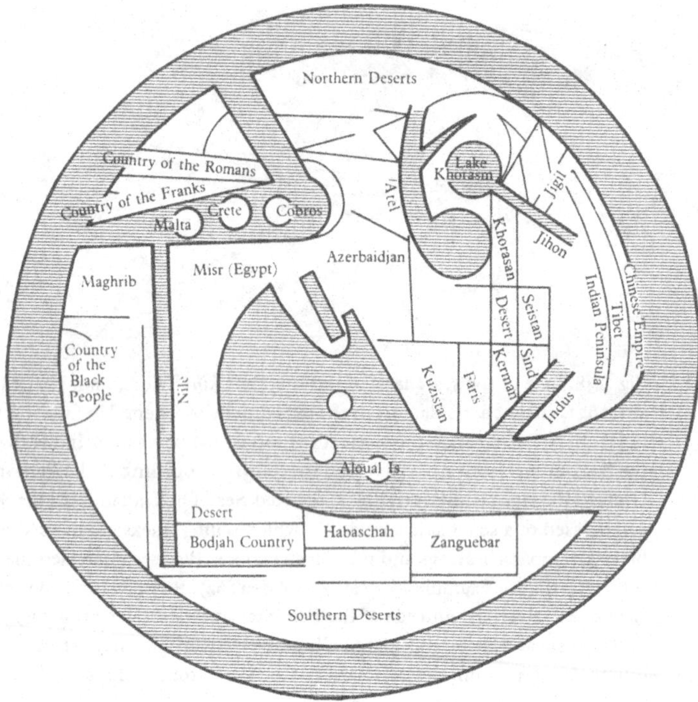

Fig. 139. An Arabic wheel-map, that of Abū Isḥāq al-Fārisī al-Iṣṭakhrī and Abu al-Qāsim Muḥammad ibn Hauqal (A.D. 950 to 970). It clearly shows the strong tendency to geometrical stylisation characteristic of the second period of Arab cartography. In the original, east was at the top, just as in the T-O maps of Latin Europe, but instead of the Earthly Paradise the Arab scholars knew enough to place in the Furthest East both China and Tibet. Note also how the tip of Africa points eastwards, a mistake which the Chinese geographers were the first to correct (from Beazley after Reinaud).

that the appearance of wheel-maps lost all resemblance to the actual contours of sea and land (Fig. 139), and they compare unfavourably with the Chinese maps we have been examining. By the fourteenth century, however, the general configuration of the world was becoming fairly clear, and in 1331 Naṣīr al-Dīn al-Ṭūsī prepared a world-map with climates for the northern half only (Fig. 140). Like the world-map of the Venetian Marino Sanuto (1306 to 1321) which it closely resembles, it shows the tip of Africa pointing east. Alone of the three contemporaries, Chu Ssu-Pên got this point correct.

In comparing the cartography of the Muslims with the Chinese, there are three chief maps to keep in mind. The first is the world-map of al-Sharīf

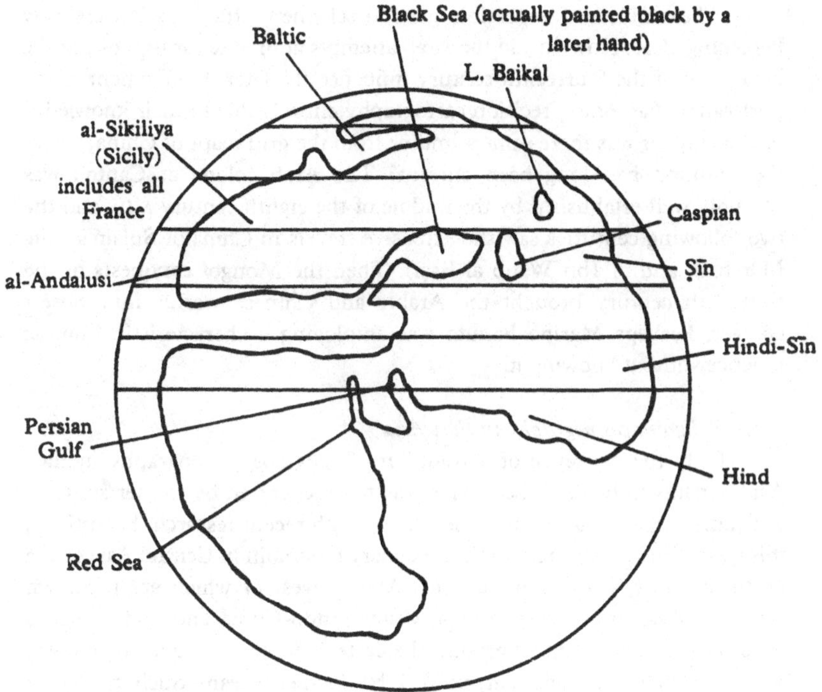

Fig. 140. The world-map of al-Ṭūsī (original sketch from autograph manuscript copy), A.D. 1331.

al-Idrīsī (1099 to 1166), made about 1150 for Roger II, the Norman king of Sicily. This was fully in the Ptolemaic tradition, using nine parallels of latitude (climates) and eleven meridians of longitude, but arranged on a projection like Mercator's and making no attempt to allow for the earth's curvature. Advantage here seems to rest with the Chinese grid map of 1137. Yet early in the next century 'Alī al-Marrākushī gave a list of 134 coordinate reference points which connected astronomy more closely with geography than was ever the case in China.

The two other sets of maps are essentially grid maps constructed not by the Chinese but by the Muslims. The first were made by al-Mustaufī al-Qazwīnī (1281 to 1349), who was certainly in touch with East Asia. His three maps are on rectangular grids, even his disc-shaped world-maps. Place-names are inserted but no symbols for physical features. Again the non-astronomical grid system has a Latin representative, Marino Sanuto, who used such a grid on his map of Palestine. But this is the solitary use of a grid by any European before the later portolans and the revival of Ptolemy. It may not be without significance that it was a map of an Arab country.

The outlines of a possible general scheme of transmission are now becoming clear. How far did the Arab attempts at precise cartography at the beginning of the fourteenth century influence the later development of the portolans? Was Arab precision cartography entirely due to their knowledge of Ptolemy, or was there some stimulus from the grid maps of China, where the tradition had long been current? The Arab colony at Canton was certainly well established by the middle of the eighth century A.D., and the two following centuries saw the extensive travels in China of Sulaimān the Merchant and of Ibn Wahb al-Baṣrī. Then the Mongol conquests of the thirteenth century brought the Arabic and Chinese worlds into closest contact. Perhaps Marino Sanuto was employing a characteristic Chinese practice without knowing it.

Religious cosmography in East Asia

Only the existence of a tradition of religious cosmography in East Asia remains to be described. This was not recognised by former students of Chinese affairs, but has become evident with recent research. Essentially, this type of map is centred on the legendary mountain in Central Asia or the north of Tibet, Mt Khun-Lun (Mt Meru), west of which are unknown regions, while to the east, Korea, China, Indo-China and India form a series of continents extending into the eastern oceans. Around these outer seas is a ring continent surrounded by further ocean. Such maps are particularly frequent in Korea where they occur as woodcuts, manuscripts and paintings on screens. No place-name later than the eleventh century A.D. has, however, been found on them.

There is no reason to doubt that the Koreans received this tradition from China, though it never seems to have been so popular there. In the *Thu Shu Pien* (On Maps and Books) of A.D. 1562 by Chang Huang there is such a map which has Mt Khun-Lun in the centre, but the great compiler says:

> This map is copied from a Buddhist work ... Although this map, taken from the *Fo Tsu Thung Chi* (Records of the Lineage of Buddha and the Patriarchs), does not clearly represent the shape of the world, I give it here. Such Buddhist attempts are not as a rule convincing.

The 'not' is of much interest, since it indicates the Buddhist–Taoist nature of the tradition, and the low opinion in which it was held by scholars who knew of the scientific tradition of Phei Hsiu.

The cosmographical content of the *Fo Tsu Thung Chi* seems to go back to A.D. 347, and it is just about this time that there is what is probably the principal literary reference to this wheel-map tradition. It occurs in the *Han Wu Ti Nei Chuan* (Inside Story of the Emperor Han Wu Ti), where the emperor obtains the secret and important 'Map of the True Topography of the Five Sacred Mountains', which is described in some detail. This description makes it clear that we are here dealing with a mixed Buddhist – Taoist tradition of religious cosmography, similar to that in Europe, but with Mt Khun-Lun as centre of the globe instead of Jerusalem. One can hardly doubt that this tradition came to China with Buddhism, perhaps joining indigenous ideas of Mt Khun-Lun as central. But it never did triumph over the tradition of scientific map-making there, as its analogue did in Europe.

CHINESE SURVEY METHODS

Let us now return briefly to the Chinese tradition of precise cartography. We may glance, first, at the probable survey methods used by Chinese map-makers; secondly at the origin of relief and special maps; and thirdly at the coming of Renaissance cartographic science to China.

It seems quite safe to assume that by the beginning of the Han the Chinese were in possession of the simple and ancient survey instruments which had been known to the Babylonians and Egyptians. We have seen that astronomical use of the gnomon goes back to Chou or Shang times, but it could not have been set up without the water-level and the plumb-line. Ropes and measuring tapes must also have been used, together with graduated poles (Fig. 141). One of the earliest literary references to calculations of surveyors in the field may be that in the *Shan Hai Ching* (Classic of the Mountains and Rivers), where Yü the Great orders two of his legendary assistants to pace out the size of the world. One of these carried counting-rods in his right hand and pointed to the north with his left. This at any rate gives a glimpse of late Chou or early Han surveyors. Similarly, the early *Chou Li* (Record of the Rites of (the) Chou (Dynasty)) speaks of the compass and square, the plumb-line and the water-level.

One of the developed forms of the plumb-line was the *groma*, associated with Hellenistic Egypt and Rome, but possibly of older origin. It consisted of two sets of plumb-lines set at right-angles and arranged to turn round a vertical axis. One pair could be used for sighting and the other to determine the direction at right-angles. It seems the Han Chinese also used this instrument, for the *Chou Li* says that the builders 'level the ground (by the use of) water(-levels), and suspend (plumb-lines)'; then 'they test the verticality of posts, gnomons or poles with the plumb-line'. In his commentary on the first phrase Chêng Hsüan remarks: 'at the four corners (of an instrument) four straight (lines) hang over the water, and (the surveyors) observe with this the high and low; when this has been decided the ground can be levelled'. As for the water-level, that in Fig. 141 is a trough with three floating sights.

Fig. 141. Chinese survey methods; apparatus figured in the *Wu Ching Tsung Yao* (Collection of the most important Military Techniques) by Tsêng Kung-Liang (A.D. 1044). To the left a graduated vertical pole and a surveyor with his sighting board; to the right a water-level with three floating sights and two plumb-lines. The text of the description of these instruments in Tsêng's book follows verbally that of a much earlier work, the *Thai Pai Yin Ching* (Manual of the White and Gloomy Planet of War) written by Li Chhüan in A.D. 759.

Besides the *groma* the other important invention of Hellenistic Egypt was the *dioptra* of Heron of Alexandria (about A.D. 65). This was the ancestor of today's theodolite. The Chinese used sighting-tubes but we have not found any ancient description of their use other than in armillary rings. If these were coming into use in the fourth century B.C. (page 159), then it is hard to believe that some azimuth mounting like that of the dioptra would not have been appreciated.

Something has been said already (page 13) of the practical geometry of Liu Hui (Fig. 142), who was a contemporary of Phei Hsiu, Têng Ai, a

Fig. 142. An illustration of survey geometry from the *Thu Shu Chi Chhêng*; measuring the height of an island crag.

general of the Wei State, was also alive at that time, and it is recorded of him that whenever he saw a high mountain or a broad marsh, he always 'estimated the heights and distances, measuring by finger-breadths, before drawing a plan of the place and fixing the position of his camp'. People used to laugh at him for being so particular, but the remark is interesting as showing how widespread in the third century A.D. were survey techniques involving, for the most part, the properties of right-angled triangles.

Fig. 143. The use of Jacob's Staff; an illustration from the *De Re et Praxi Geometrica* by Oronce Finé (Paris, 1556).

Perhaps the most important survey instrument of the European Middle Ages was the cross-staff or Jacob's staff (Fig. 143). The accepted view is that this graduated rod 1·2 m long, with a sliding cross-piece, was first described by the Provençal Jewish scholar Levi ben Gerson (1288 to 1344) in 1321. It would be natural to think that knowledge of it reached the Chinese by way of the Arabs in the fifteenth century, if not indeed later through the Jesuits. Similarly it has been supposed that the use of cross-wires in sighting-tubes was introduced to China by Jamāl al-Dīn in 1267. The following remarkable passage from Shen Kua, however, shows that the cross-staff was already known in China in the eleventh century, in the Sung, and that the cross-wire grid had an ancestry as far back as the Han:

> When I once dug in the garden of a house at Haichow, I unearthed a crossbow trigger-mechanism (or, a crossbow-like instrument). On looking at the whole breadth of a mountain, the distance on the instrument was long; on looking at a small part of the mountain-side, the distance on the instrument was short (because the cross-piece had to be pushed further away from the eye, and the graduation started from the further end). The (stock of the crossbow) was like a rule with graduations in inches and tenths of an inch. The idea of it was that by (placing) an arrow (across it at different points) and looking past the two ends of the arrow one could measure the degree of the mountain on the instrument and thus calculate its height. This is the same as the method of similar right-angled triangles of the mathematicians.

Shen Kua then discusses the prowess of the Han prince of Chhen with the crossbow and of his technique says:

> I once arranged such a 'three lengthwise and three crosswise' (grid) on a crossbow, and also sighted the target with the cross-piece arrow, and the result was that my shots were successful seven or eight times out of ten.

The significance of this entire passage, written about 1085, is that in Europe the cross-staff was also known as the crossbow or arbalest, and may plausibly be supposed to have developed from that propulsive mechanism. But the crossbow was much more ancient and widespread in China than in Europe and was introduced into Europe twice, once in Hellenistic and early Byzantine times, and then in the eleventh century. The European application to surveying around Levi ben Gerson's time would be natural. But in China, on the other hand, it formed the standard weapon of the Han armies, and was in continuous use from the fourth century B.C. onwards. Could this fact have had something to do with the superiority in survey methods which assisted the continuity of grid cartography after Phei Hsiu? In any case, Shen Kua certainly had in his hands a crossbow-like instrument with graduated divisions for measurement of heights, breadths and distances. Moreover there is evidence for transmission from China by Hebrew travellers, and merchants carrying ideas and techniques between East Asia and Western Europe, and this may well have been how the association with Levi ben Gerson came about.

A passage on the application of survey methods to map-making was also written by Shen Kua in the *Mêng Chhi Pi Than* (Dream Pool Essays):

> As roads and paths are sometimes winding and sometimes straight without any definite rule, if a walker starts out in any one of the four directions from a point along a path, his pacing will not help us get the direct distance. Therefore what we call 'straight lines in the four directions' have to be measured by other methods, ...
>
> I recently made a map of the counties and prefectures on a scale of 5 cm for 100 *li*. I used the methods of rectangular grid, mutual inclusions (similar right-angled triangles?), checking from the side, heights and depths, right angles and acute angles, curved and straight lines. With these seven methods you can work out the distances as a bird would fly them. The finished map had four-cornered, square, divisions strictly to scale. Then the four (azimuth) directions and the eight positions may be increased to twenty-four, ... Thus later generations with the help of my recorded data, and using the twenty-

> four directions, will be able to reconstruct the map showing the positions of districts and towns without the slightest mistake, even if the original copies should be lost.

This statement is of particular interest, because it suggests rather strongly that towards the end of the eleventh century Chinese cartographers were recording compass-bearings, as in modern ordnance surveys. For it should be remembered that elsewhere in the same book we have the earliest definite reference to the magnetic needle in any literature. Not until three centuries later did European portolans show compass roses.

For determining the lie of sloping ground, graduated poles were used in Shen Kua's time, as we know from one of his autobiographical sections where he describes work he once did as a government hydraulic engineer (between 1068 and 1077). This serves to remind us of a point of social importance, namely, the stimulus which the characteristic water-conservancy works of China must have given to the art of cartographic survey. Lastly there is the question of whether the ancient and medieval Chinese map-makers used a wheel-driven device for measuring distance – an hodometer. Since the first Chinese references to hodometers are certainly as old as those in Europe, going back to Prince Tan of Yen (240 to 226 B.C.) and since Chinese map-making must have involved so much dead-reckoning, one cannot help wondering whether men like Chang Hêng and Phei Hsiu did not make use of it. The nature of such vehicles will be discussed in a later volume.

RELIEF AND OTHER SPECIAL MAPS

The idea of representing the surface of the earth in modelled form on an exaggerated scale may seem to us obvious enough. And indeed it has a much longer history than is usually suggested by historians of geography. In 1665 John Evelyn published in England a description of '... New Kind of Maps in Bas-Relief' which were made in France, but some give priority to Paul Dox who in 1510 represented the neighbourhood of Kufstein in this way. Nevertheless their origin is earlier; Ibn Baṭṭūtah (1304 to 1377) described such a map he saw at Gibraltar, but in fact relief maps of a scientific character were well known in China in the eleventh century.

The main evidence for this comes once more from Shen Kua in his *Mêng Chhi Pi Than* (1086). He wrote:

> When I went as a government official to inspect the frontier, I made for the first time a wooden map upon which I represented the mountains, rivers, and roads. After having explored personally the mountains and rivers (of the region), I mixed sawdust with wheat-flour paste (modelling it) to represent the configuration of the terrain

Fig. 144. Two kinds of vessel from which the earliest relief maps may have originated. On the left a pottery 'hill-censer' (*po shan hsiang lu*) of the Han period, its cover moulded so as to represent one of the magic mountain islands of the Eastern Sea. On the right a pottery mortuary jar the lid of which also represents modelled mountains. After Laufer, *Chinese Pottery...*

> upon a kind of wooden base ... The emperor invited all the high officials to come and see it, and later gave orders that similar wooden maps should be prepared by all the prefects of frontier regions. These were sent up to the capital and conserved in the imperial archives.

Shen Kua must therefore take an important place in the history of relief maps, as of other aspects of geography. But there are further records of Sung scholars taking an interest in the relief techniques.

How far back the idea of relief maps goes before the time of Shen Kua is not clear, but there are several features in Chinese art which may well have given rise to it. One of these is the custom of representing sacred mountains in sculptured relief on incense-burners and jars. One is shown in Fig. 144*a*. The cover of such incense-burners, made both in bronze and in pottery, is always shaped in the form of a realistic hill or mountain, with holes through which the perfume escaped, and some arrangement by which the mountain is surrounded by water. The art motif must be at least as old as the Former Han, for the censers are referred to by Chang Chhang, who died in 48 B.C., in his *Tung Kung Ku Shih* (Stories of the Eastern Palace), as well as in other Han books. In later times, these 'hill censers', fitted with long handles, were adopted in Chinese Buddhism for liturgical purposes, but as Chinese archaeological books often suggest a Taoist identification, the usual view is that the hill censers were supposed to represent the sacred island-mountains of the Eastern Sea. The representation of these mountains long persisted in other materials such as jade.

There is, however, another possibility. The earliest Chinese maps were carved on wood, and the Fu Pan Chê – the Bearers of the Tables of Population – to whom Confucius always used to bow, has been cited as an example. There seems here a very old connection between geographical maps and population statistics. Among the many meanings of the word *pan* is that of a mould with branching channels used for casting coins, and also, of course, the carved board from which each page of a book was printed. No doubt in China the carving and incising of characters was particularly important from the oracle-bones onwards, and so might have given rise to the representation of the earth's surface in relief. Of course carved inscriptions were also a feature of the Babylonian and Egyptian civilisations. In any case there is one strange account of what may have been a relief map in the third century B.C. which cannot be omitted. In describing the tomb of Chhin Shih Huang Ti, the *Shih Chi* says:

> In the tomb-chamber of the hundred water-courses, the Chiang (the Yangtze River) and the Ho (the Yellow River) together with the great sea, were all imitated by means of flowing mercury, and there were machines which made it flow and circulate. Above (on the roof) the celestial bodies were all represented; below (presumably on the floor or on some kind of table) the geography of the earth was depicted.

Here, at least, there must have been channels for the mercury to flow in, so that a relief map is implied. This was in 210 B.C. Then, in the Han, we find in A.D. 32 a mention of strategic maps on which mountains and valleys were modelled in rice.

The use of wooden maps also led to one rather remarkable development in the fifth century A.D., the making of what seems to have been a 'jig-saw' map. The official history of the Liu Sung dynasty says Hsieh Chuang (421 to 466) made a wooden map some 3 m square. 'When one separated (the parts of the map) then all the districts were divided and the provinces isolated; when one put them together again, the whole empire then once more formed a unity'.

THE COMING OF RENAISSANCE CARTOGRAPHY TO CHINA

In 1583 Matteo Ricci, the first leader of the Jesuit mission, established himself at Chao-chhing, and it was there that he was asked by Chinese scholarly friends to prepare for them a map of the world. This was the beginning of his famous world-map of 1602. It was on a flattened sphere projection with parallel latitudes, showed America, and was certainly based on the 1570 world-map of Abraham Ortelius. Great geographical activity followed in the Khang-Hsi reign-period (1662 to 1722) when the emperor was personally interested in extending scientific knowledge of his vast dominions. An elaborate programme of work led to the Khang-Hsi Jesuit Atlas which, when complete (1717), was not only the best map ever made in Asia till then, but better and more accurate than any European map of the time. Moreover Western geographers now obtained new access to Chinese sources and so improved their own work. For example, there was the famous *Atlas Sinensis* of Martin Martini (1655), largely based on the *Kuang Yü Thu* (page 267 above).

Chinese geographical scholarship continued in the eighteenth century and many fine contributions were made, including a very great work, the Chhien-Lung Atlas of China (1769 and 1775), to a scale of 1 : 1,500,000 based on Jesuit surveys some ten years earlier. It is interesting, however, that there was much reluctance on the part of the Chinese to give up their rectangular grid system, and during the nineteenth century many atlases showed grid lines as well as lines of longitude and latitude.

COMPARATIVE RETROSPECT

We can now gain a comprehensive look at the development of geography in East and West (Table 36). We must distinguish between scientific cartography and religious cosmography; the former tradition began with the Greek and Hellenistic geographers but then died out completely, leaving the field to the second tradition. But just as this occurs in Europe, we find the Chinese initiating a long line of scientific map-making which is never interrupted until it blends with the cartography of the Renaissance.

Could there have been any transmission of the idea of meridians and parallels eastwards from Marinus and Ptolemy to China? The answer is uncertain. Such transmission would have been possible through the early travels of traders and 'ambassadors' from Roman Syria. But even if it could be proved an idea was transmitted, it would not help us much with regard to the detailed local sources of either Chang Hêng or Phei Hsiu, and we have no need to assume any lack of originality on their part.

In the West, Hellenistic precision map-making was conserved in a dormant state only in the Byzantine civilisation, but after it became available to the Arabs in the ninth century A.D., it led to much greater geographical achievements in their culture. Meanwhile the Latin world was dominated by religious wheel-maps until 1300 when an awakening came not from scholarly geographers but from practising sailors. That this was at least partly dependent on the transmission of the magnetic compass from China cannot be questioned. But soon the sailors' rhumb-lines became replaced by a rectangular grid. Did this also come from the East?

Table 36. *Chart to show comparative development of cartography in East and West*

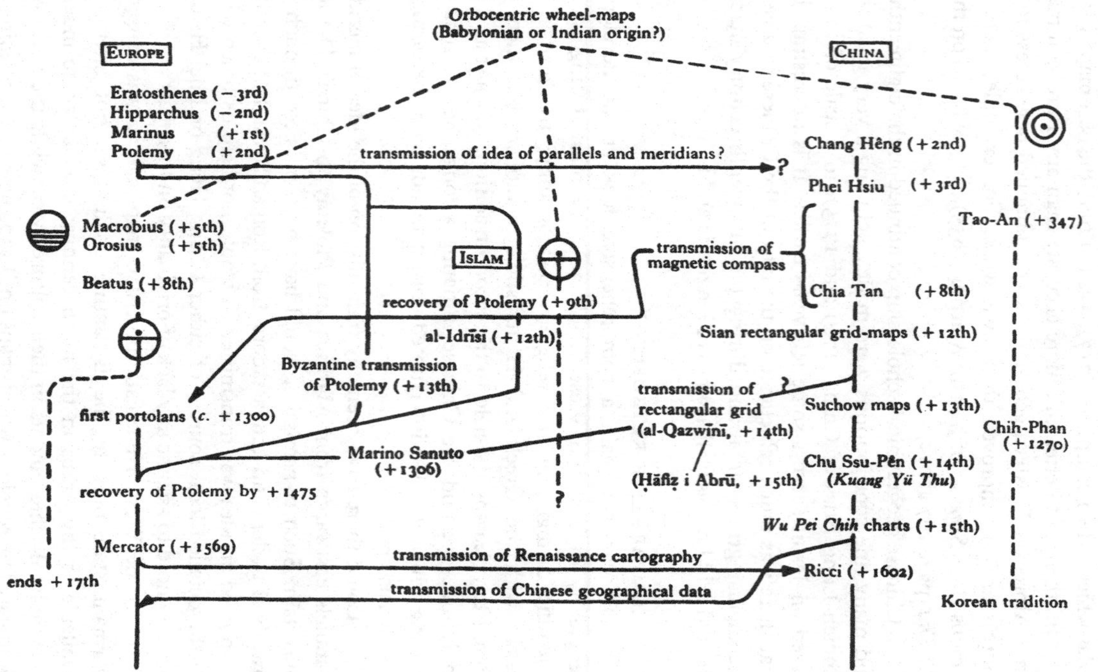

Rectangular grids were being used from the beginning of the portolan period, and since the Arabs were well established in Canton from the eighth century onwards, it would be surprising if Arabic travellers from there had not brought back knowledge of Chinese precision cartography and their rectangular grid.

The religious cosmographic tradition too was present among both the Arabs and the Chinese, though with the latter, at any rate, it never dominated to the extent that it did in medieval European geography. Perhaps Confucian good sense and Taoist skill prevented its enthusiastic reception. In India it seems to have been very ancient, but as to its ultimate origins it may have been Babylonian; maps from there between 2500 B.C. and 500 B.C. are known, and all are disc-like. In any case there is little doubt that the Chinese received it from India.

Lastly it only remains to add that while the transmission of Renaissance cartography to China at the time of Matteo Ricci must not be under-estimated, the reverse transmission of information about East Asia to the eighteenth-century geographers of Europe must also be remembered. It was owing to the solid work of generations of Chinese map-makers that knowledge of this part of the world became incorporated in modern geography.

The return of the rectangular grid to Europe. At the beginning of the present century cartographers had become so adept at representing on paper the curvature of the earth, and the astronomical determination of latitude and longitude had become so refined, that no one would have taken anything but a very superior attitude to the traditional Chinese rectangular grid. Yet in the First World War it was found, about 1915, that in the cooperation of artillery with survey engineers for accurate gunfire, rectangular coordinate mapping was most convenient. More recently still it has become essential in trans-polar flying where any course taken (except in the north–south direction) rapidly crosses the converging meridians at a constantly changing angle. We do not suggest, of course, that the similarity with the age-long Chinese tradition was known to those who made these more recent innovations in Europe.

5

The sciences of the earth: (ii) Geology and related sciences

The study of the Chinese contribution to geology is difficult because no history of it has ever been written, either in Chinese or in a Western language, so we can give, therefore, only a rough first approximation. Moreover, geology is a largely modern, post-Renaissance science. Scientific geology really begins with the *Prodromus* of Niels Stensen (Nicolaus Steno) of A.D. 1668, which introduced ideas such as folded strata, faulting, volcanic intrusions, eroded forms, etc. Yet not until the latter half of the eighteenth century did geology as we know it take shape, with the opposing ideas of those – the Neptunian school – who thought that all rocks were formed under water and the Plutonians, who believed that granite and many other rocks were forged in subterranean heat. This earlier general backwardness in investigating the structure of the earth's crust may broadly be said to have been shared by the Chinese, though they were responsible for several discoveries and inventions of great interest.

GENERAL GEOLOGY

Pictorial representations

The capacity for accurate observation and representation of geological forms as shown in Chinese pictures and book illustrations should not be under-estimated. It is probable that a whole textbook of geology could be illustrated largely from such sources. Numerous illustrations of this kind are to be found in the late encyclopaedias, such as the *Thu Shu Chi Chhêng* (Imperial Encyclopaedia) of A.D. 1726, and in the late geographies; Figs. 145 and 146 afford two examples, but perhaps the most striking of all such pictures is that which the Sung painter Li Kung-Lin (worked about A.D. 1000) made of an exposed arch of strata (Fig. 147). The fact that geological identifications can readily be made is a remarkable testimony to the faithfulness to Nature with which Chinese artists worked. To illustrate geological structures was indeed far from their thoughts, and the accurate

Fig. 145. Geology in Chinese art: (*a*) the rejuvenation of a valley at Li-Shan, near Fei-hsien, in Shantung. From *Thu Shu Chi Chhêng*.

Fig. 146. Geology in Chinese art: (*b*) deposit of water-rounded boulders at I-Shan in southern Shantung. From *Thu Shu Chi Chhêng*.

Fig. 147. Geology in Chinese art: (*c*) an exposed anticlinal arch at Lung-Mien Shan near Thung-chhêng in Anhui, just north of the Yangtze between Hankow and Nanking; a painting by Li Kung-Lin (*c*. A.D. 1100).

Fig. 148. Geology in Chinese art: (*d*) the eroded cliffs of Yen-Tang Shan near Wênchow on the coast of southern Chekiang (*Thu Shu Chi Chhêng*). These were some of the mountains which stimulated Shen Kua in the eleventh century A.D. to consider erosion and sedimentation, and to state the basic geological principles concerning them.

description of such forms may well have little to do with the aesthetic appreciation of Chinese painting itself, yet surely it is the old empirical closeness to Nature of the Taoists which was still there, and the world they depicted was the real world (Fig. 148).

Indeed, for many centuries, Chinese literature has contained handbooks or guides for painters which include a great variety of standard components of pictures, and among these, geological structures (hills, mountains, rocks, waters) naturally have a conspicuous place. Among the most famous of these is the *Chieh-Tzu Yuan Hua Chuan* (Mustard-Seed Garden Guide to Painting) of 1679, a book which in some editions included examples of very early colour block-printing. Many general terms are used in its descriptions, such as 'outline mountain ranges like a pile of cut-off wheel-rims', 'the last enveloping outline of a peak far away' and 'mountain wrinkles' (of which rather more than twenty variations of what was possible for a brush were recognised), but they show a live awareness of geological features.

The origin of mountains: uplifting, erosion, and sedimentary deposition

As regards statements concerning the origin of mountains, Chinese literature proves unexpectedly rich. Here the most famous text is that of the Neo-Confucian Chu Hsi (A.D. 1130 to 1200). In the *Chu Tzu Chhüan Shu* (Collected Works of Master Chu [Hsi]), the following passage appears after he has been talking about the periodical destruction of the world and its re-creation after very long intervals of time. Beginning with primeval fire and water, he suggests that the earth formed as a deposit from the water; he goes on to say:

> The Five Peaks are all but one Chhi, (blown by) a vast breathing. The waves roar and rock the world boundlessly, the frontiers of sea and land are always changing and moving, mountains suddenly arise and rivers are sunk and drowned. Human things become utterly extinguished and ancient traces entirely disappear; this is called the 'Great Waste-Land of the Generations'. I have seen on high mountains conchs and oyster shells, often embedded in the rocks. These rocks in ancient times were earth or mud, and the conchs and oysters lived in water. Subsequently everything that was at the bottom came to be at the top, and what was originally soft became solid and hard. One should meditate deeply on such matters, for these facts can be verified.

The great importance of this passage for life in past times – the history of palaeontology – will be mentioned shortly, but the main geological interest is that Chu Hsi recognised the fact that the mountains had been elevated since the day when the shells of the living animals had been buried in the soft mud of the sea-bottom. Three centuries later in the West in the time of Leonardo da Vinci, it was still supposed that the fact that shells were found in the Apennine mountains indicated that the sea had once stood at that level, although Leonardo himself attributed the origin of mountains to elevation and distortion after the formation of their fossil-containing rocks.

It might be thought that this was a flash of insight on the part of the great Sung philosopher, and stood alone. In fact, however, it was only part of a train of thought which had begun centuries before and was to continue long afterwards. The pursuit of this leads us to a peculiar technical term of the Taoists – *sang thien*, the 'mulberry-grove'. Originally a place-name and perhaps a constellation name, by the Thang it had become used for land which had once been covered by the sea, or would be in the future. Li Pai referred to it in poems, and at the same time Yen Chen-Chhing wrote an essay, *Ma-Ku Shan Hsien Than Chi* (Notes on the Altars to the Immortals on Ma-Ku Mountain), in which he quoted the legendary Taoist woman immortal Ma Ku as remarking that a place previously sea had turned into land and that she had seen such a change occurring. And Yen Chen-Chhing goes on to say:

> Even in stones and rocks on lofty heights there are shells of oysters and clams to be seen. Some think they were transformed from the groves and fields once under water (*sang thien*).

This essay must have been written about A.D. 770.

But Yen Chen-Chhing's views on the *sang thien* were not new in the eighth century. Yen was quoting from a book attributed to Ko Hung, and if the attribution is right, that book would date from about A.D. 320. It may not be as old as this but it is certainly pre-Thang. Yet we can go back earlier. In the mathematics section we met with the *Shu Shu Chi I* (Memoir on Some Traditions of the Mathematical Art), supposedly written in A.D. 190 but which cannot be later than about 570. Here we find:

> If one does not know the length or shortness of time, how can one appreciate what Ma Ku meant by 'the *sang thien*' (i.e. the long periods of centuries during which the sea is turned into dry land)?

There are other examples and it is therefore safe to say that sometime between the second and sixth centuries A.D., the expression *sang thien* had a familiar meaning equivalent to what we should now call 'geological time'.

Though the ideas about mountain formation so far described have a distinctly Taoist and Neo-Confucian flavour, there can be little doubt that they were originally connected with the Indian notion, brought to China probably by the Buddhists, of periodical cataclysms in which the world was destroyed and formed anew. Hence some interest attaches to a curious story preserved in the *Yeh Kho Tshung Shu* (Collected Notes on the Rustic Guest) of A.D. 1210, which relates to an incident alleged to have taken place in 120 B.C. During digging near the Khun-ming Lake 'black ashes' (*hei hui*, perhaps asphalt, coal, peat or lignite) were found, and to a question of the emperor Han Wu Ti regarding these, his adviser Tungfang Shuo replied 'You had better ask the Taoists from the Western countries' i.e. the Buddhist monks. A book from the eighteenth century quoting a source from the first century A.D., reports that the Buddhist monks replied that 'These are the remains of the last cataclysm of heaven and earth'. Modern research on the date of the entry of Buddhism into China does not allow us to accept the story as it stands, since monks were certainly not there in the Former Han dynasty. But the story must be quite old for it is to be found in other books and probably dates from the earlier years of Buddhism in China, i.e. from the late first century A.D. onwards.

References to mountain-building in European classical times are extremely scarce. European medieval views were dominated by the legend of Noah's Flood, essentially an echo of very early Babylonian happenings. This was the background of the explanation of fossils on high mountains given by Leonardo and already mentioned. Long before him, however, there had been advanced geological thinking in the Islamic civilisation in the tenth century A.D., when changes in sea-level, denudation, river evolution, etc. had been discussed. Ibn Sīnā (Avicenna) in the eleventh century ascribes the formation of heights to natural convulsions such as earthquakes, and describes erosive action and other matters which distinctly resemble the ideas of Chu Hsi a century later. A true explanation of fossils would be expected to follow, and Avicenna gives it.

The slow erosion of uplifted strata was quite clearly conceived in medieval Chinese thinking. There was an old expression *ling chhih* or *ling i* which probably meant originally the gradual slope of hills, but which commentators in the Chin and later interpreted in the sense of reduction in course of time. River valley erosion is referred to by the Han poet Chia I and by Emperor Yuan of the Liang (about A.D. 550). About the year 1070 in the Sung, Shen Kua wrote:

> Now I myself have noticed that Yen-Tang Shan is different from other mountains. All its lofty peaks are precipitous, abrupt, sharp and strange; its huge cliffs, 300 metres high, are different from what one finds in other places ... Considering the reasons for these shapes, I think that (for centuries) the mountain torrents have rushed down, carrying away all sand and earth, thus leaving the hard rocks standing alone.
>
> In places like Ta Lung Chhiu, Hsiao Lung Chhiu, ... one can see in the valleys whole caves scooped out by the forces of water.

He also describes sedimentary deposition:

> Now the Great (i.e. the Yellow) River, the Chang Shui, ... and the Sang Chhien are all muddy silt-bearing rivers. In the west of Shensi and Shansi the waters run through gorges as deep as 30 metres. Naturally mud and silt will be carried eastwards by these streams year after year, and in this way the substance of the whole continent must have been laid down.

Thus in the eleventh century Shen Kua fully understood those conceptions which, when stated by James Hutton in 1802, were to be the foundation of modern geology.

Caves, underground waters, and shifting sands

The study of caves (*tung*) and the formations found in them has been pursued throughout Chinese history; they were of particular interest to the early Taoist hermits, and the word continued always to have a numinous significance in the Taoist religion. Stalactites (Fig. 149) are included in the earliest lists of inorganic chemical substances and drugs which have come down to us, namely the *Chi Ni Tzu* (Book of Master Chi Ni), probably of the fourth century B.C., and the *Shen Nung Pên Tshao Ching* (Pharmacopeia of the Heavenly Husbandman), in date either first or second century B.C.

The existence of underground streams was well recognised. There is a mention of one in a Sung book, while a connection of some wells with the sea was suspected in certain cases where the water-level seemed to rise and fall with the tides – hence the name *hai yen*, eye of the sea. As for springs, a great deal of information is to be found in the encyclopaedias and geographical collections. In these, petrifying springs, so important in the history of geological and mineralogical thought, play their part, for they attracted much attention in China and were described with great clarity.

Ancient Chinese writings refer constantly to two place-names which indicate phenomena of geological interest. In the time of the Warring States and the early Han, just as there are always references to Mt Khun-Lun, so there are to the *liu sha* (the shifting sands) and the *jo shui* (weak water), both in the far west or north-west. But for our purposes identification of the places is of secondary importance; the interest of the names lies in the phenomena concerned. The shifting sands or quicksands were surely just that, and Shen Kua wrote in his *Mêng Chhi Pi Than* (Dream Pool Essays):

> At the Wu Ting river, going over with a horse, the sand shifted 30 metres away from where one was, like people treading on curtains, although where we trod was firm. In some places, however, they would sink in. Horses, camels and wagons, have all been engulfed, and hundreds of soldiers have disappeared there. This is indeed *liu sha* – flowing sand.

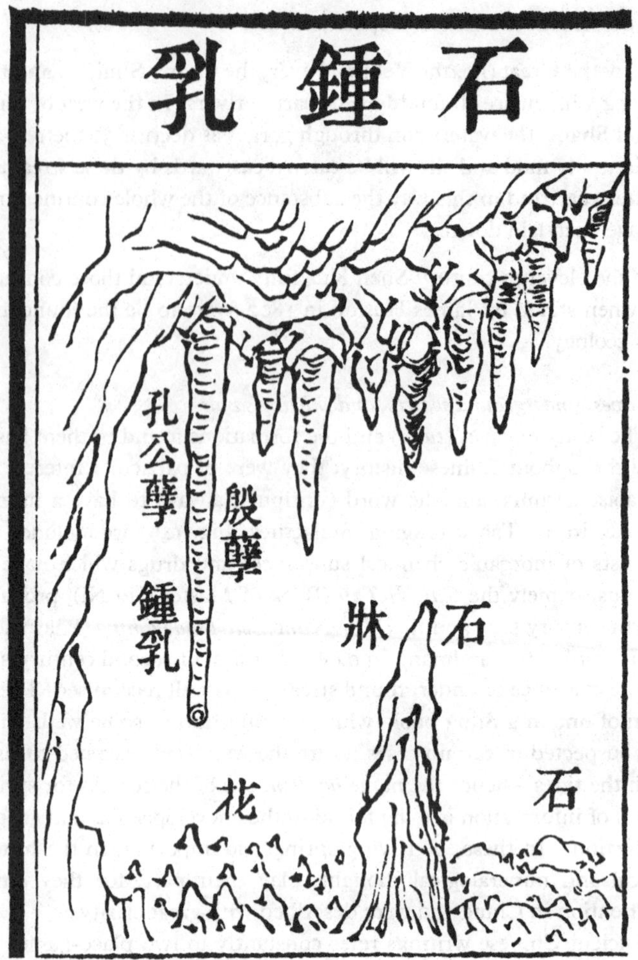

Fig. 149. Stalactites, stalagmites, and crystalline deposits; from Li Shih-Chen's *Pên Tshao Kang Mu* of A.D. 1596.

There are certainly localities in north-west China today which still present the same difficulties. The phenomenon of 'singing sands' is also referred to in Chinese texts; it is particularly associated with the sand dunes surrounding the temple and lake of Yüeh-Ya Chhüan, some kilometres west of Tunhuang.

Petroleum, naphtha, and volcanoes

The second ancient term of interest, *jo shui*, is often taken simply as the name of a river. But ancient authors repeatedly state that wood will not float in it, indeed Kuo Pho, about A.D. 300, said it would not bear the weight even of a goose feather. While some may prefer to believe that these were fancies derived from an ancient name, the suggestion may be put forward that in all these cases the foundation for these stories was the natural occurrence of petroleum seepages. The reference is often to various parts of Central Asia now hard to identify, but such seepages certainly exist and have existed in the western provinces, especially Kansu and Szechuan. For instance, the natural petroleum which occurred at Lao Chün Miao in Kansu was locally known and used for greasing cart-axles many years before the present oil field with its refineries was established there. It is mentioned in many old records. For instance, Thang Mêng in his *Po Wu Chi* (Record of the Investigations of Things), written about A.D. 190, says:

> In the mountains south of Yen-shou there are certain rocks from which springs of 'water' arise ... This liquid is fatty and sticky like the juice of meat. It is viscous like uncongealed grease. If one sets light to it, it burns with an exceedingly bright flame...

Shen Kua wrote about it, as one might expect, and explains that his samples from Fu and Yen burned with a thick smoke. He collected the deposit for making ink. It was, he said, superior to pine-wood resin ink and he thought it might become widely adopted, for:

> The petroleum is abundant, and more will be formed in the earth while supplies of pine-wood may be exhausted ... All the woods south of the Yangtze and west of the capital are going to disappear in time if this goes on, yet the ink-makers do not yet know of the benefit of petroleum smoke.

These are striking remarks about deforestation from anyone writing as early as A.D. 1070. Also very striking, and characteristic of Chinese literary culture, is the fact that Shen Kua could think of nothing better to do with petroleum than to make ink from its soot.

As far as volcanoes are concerned, all information had to come from outside since there are no volcanoes in China itself. One of the first references is in the third century A.D., but it does not seem to be an eye-witness account; it was perhaps a description brought by Syrian traders or from reports of travellers to the south seas like Khang Thai, who was envoy to Indo-China about A.D. 260. On the other hand there are many hot springs in China, and in the Thang, Hsü Chien in his encyclopaedic *Chhu Hsüeh Chi* (Entry into Learning) noted that if the upper reaches of a stream smell sulphurous, its springs are likely to be warm, or hot. In the Sung it was suggested that the heat was caused by the underground combustion of alum and sulphur, and in the Ming, Wang Chih-Chien listed many kinds of mineral-laden waters.

PALAEONTOLOGY

Though the term fossils is now reserved for the remains of plants and animals embedded in the strata of the earth's surface, its original meaning was much wider and included everything of interest which could be obtained by digging. Nevertheless, taking the modern definition, it is clear that the Chinese recognised fossils at a comparatively early date and certainly before any such understanding appeared in Europe or Islam.

Fossil plants

China certainly contributes to the beginnings of palaeobotany. Knowledge of the petrification of pine-trees may go back to the third century A.D. since encyclopaedias quote Chang Hua as saying that all trees turn into stone after three thousand years. In the Thang great interest was taken in the matter. In A.D. 767 the painter Pi Hung made a famous fresco depicting fossilised pine-trees, and at the end of the ninth century petrified pieces of pine-tree were described, while fifty years later the Taoist poet Lu Kuei-Mêng wrote two poems on them. In the Sung, in the *Mo Kho Hui Hsi* (Fly-Whisk Conversations of the Scholarly Guest), Phêng Chhêng wrote (about 1080):

> In Hu Shan there are Pu tree(-trunks) some metres in length, of which half has been changed into stone and half is still the hard wood. Tshai Chün-Mo, seeing these and thinking them particularly strange brought one to his house and I myself saw it there.

The observation that only part of a tree-trunk may petrify was a good one, but one has to be careful for there could have been some confusion with another kind of stone called *sung lin shih* (pine-forest stone), which was found in Szechuan; this simply has tree-like markings due to crystallisation of manganese dioxide. One of the first descriptions of it is in the thirteenth century. And there are some other doubtful cases of identification. Nevertheless, Shen Kua gives a clear reference to fossil plants:

> In recent years [about 1080] there was a landslide on the bank of a large river in Yung-Ning Kuan near Yenchow. The bank collapsed, opening a space of several metres, and under the ground a forest of bamboo shoots was thus revealed. It contained several hundred bamboos with their roots and trunks all complete, and all turned to stone...

Fossil animals

With the fossils of animals we are on much surer ground. The brachiopods are a very ancient group of invertebrate animals which have a superficial resemblance to bivalve molluscs such as mussels. Because many extinct species known as *Spirifer* had shells which looked like the out-stretched wings of a bird, the Chinese name for them was *shih yen* (stone-swallows). In the West they were identified as fossils in 1853, but the first important reference to them occurs in China towards the end of the fifth century A.D., where they are spoken of as '... stone oysters which look like swallows ... During thunderstorms, these stone-swallows fly about as if they were real swallows.' But this is a later quotation, and it seems likely that they had first been recognised in the fourth century. From the sixth century they were incorporated into the pharmaceutical compendia and a drawing of one of them is included in Fig. 150; they were also used as imperial tribute for a time. They were recognised in 1176 by Tsêng Min-Hsing as having once been in the sea.

It was a curious idea that stone-swallows came out of the rocks and flew about in windy and stormy weather. But the legend at any rate stimulated Tu Wan, a Sung scholar, to disprove it experimentally in the twelfth century. He also gave the reason for the belief.

> It was said in ancient times that when it rained they flew about. In recent years, however, I have climbed up high cliffs and found many of these stone shapes with the form of swallows. Some of them I marked with my pen. As the rocks were exposed to the blazing sun they cracked and weathered when the thunder showers came, and the ones which I had marked fell to the ground.... It was because of the expansion in the heat and contraction in the cold that they fell flying through the air. They cannot really fly.

The texts also explain that the stone-swallows were dissolved or preserved in vinegar for use as a medicine. The traditional Chinese diet was always lacking in calcium (owing to the absence of milk products from the diet) and a good source of assimilable lime was needed. Indeed the *Pên Tshao Kang Mu* (The Great Pharmacopoeia) prescribes fossils for dental and other troubles likely to be due to a low calcium content in the blood.

Other fossils are mentioned – the 'stone oyster' was one, the 'conch' (spiral-shelled mollusc) another. Some very impressive coiled fossil shells of soft-bodied animals which possessed an external skeleton were also recognised by the Chinese, at any rate from the Sung onwards, when the first pharmaceutical natural history in which they appear was published in A.D. 1070. Besides these 'stone serpents' there were the straight-shelled nautiloids, and limestone containing these was known by the time of the

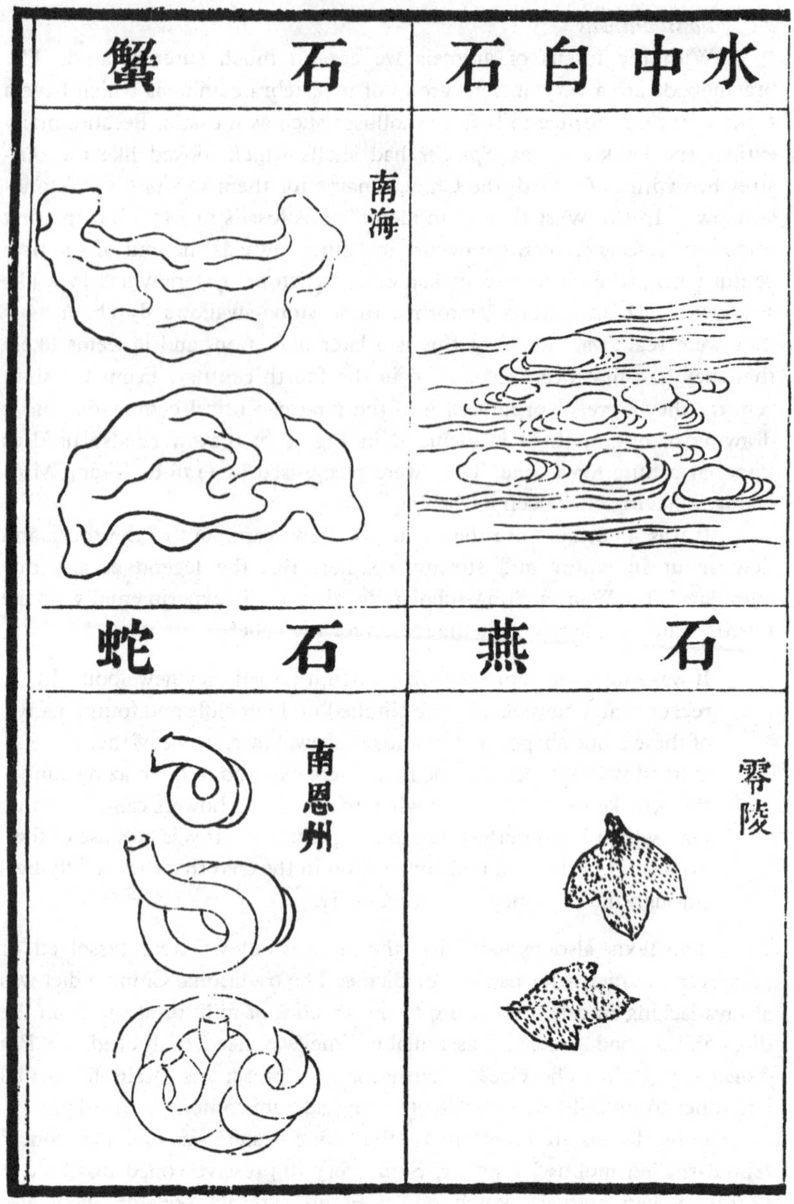

Fig. 150. Drawings of fossil animals from Li Shih-Chen's *Pên Tshao Kang Mu* of A.D. 1596. At the top of the left, *shih hsieh*, i.e. stone crabs. Below, to the left, *shih shê*, i.e. stone serpents; and to the right, *shih yen*, i.e. brachiopods such as *Spirifer* and related genera.

Ming. Fossil arthropods had more success in China than the fossil cephalopods just mentioned – stones containing trilobites were called 'bat-stones' because of a superficial resemblance to bats' wings, and crabs from the Pleistocene were widely known. They, too, were introduced into medicine in the Sung.

The Chinese were also interested in fossilised and partially fossilised vertebrates – fishes, reptiles and mammals. The first reference to stone fishes comes in the *Shui Ching Chu* of Li Tao-Yuan (died A.D. 527), and then in 1133 in the *Yün Lin Shih Phu* (Cloud Forest Lapidary), which not only describes the fossils which 'are all as perfect as if drawn with ink', but also the precise location and depth of the geological layers among which they are found. The bones and teeth of fossil reptiles, birds and mammals were always known in China as 'dragon's bones' and 'dragon's teeth', and the great esteem in which such bones were held medicinally aided modern palaeontologists to discover fossil man in China (*Sinanthropus pekinensis*). Moreover, it was through examination of drug-store material that the first oracle-bone inscriptions came to light. It is interesting, however, that the first incorporation of vertebrate fossils into the pharmaceutical natural histories occurred earlier than for the other fossils mentioned. The common name 'dragon's teeth' shows, of course, that the fossils were always regarded as the remains of animals long dead, even if somewhat mythological, but the existence of prehistoric monsters is clearly stated by Khou Tsung-Shih early in the twelfth century.

Compared with the pre-Renaissance West, the Chinese from about the first century B.C. defined and recognised their material with a good deal more precision and comprehension. As for the ancient Mediterranean civilisations, the oldest Greek reference is that ascribed to Xenophanes (sixth century B.C.) but known only through a third-century A.D. quotation. Here he is said to have believed that the land and sea 'were once mixed up' and as proof of this quoted the existence of shells of sea-animals found far inland and on mountains, as well as imprints of fish and seaweed. The quotation clearly contains an echo of the Indian belief in periodical cataclysms. Later Greek authors make some references to fossils but conclusions of a geological character were not usually drawn; the facts were simply noted as curiosities.

The picture thus presented is very reminiscent of the Great Interruption which was so prominent in precise map-making. Brilliant insights or great achievements were attained by the Greeks, but from about the end of the second century A.D. to the fifteenth, China was much more advanced than Europe, until modern science begins to appear. It does not seem that any stimulus was received by the Chinese from the West at the beginning of their best period.

Seismology

EARTHQUAKE RECORDS AND THEORIES

If China had no volcanoes it nevertheless formed part of one of the world's greatest areas of seismic disturbance from the earliest times. It was natural, therefore, that the Chinese should have kept extensive records of earthquakes, and these indeed constitute the longest and most complete series which we have for any part of the earth's surface. They tell us that up to A.D. 1644 there had been 908 recorded shocks (for which we have precise data), one of the earliest being that of 780 B.C. when the courses of three rivers were interrupted. At Nanking between A.D. 345 and 414 there were 30 shocks, and between 1372 and 1644 no less than 110, but the main area always lay north of the Yangtze and in all the western provinces. From the records there emerge twelve peaks of frequency between the end of the Sung and the beginning of the Chhing, showing a 32-year period, but although earthquakes sometimes affected several provinces, different regions did not usually have them simultaneously. There seems to have been no relationship between Chinese and Japanese earthquakes. One of the worst Chinese ones was that of 25 September in A.D. 1303 in Shansi, while that of 2 February 1556 is said to have killed more than 800,000 people in Shansi, Shensi and Honan. A famous siege of Sian in A.D. 1128 was ended by an earthquake.

No great progress was made in ancient or medieval China regarding the theory of earthquakes, which in Europe had also to await post-Renaissance theories about the nature of the earth's crust. However, in connection with the eighth-century B.C. earthquake, we find an early statement of ideas about them. The *Shih Chi* (Historical Records) of about 90 B.C. state:

> In the second year of the reign of King Yu (of Chou), the three rivers of the western province were all shaken and their beds raised up. Poyang Fu said: 'The dynasty of the Chou is going to perish. It is necessary that the *chhi* of heaven and earth should not lose their order; if they overstep their order it is because there is disorder among the people. When the Yang is hidden and cannot come forth, or when the Yin bars its way and cannot rise up, then there is what we call an earthquake. Now we see that the three rivers have dried up by this shaking; it is because the Yang has lost its place and the Yin has overburdened it. When the Yang has lost its rank and finds itself (subordinate to) the Yin, the springs become closed, and when this has happened the kingdom must be lost...'

Let it not be thought that theories of this kind were more primitive than those which the Mediterranean world entertained on the subject of earthquakes. In Greece there were suggestions that earthquakes were caused by excess of water from the upper regions bursting into the under parts and hollows of the earth, or by masses of earth falling into cavernous places during the process of drying, or even, according to Aristotle, to the instability of the vapour generated by the drying action of the sun on the moist earth, and to difficulties met with by the vapour in escaping. This last is closely parallel to Chinese ideas of the imprisonment of *chhi*; indeed, the importance of pneumatic theories in both civilisations is striking. However, no further progress was made in seismic theory until modern times.

THE ANCESTOR OF ALL SEISMOGRAPHS

But if theory did not advance, China has unquestionably the credit of having produced the ancestor of all seismographs. This was due to the brilliant mathematician, astronomer and geographer, Chang Hêng (A.D. 78 to 139), whom we have already met in other connections.

The basic text which describes Chang Hêng's instrument is contained in the biographical chapter devoted to him in the *Hou Han Shu* (History of the Later Han Dynasty). The information which is given is fortunately rather detailed.

> In the first year of the Yang-Chia reign-period (A.D. 132) Chang Hêng also invented an 'earthquake weathercock'.
>
> It consisted of a vessel of fine cast bronze, resembling a wine-jar, and having a diameter of eight *chhih* (2 metres).
>
> It had a domed cover, and the outer surface was ornamented with antique seal-characters and designs of mountains, tortoises, birds and animals.
>
> Inside there was a central column capable of lateral displacement along tracks in the eight directions, and so arranged (that it would operate) a closing and opening mechanism.
>
> Outside the vessel there were eight dragon heads, each one holding a bronze ball in its mouth, while round the base sat eight (corresponding) toads with their mouths open, ready to receive any ball which the dragons might drop.
>
> The toothed machinery and ingenious constructions were all hidden inside the vessel, and the cover fitted down closely all round without any crevice.
>
> When an earthquake occurred the dragon mechanism of the vessel was caused to vibrate so that a ball was vomited out of a dragon-mouth and caught by the toad underneath. At the same

> instant a sharp sound was made which called the attention of the observers.
>
> Now although the mechanism of one dragon was released, the seven (other) heads did not move, and by following (in azimuth) the direction (of the dragon which had been set in motion), one knew (the direction) from which the earthquake (shock) had come. When this was verified by the facts there was (found) an almost miraculous agreement (i.e. between the observations made with the apparatus and the news of what had actually happened).
>
> Nothing like this had ever been heard of before since the earliest records of the *Shu* (*Ching*).
>
> On one occasion one of the dragons let fall a ball from its mouth though no perceptible shock could be felt. All the scholars at the capital were astonished at this strange effect occurring without any evidence (of an earthquake to cause it). But several days later a messenger arrived bringing news of an earthquake in Lung-Hsi (Kansu). Upon this everyone admitted the mysterious (power of the instrument). Thenceforward it became the duty of the officials of the Bureau of Astronomy and Calendar to record the directions from which earthquakes came.

The central column of Chang Hêng's seismograph was essentially a pendulum which could have been either suspended or inverted. Fig. 151 shows a reconstruction with a suspended pendulum, and Fig. 152 with an inverted pendulum. All seismologists have appreciated the technical difficulty of constructing an apparatus in which only one of the balls should drop out (and thus 'write' its record of the earthquake shock) since there are always, besides the main longitudinal (to-and-fro) shock-wave, other components, mainly lateral, some of which may be of great force. It was necessary therefore to include some arrangement whereby the apparatus would be immobilised immediately after giving its first response. In the reconstructions shown methods of immobilisation are suggested. A third reconstruction, again using an inverted pendulum, is illustrated in Fig. 153. That people of the Han could have constructed such a mechanism using levers as Figs. 151 and 153 suggest is proved by the levers and cranks used in contemporary devices such as the crossbow trigger. However, it may be that the hanging pendulum device of Fig. 151 would not be immobilised satisfactorily from additional lateral shocks. At all events an instrument of the kind illustrated in Fig. 152 has been constructed at Tokyo University, and it was found that in such a device the ball was usually released by the secondary lateral waves, not by the first longitudinal one unless this was very strong. This might therefore mean that Chang Hêng would have had to calibrate his instrument by trial and error.

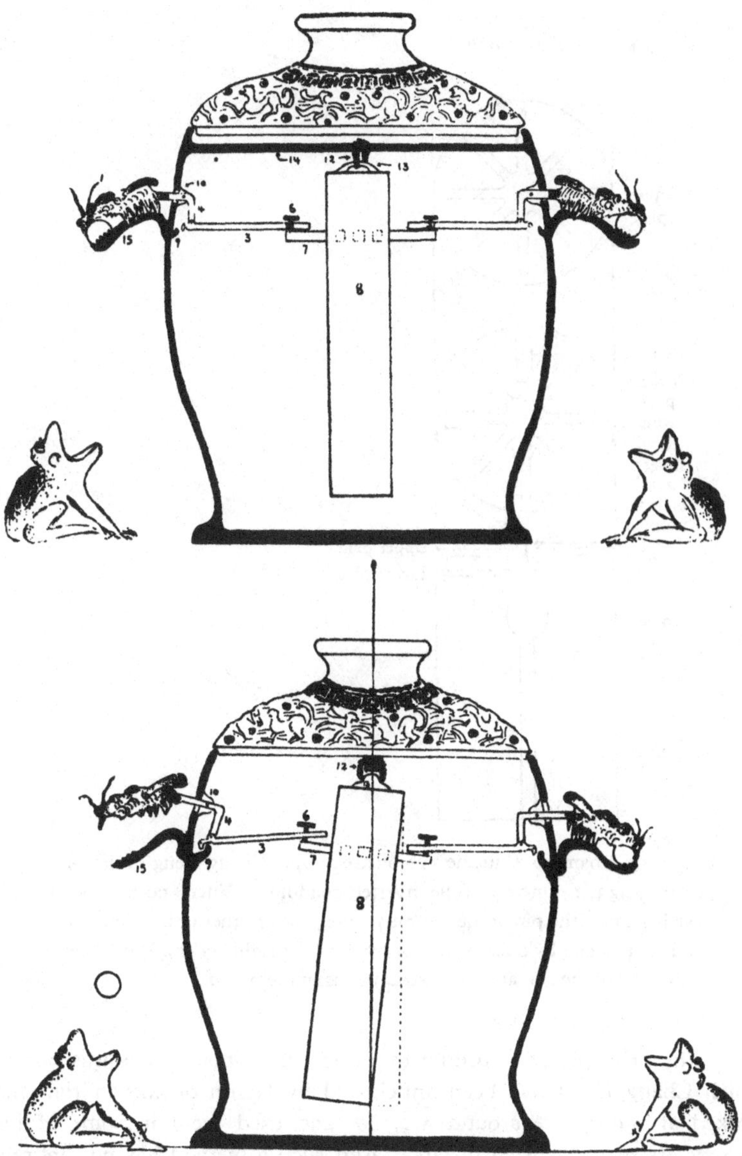

Fig. 151. An attempted reconstruction of the mechanism of Chang Hêng's seismograph by Wang Chen-To. The pendulum carries eight mobile arms radiating in as many directions and each connected with cranks which are provided with catch mechanisms at the periphery. Any one of the cranks which raises a dragon head and so releases a ball is thus at the same time caught and held, thus immobilising the instrument. The parts are as follows: 3, crank; 4, right-angle lever for raising the dragon's head; 6, vertical pin passing through a slot in the crank; 7, arm of the pendulum; 8, pendulum; 9, catch; 10, pivot on a projection; 12, sling suspending the pendulum; 13, attachment of sling; 14, horizontal bar supporting the pendulum; 15, lower jaw of dragon supporting ball.

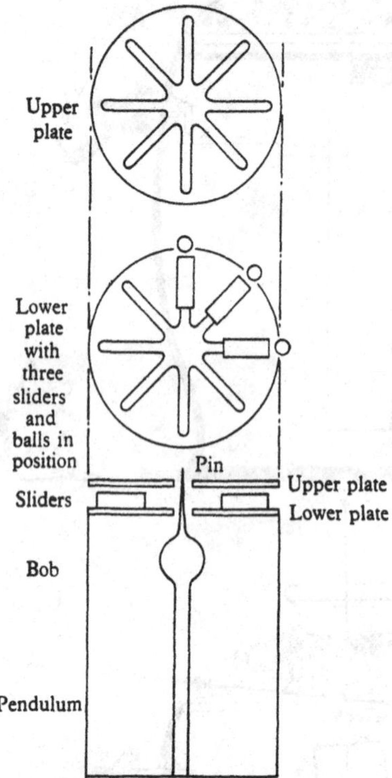

Fig. 152. Imamura Akitsune's reconstruction of Chang Hêng's seismograph employing the principle of the inverted pendulum. When knocked over by an earth tremor, the pin at the top must enter one or other of the slots provided and expel a ball by pushing one or other of the eight sliders. It can then no longer leave the slot and the instrument is immobilised.

The principle of recording by means of dropping balls was one in which Chang Hêng had been anticipated by Heron of Alexandria, who worked at some time around A.D. 60, and used them in some of his hodometers for measuring distance. And such a method has not entirely vanished, having been used in modern times in an instrument for recording water currents at sea. It is clear from other Chinese literature that Chang Hêng's instrument was considered of much importance at the time. This is not surprising, for it would enable officials of the central government to have early warning of an earthquake in a distant province and so be able to take measures to deal with needs or disturbances which might ensue.

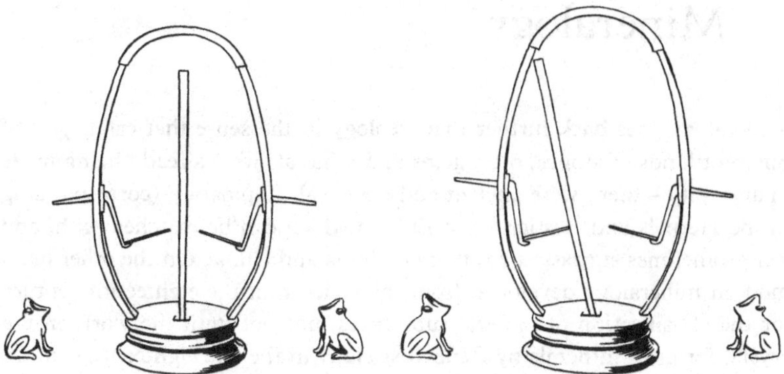

Fig. 153. Another reconstruction of Chang Hêng's seismograph by Wang Chen-To, accepting Imamura's principle of the inverted pendulum. Only the skeleton instrument is shown here, Wang Chen-To still favouring lever systems rather than slots and sliders.

It is therefore parallel with the rain-gauges and snow-gauges previously mentioned.

Some have suggested that Chang Hêng's seismograph was a chance achievement with no subsequent history, but this does not seem to be correct. We can find at least two references to similar devices in later centuries. Hsintu Fang described such an instrument, along with others, in the sixth century, and if he could describe it and find imperial favour, the presumption that he made one is not unreasonable. The second reference is to Lin Hsiao-Kung, who took up the seismograph a century later, between 581 and 604, wrote a book on it, and being in charge at court of all matters concerned with predicting fortunes and misfortunes, again probably made such a device.

Possibly the seismographic pendulums of sixth-century China found their way somehow to the West, for we hear of seismographic instruments at the famous Marāghah observatory in thirteenth-century Persia. Then again there is a great gap until 1703 when de la Hautefeuille set up the first modern seismograph. The principle he adopted was the spilling of an overfilled dish of mercury, and this was used throughout the eighteenth century, even as late as 1848. The fact that the method was so different may suggest that it was a new development, possibly a case of stimulus diffusion (this abridgement, volume 1, page 75). Only since the mid-nineteenth century have the extremely delicate and complex devices current today been developed.

Mineralogy

Mineralogy goes back further than geology in the sense that catalogues of different kinds of stones, ores, gems and minerals were already being made in antiquity – there were ancient and medieval 'Lapidaries' (corresponding to the Herbals and Bestiaries) of Europe which can be matched with, and were sometimes surpassed by, those of India and China. On the other hand modern mineralogy developed from the concern of the eighteenth century for the classification of mineral substances, but not until the work on the crystal forms of minerals by René-Just Haüy in the late eighteenth century and early nineteenth did today's systematic mineralogy arise.

It is at this point that we come, for the first time in this book, to a science which was, in its earlier phases at least, purely descriptive – one in which the Chinese contributed at least as much as the Europeans to what there was of scientific and dispassionate study of stones and mineral ores. Such a descriptive science confronts us with special difficulties, since we cannot here deal with all the individual members of the clan of inorganic mineral substances which were studied. We therefore select a certain number of substances for brief discussion, choosing those on which historians of science have thrown most light.

THE THEORY OF 'CHHI' AND THE GROWTH OF METALS IN THE EARTH

In the West, Aristotle recognised two emanations of the earth, given off under the influence of the sun (page 226). One, derived from the moisture within the earth and on its surface, was a moist vapour (i.e. that which forms clouds and returns as rain). The other, which came from the earth itself, was hot, dry, smoky and highly combustible, like a fuel. Metals, he believed, were congealed from the moist exhalation, and all other minerals and rocks by the dry. The exact processes, as Aristotle describes them, are obscure. However, the important point for us is the 'pneumatic' character of the doctrine, for the *chhi* of Chinese writers seems a conception quite parallel to Aristotle's 'exhalations'.

The *Pên Tshao Kang Mu* (The Great Pharmacopoeia) of A.D. 1596 puts the matter thus:

> Stone is the kernel of the *chhi* and the bone of the earth. In large masses it forms rocks and cliffs, in small particles it forms sand and dust. Its seminal essence becomes gold and jade, its poisonous principle becomes arsenolite and arsenious acid. When the *chhi* becomes congealed it forms cinnabar and green vitriol. When the *chhi*

> undergoes transformation, it becomes liquid and gives rise to alums and mercury. Its changes (are manifold) for that which is soft can become hard, as in the case of milky brine which sets to rock (-salt); and that which moves can become immobile as in the case of the petrifaction of herbs, trees, or even of flying or creeping animals, which once had animation yet turn into that which has not. Again, when thunder or thunderbolts turn to stones, there is a transformation of the formless into that which has form...

This is surely the same doctrine as that of the Greek exhalations. Probably both derive from sources more ancient still, perhaps Babylonian, and finding expression in Sanskrit texts earlier than anything in either Greece or China. The quotation just given was of the sixteenth century, but it is possible to trace the view back through the eleventh to the fifth century A.D., and even to the second B.C.

In so far as the ancients recognised what we should now call processes of slow chemical change in the constituents of the earth's crust, they were not so far wrong; where they erred was in the assumption that metallic and other elements themselves changed into one another. All the same, this conviction that metals and minerals did change into one another while slowly growing in the earth was powerfully effective in encouraging the belief that, by suitable methods, alchemists could succeed in accelerating similar changes under laboratory conditions with the formation of gold itself. The Taoists pursued this idea, and on the way concluded that there was a series of chemical changes in the earth whereby minerals and metals were produced in a definite order. It has been suggested that the Chinese obtained their ideas from Western sources, but the following quotation from the *Huai Nan Tzu* (Book of (the Prince of) Huai Nan) of the second century B.C., which is earlier than most Greek chemical writings, makes it evident this was not so:

> When the *chhi* of the central regions (lit. the 'main-lands') ascends to the Dusty Heavens, they give birth after 500 years to *chuëh* (an unknown mineral, perhaps the orange-red arsenic sulphide known as realgar). This in its turn produces after 500 years yellow mercury, yellow mercury after 500 years produces yellow metal (gold), and yellow metal in 1000 years gives birth to a yellow dragon. The yellow dragon, penetrating to the treasuries (of the earth) gives rise to Yellow Springs. When the dust from the Yellow Springs ascends and becomes yellow clouds, the Yin and Yang beat on one another, produce peals of thunder, and fly out as lightning. The (waters) which were above thereupon descend (as rain), and the running streams flow downwards uniting in the Yellow Sea.

> When the *chhi* of the eastern regions (lit. the 'edge-lands') ascends to the Caerulean Heavens, they give birth after 800 years to *chhing tshêng* (azurite or malachite, i.e. copper carbonate). This in its turn produces after 800 years green mercury, green mercury after 800 years produces blue metal (presumably lead), and blue metal in 1000 years gives birth to a Caerulean dragon. The Caerulean dragon, penetrating to the treasuries (of the earth) gives rise to the Green Springs. When the dust from the Green Springs ascends and becomes blue clouds, the Yin and Yang beat on one another, produce peals of thunder,...

It would be tedious to prolong this passage, which continues its parallelism to the end. So far we have had the yellow Centre and the blue-green East: the rest of the pattern of symbolic relationships is as follows:

South	'Bull-lands'	Red	700 yr *chhih tan* (red cinnabar)	Red mercury	Copper
West	'Weak lands'	White	900 yr *pai yü* (arsenolite)	White mercury	Silver
North	'Cow-lands'	Black (or dark)	600 yr *hsüan chih* (dark grindstone)	Black mercury	Iron

The passage is certainly archaic. The association of colours, compass-points and so on is an example of the symbolic correlations devised early on in Chinese science (volume 1 of this abridgement, pages 531–537) and should not surprise us, but what is quite surprising is to find the same juxtaposition of meteorological and mineralogical exhalations which was characteristic of Aristotle. The text also mentions many of the inorganic substances which later became so important in alchemy, and strongly indicates that the alchemical-mineralogical doctrine is as early as the school of Tsou Yen in the fourth century B.C. Certainly, the theory of the metamorphosis of minerals was fully developed by 122 B.C. It is very difficult to believe that Tsou Yen and his school derived it from Aristotle (384 to 322 B.C.) or from the philosophers before Socrates, in view of what we know of the difficulties of contacts between the cultures at that time. Perhaps it originated from some central source whence it radiated in both directions.

PRINCIPLES OF CLASSIFICATION

The two radicals in Chinese which were most important when minerals were being classified were *shih* (石) for all stones or rocks, and *chin* (金) for all metals and alloys. Oracle-bone forms for *shih* show what may be a mouth (signifying a person) under an overhanging projection, perhaps a man sheltering in a rock cave (石). The second has often been taken as a drawing of a mine, with a cover over the shaft, and two nuggets of ore (金), but this seems perhaps rather sophisticated for the Shang. A third radical, *yü* (玉), was applied to jade and all kinds of precious stones. A fourth, *lu* (鹵), was dedicated to salt, but unfortunately was not given any general use until modern times. The Shang graph is 鹵 and it may be suggested, perhaps, that it was a bird's-eye view of the salt pans in which brine was evaporated, or an attempt to draw a large crystal of salt.

Colours played an important part in the classification of minerals, as was natural. *Tan* (丹) which came to mean red, was a great witch-word in Chinese alchemy and mineralogy, having had the significance of cinnabar (mercuric sulphide) as far back as it can be traced. Later, due to its use by Taoists as a drug for immortality, it came to mean any medicine, pill or prescription. Similarly other minerals and salts were given names ending in yellow (*huang*) or blue-green (*chhing*), for example 'male yellow', somewhat in the same way as in our own time dyes of complex organic structure are known as Nile Blue or Brilliant Green.

Another word which should have been capable of extended use in classification was *fan*, used for alum and, with qualifiers, all related substances. The character (礬) with its hedge-like upper part is probably derived from the traditional Chinese method of evaporating brine and similar substances by pouring them continuously in the open air over a large structure like a hedge made of dry thorn branches (Fig. 154). Powders were known as *sha* ('sands', lit. small particles of rock), *hui* ('ashes'), *fên* (from finely divided rice meal), *ni* ('muds'), *shuang* ('frosts', if white in colour), and *thang* 'sugars' (for example 'sugar of lead'). Inorganic substances were termed 'fats' if greasy or viscous like clays or soapstone, and there was, of course, the belief that in the earliest stages of growth, ores were soft, plastic and of viscous consistency, an idea which was echoed in the West by medieval and Renaissance writers.

MINERALOGICAL LITERATURE AND ITS SCOPE

First and foremost, Chinese mineralogical literature is to be found in that remarkable series of compendia stretching from the Han to the Chhing, comprised in the term *pên tshao* – pharmaceutical natural histories. There are some two hundred collections, all of them large. Most of the objects they describe are plants and plant drugs but mineral drugs were included from the beginning. Here we can only give the briefest review of them. More detail will come in a later volume.

The *Shen Nung Pên Tshao Ching* (Pharmacopeia of the Heavenly Husbandman) is certainly a work of Han date; it no longer exists separately (only in reconstructions and commentaries) but is fully incorporated into

Fig. 154. Evaporator of the traditional type at the Tzu-liu-ching brine-field (original photograph). Solutions of salts are continuously poured over these large thatch structures in the open air; the air–water surface is thus enormously increased and the brine concentrated. If this method is ancient it may explain the structure of the character, *fan*, alum.

other pharmaceutical natural histories. Important is the fact that mineral remedies were included in the pharmaceutical natural history from its earliest beginnings, and this first one described 46 inorganic substances. This was in contrast with Europe, where the Roman physician Galen imposed a doctrine of reliance on plant drugs alone for twelve hundred years or so. The *Pên Ching* list, which dates from either the second or first century B.C., is not the oldest we have. The *Chi Ni Tzu* book, which may be of the fourth century B.C., is certainly pre-Han, and lists 24 substances, most of which are included in the later pharmaceutical natural history, although it seems to be more a record of economic geography than medical in aim.

All subsequent pharmaceutical natural histories list inorganic mineral substances, but the climax of the series is the *Pên Tshao Kang Mu* (The Great Pharmacopoeia) of A.D. 1596, edited by Li Shih-Chen. This gives an elaborate treatment of 217 substances. However, there were several Chinese books in which minerals figured more as chemicals than as materia medica. From the Thang there is the *Shih Yao Erh Ya* (Synonymic Dictionary of Minerals and Drugs) by Mei Piao, prepared about A.D. 806. This is a veritable key to the language of the Thang alchemists, for it lists 335 synonyms of 62 chemical substances. Then, with the Sung flowering of scientific monographs on special subjects, comes a whole group of books devoted to stones and minerals, the starting point of which was aesthetic rather than medical. Some have only survived in part, but the *Yün Lin Shih Phu* (Cloud Forest Lapidary) of 1133 by Tu Wan is complete. If, as in Western lapidaries, each mineral or stone is associated with magical-medical formulae and improbable fables, the spirit of observation and analysis totally lacking in the European treatises is certainly present. Unfortunately the Ming and Chhing hardly lived up to this standard.

There is a further point. From the time of the introduction of carbon-black ink and brushes it was the delight of Chinese scholars to select stones suitable for grinding their ink-blocks with water. This led to the description, at least, of a considerable variety of rocks. The most famous treatise on the subject is that of Mi Fu, a high official of the Sung, who in about 1048 wrote the *Yen Shih* (On Inkstones); though it is not the first, which was the *Yen Phu* (Inkstone record) of the late tenth century, perhaps by Shen Shih or So I-Chien, which describes 32 sorts of stones. There seems to be no close parallel to this literature in the West. Nevertheless, it is possible to construct a comparative table showing the coverage of the successive lapidaries in East and West (Table 37).

GENERAL MINERALOGICAL KNOWLEDGE

The oldest classification of minerals, found already in the Greek Theophrastus (372 to 287 B.C.), was simply whether or not they were changed by heat. Fusibility, therefore, distinguished stones from minerals, and this we duly find in the fourth-century A.D. *Pao Phu Tzu*:

> Other substances decay when buried under the ground, or are melted when subjected to fire. But if the five kinds of mica are put right into a blazing fire, they will never be destroyed, and if buried they never decay.

The term for mica throughout Chinese history was 'cloud-mother', *yün mu*. A description of it given by Su Sung in the *Pên Tshao Thu Ching* of A.D. 1070 is not bad. It runs:

> Mica grows in between earth and rocks. It is like plates and layers which can be separated, bright and smooth. The best kind is white and shining ... Its colour is like purple gold. The separate laminae look like the wings of a cicada. When they are piled up they look like folded gauze. It is said this belongs to the category of glass. It can be used in the preparation of medicines.

Table 37. *The coverage of stones and substances in Western and Eastern lapidaries*

	Western lapidaries	No. of stones or substances described	Eastern lapidaries	No. of stones or substances described
c. 350 B.C.			*Chi Ni Tzu* (Chi Jan)	24
c. 300 B.C.	*De Lapidibus* (Theophrastus)	*c.* 70		
c. 1st century B.C.			*Shen Nung Pên Tshao Ching*	46
A.D. 60	*De Mat. Med.* (Dioscorides)	100		
c. A.D. 220	*Rivers and Mountains* (pseudo-Plutarch)	24		
A.D. 300	*Cyranides* (Hermetic)		*Pao Phu Tzu* (Ko Hung)	*c.* 70
A.D. 818			*Shih Yao Erh Ya* (Mei Piao)	62
			(total synonyms)	(335)
A.D. 1022	*Kitāb al-Shifā'* (Ibn Sīnā)	72		
c. A.D. 1070	*Lapidarium* (Marbodus)	60	*Pên Tshao Thu Ching* (Su Sung)	58
A.D. 1110			*Chêng Lei Pên Tshao* (Thang Shen-Wei)	215
A.D. 1120			*Hsüan-Ho Shih Phu* (Tsu-Khao)	63
A.D. 1133			*Yün Lin Shih Phu* (Tu Wan)	110
c A.D. 1260	*De Mineralibus* (Albertus Magnus)	70		
c A.D. 1278	*Lapidarium* (Alfonso X)	280		
A.D. 1502	*Speculum Lapidum* (Camillus Leonardus)	279		
A.D. 1596			*Pên Tshao Kang Mu* (Li Shih-Chen)	217

Su Sung then goes on to classify eight different types of mica according to colour, and gives a warning about eating two particular types. Here already is not only a clear description of thin transparent plates, but also a systematic attempt to distinguish between the many kinds which are to be found.

The descriptions in the pharmaceutical natural histories were, of course, therapeutic in orientation (and will be mentioned in a later volume when Chinese pharmacology is described), but what is worth noting here is that all the old Chinese mineralogical works paid attention to crystal form, noticing which substances had crystals of hexagonal, needle-like, pyramidal or other types. As early as the eleventh century Su Sung had written about the particular kinds of fractures found when one breaks minerals and caused by their crystal structure. In describing cinnabar he says:

> Cinnabar is found several metres deep in the ground. The local people find it by means of the 'sprouts' (i.e. signs drawn from the presence of other kinds of stone or even herbs).
>
> Cinnabar is thus found in association with a kind of white stone which is known as the 'cinnabar bed'. The mineral grows on this stone...
>
> Upon breaking the lumps of the mineral, it is seen to form precipitous slopes (with surfaces) like walls, as smooth inside as plates of mica...
>
> The *Lei Kung Yao Tui* (Lei Kung's Answers concerning Drugs) says one can find lumps (crystals) ... with fourteen surfaces, each looking as bright as a mirror. On gloomy or rainy days, humidity like a red juice forms on the broken surfaces.

This is interesting for several reasons. We catch a glimpse of the great attention paid to signs of ore beds, and to the special characteristics of the country rock, while the characteristic feature of the mineral is noted together with an attempt to describe the crystal form.

Effects due to the closeness of mineral deposits are not infrequently noted. For instance Tu Wan, speaking of the green colour of a certain stone, says that it arises from a distillation of the vapours of the 'sprouts' of copper which are not far off, since this stone always occurs in the neighbourhood of copper ores. This is not very different from saying that the stone contains inclusions of malachite or other copper-bearing mineral. It is reminiscent of the views put forward three centuries later by the German metallurgist and mining engineer Georgius Agricola who is often considered the founder of modern theory, according to which ore channels were formed by the circulation of ground waters in fissures following the deposit of the surrounding rocks. Similar statements concerning copper sulphate and copper carbonate occur in other Chinese books of the tenth and eleventh centuries.

Perhaps the most vigorous description of the idea of the deposition of ore beds from the circulation of ground waters in rock fissures comes from the pen of Chêng Ssu-Hsiao, who died in 1322, two centuries before Agricola:

> In the subterranean regions there are alternate layers of earth and rock and flowing spring waters. These strata rest upon thousands of vapours (*chhi*) which are (distributed in) tens of thousands of branches, veins and thread (-like openings). (There are substances there) both soft and firm, ever flowing back and forth, and undergoing transformations. (The veins are) slanting and delicate, like axles interlocking and communicating. (It is like a) machine rotating in the depths, (and the circulation takes place as if the veins had) intimate mutual connections (and as if) there were piston-bellows (at work). The mysterious network spreads out and joins together every part of the roots of the earth...
>
> Now if the *chhi* of the earth can get through (the veins), then the water and the earth (above) will be fragrant and flourishing ... and all men and things will be pure and wise ... But if the *chhi* of the earth is stopped up, then the water and earth and natural products (above) will be bitter, cold and withered ... and all men and things will be evil and foolish...'

Though the idea of some kind of circulation going on in the earth was quite common in geomantic circles (it fitted in with the belief that the disposition of houses and tombs should harmonise with the local currents of 'cosmic breath'), Chêng applies it here with clarity to the deposition of minerals by evaporation of (or precipitation from) ground waters in ore channels. It is interesting, too, that he adapts to his purpose the ancient medical theory that diseases were caused by stopping up the pores.

The old books mention innumerable practical uses of the various minerals. At a time when very few Europeans were clear as to the distinction between sal ammoniac, saltpetre, alum and the like, the industrial processes of the Chinese, such as tanning, dyeing, painting and firework-making, had led them to make the necessary identifications. However, the applications of minerals were often broad: for instance, copper sulphate found employment as a fungicide as well as in medicine, and white arsenic, obtained as a by-product from copper-smelting, was applied to the roots of rice-plants during replanting to protect them from insect pests.

NOTES ON SOME SPECIAL MINERALS

We give here some brief notes on particular mineral substances. It will be convenient to postpone mentioning some until later chapters: for example, coal will be discussed under mining and metallurgy, and fuller's earth under textiles.

Alum

A substance of very great industrial importance, its main application was as a mordant (i.e. a substance for fixing a colour) in dyeing. For this it had to be pure. It was also used for making hides supple, for sizing paper, glass-making, clearing natural waters and fire-proofing wood. There were medical uses too – for stopping bleeding, to cause vomiting, and as an astringent. The Chinese distinguished native alum from that prepared by roasting alunite, although roasting does not seem to have been carried out in Asia until the tenth century A.D. A third method appeared in Arabia in the twelfth century A.D.; it consisted of boiling the rocks containing aluminium sulphate with urine, so as to form ammonium alum, but this is a method which does not appear to have been used in China. Sources seem to have been somewhere to the west of China, in Po-Ssu (which meant the south seas as well as Persia); and the tenth-century A.D. *Hai Yao Pên Tshao* (Drugs of the Southern Countries beyond the Seas) states that there were two kinds of alum, one from the Malayan Po-Ssu, the other from Arabia.

Sal ammoniac

Ammonium chloride (sal ammoniac) was also of importance both medically (as a stimulating expectorant and a mild medicine for stimulating the production of bile) and chemically. It is not the same substance as the 'hammoniac' mentioned by Pliny, nor was it known in Alexandria. It was introduced into chemistry by Arabic alchemists in the tenth century A.D. but they got it from their Chinese colleagues, for it occurs naturally in Central Asia, and was probably collected there at an early date. It is mentioned in Wei Po-Yang's *Tshan Thung Chhi* (The Kinship of the Three; or, The Accordance (of the *Book of Changes*) with the Phenomena of Composite Things) of the second century A.D., which bears a correct reference to its cooling effect on boils; and its next main reference is in a natural history of 660 A.D. All Chinese books agree that sal ammoniac came from the western provinces, i.e., Szechuan, Kansu, Sinkiang and Tibet, where it was collected from active volcanic areas or perhaps from burning coal seams.

Asbestos

Asbestos is unusual among minerals in having a fibrous structure; it is composed of calcium magnesium silicate. Though it occurs in many forms and with various admixtures of other elements, the fibres may usually be separated and look like flax. Chrysolite, or fibrous serpentine, is a hydrous magnesium iron silicate which resembles asbestos and has similar properties. From very early times it was discovered that these fibres could be woven so as to form a kind of cloth which was indestructible by fire. It was natural that such a thing should be considered a wonder and a work of great art, and in fact there are numerous references to asbestos cloth from late B.C. times onwards. In the West, Strabo (63 B.C. to A.D. 19) speaks of 'fire-resisting napkins', and he was followed by many other authors. It has been said that the Chinese first knew of asbestos through trade with Roman Syria in the fourth century A.D., and certainly it is listed in fourth-century Chinese literature as a product of Arabia. However, the *Lieh Tzu* book reports:

> When King Mu of the Chou dynasty made a great expedition against the Western Jung people, they presented (to propitiate him) a Khun-Wu sword and some fireproof cloth. The sword was 51 centimetres long, a red blade of fire-transformed (or refined) steel which would cut jade like clay. The fireproof cloth was cleaned by being thrown into a fire, when the cloth became the colour of the fire, and the dirt assumed the colour of the cloth. When taken out of the fire and shaken, it became as white as snow. A certain prince did not believe it and thought those who brought news of it must be mistaken, but Hsiao Shu said, 'Must the prince insist on maintaining a preconceived idea, and deny the (demonstrable) truth?'

As it is hard to be sure which parts of this book are of pre-Han date and which are not, we cannot affirm that the passage was written in the third century B.C. or thereabouts, but this does not seem unlikely. It strikes a keynote for many subsequent discussions, and Hsiao Shu appears as the prototype of all those Taoists who used the example of fireproof cloth to convince Confucian sceptics that there was more in heaven and earth than was dreamt of in their philosophy. There are other uncertain references, but in A.D. 300 Wang Chia quoted a story referring to King Chao of Yen having lamps lit by 'dragon blubber' and using asbestos wicks. There were two rulers of Yen with the same name but, if the story be correct, it must refer either to 598 B.C. or 308 B.C., while there is other evidence of royal gifts made during the third century A.D.

Explanations about the origin and nature of asbestos are various. In

A.D. 300 Ko Hung gave an account which is an example of the salamander and phoenix legends, natural perhaps at a time when it was thought that all cloth fibres must have a vegetable or animal origin. In the West, Strabo, Dioscorides and Plutarch of the first century A.D. knew it was a mineral fibre, but the earliest Chinese text which is clear on this is of the fifth or sixth century A.D., referring to it as 'stone hemp' or 'stone veins' which make string or cord. The idea of an animal or vegetable source lasted for a time among some writers but it was finally rejected. Once again, then, we meet with the same pattern of advance; the Greeks are the first to record accurate information, but then between Hellenistic times and the Renaissance the Chinese are more advanced than the Europeans.

Borax

Borax (native sodium tetraborate, or tincal) is not mentioned in ancient Chinese writings, and first appears in a pharmaceutical natural history of the tenth century A.D. This is about the same date as its first mention in the West. It probably came then, as later, from natural deposits in what is now Chhinghai province (north-eastern Tibet), and may also have been produced, as it is today, from certain of the brines at Tzu-liu-ching. Its use as a preparatory agent for soldering and brazing (in the molten state it cleans metal surfaces by dissolving metal oxides) goes back in China to the eleventh century A.D. Its mild non-irritant antiseptic qualities were appreciated by pharmacists in China as well as in the West.

Jade and abrasives

Jade is no ordinary mineral. The love of it was one of the most characteristic features of the Chinese civilisation and its texture, substance and colour gave inspiration to carvers, painters and poets for more than three thousand years. Jade generated a vast literature both in Chinese and Western languages, but this is mostly concerned with aesthetic appreciation and social uses. We are here concerned rather with the technological aspects of the mineral, its mining and, above all, the manner in which it was worked, no easy achievement in view of its great hardness.

The word which we use, jade, is a corruption of the Spanish *ijada*, meaning flank or loins, the full form being *piedra de ijada*. During the course of the Spanish conquests, green stones much prized by the Mexicans as amulets against diseases of the kidneys were brought back to Europe with repute of their worth. The ancient Chinese word had none of these connotations. It is a remarkable fact, however, that the appreciation of jade and the art of working it was not confined to China; it was a feature also of ancient Central American civilisation, and of the New Zealand Maoris.

This is one of the features which suggest certain common origins for the jade-loving cultures on both sides of the Pacific, or possible contacts between them.

True jade, or nephrite, is a silicate of calcium and magnesium with a concealed crystalline structure. Jadeite, though similar in appearance, is a silicate of sodium and aluminium and a member of a different group of minerals. It is generally composed of small grains rather than minute fibres like jade, hence, although slightly harder than true jade, it does not offer quite such a tough resistance to the tools of carvers. The colours of jade are mainly due to the presence of compounds of iron, manganese, and chromium. There are also other minerals which resemble jade fairly closely, which the Chinese called 'false jade', and all are considerably softer than true jade.

The rivers and mountains of Khotan and Yarkand in Sinkiang were the principal, perhaps the only, centres of production of true jade for over two thousand years, and as early as the fourth century B.C., the Yüeh-chih people were intermediaries in the trade. The jade was found in the valleys of two rivers, the Karakash and the Yurungkash, being either mined or collected as lumps in the river-bed (Fig. 155). Jadeite was not known in China before the eighteenth century, at which time it began to be imported from Burmese deposits.

As for the ancient history of jade in China, we know that it was carved already by the Shang people (thirteenth century B.C.), as objects have been found at Anyang. Neolithic jade tools are known, and give evidence of having been made without the help of metal, but abrasive sand must have been used, probably with thin sheets of sandstone or slate as cutters. Indeed, abrasion is the secret of all subsequent jade working, as it was in the earliest periods of ancient Egypt when vases of diorite and bottles of rock-crystal were fashioned. Hence the interest in the Khun-Wu sword (page 316), which comes into prominence about the third century B.C. One feels that just as asbestos was a real thing so also there was something behind this curious term which, later, became proverbial for any steel knife of fine quality. The central invention for abrasive working was, however, the rotary disc knife (Fig. 156). The oldest reference to it is indirect and in the twelfth century A.D., and the first specific reference is in 1500, but we should remember that the methods used in ancient times were kept secret by the technicians, in this case as in so many others. Indeed the possibility arises that what lay behind the Khung-Wu story was the first application of a treadle-driven rotary steel knife and abrasive sand. The microscope has shown traces of the use of rotary tools on late Chou period jade discs, and this would agree with such a suggestion, which is also consistent with the rather late date for the general use of iron in China.

Fig. 155. Collection of water-worn jade nuggets by women and girls in the Karakash and Yurung-Kash rivers at Khotan (Sinkiang); a picture from Sung Ying-Hsing's *Thien Kung Khai Wu* of A.D. 1637.

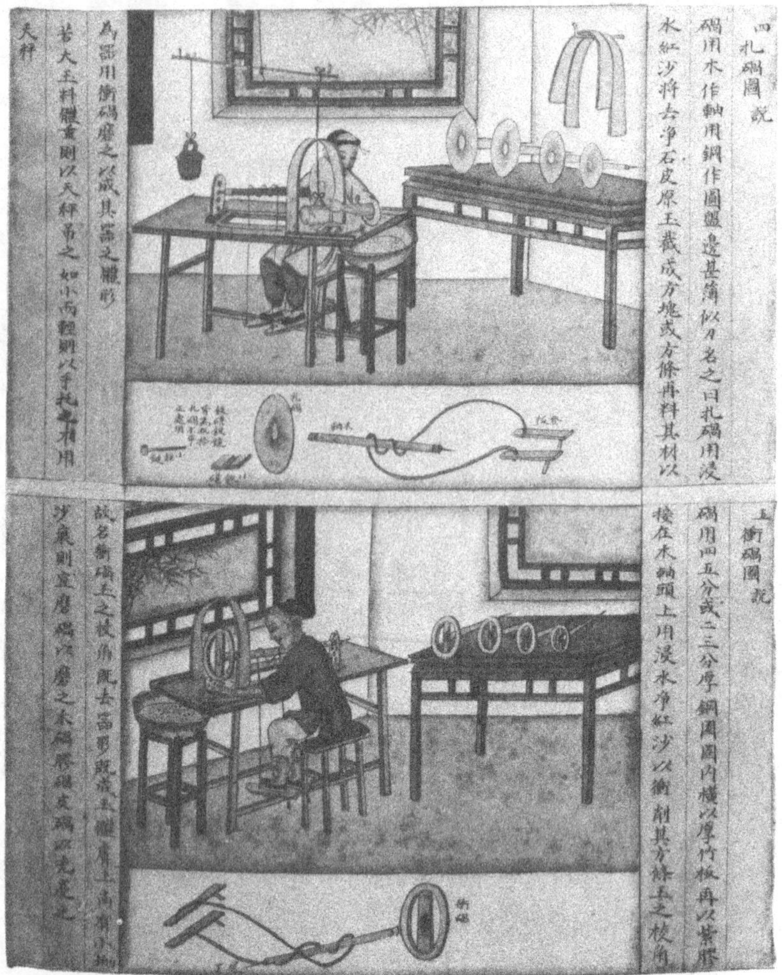

Fig. 156. Rotary tools for working jade (from Li Shih-Chhüan). Above, the steel disc knife (*cha tho*, here called *cha thuo*) with its treadle mounting and protective shield; the accompanying text mentions that crushed almandine garnets should be used with it. It is also said that when the piece of jade being worked upon is heavy, a balance suspension should be rigged up, and indeed one is shown in the picture. Below, the steel grinding wheel (*mo tho*, here called *chhung thuo*) also rotated by a treadle and operating with crushed garnets as abrasive. The text adds a reference to the polishing wheels such as the *chiao thuo*, made of a mixture of shellac and carborundum, and the leather buffing wheels.

Grease was used as a medium for the abrasives. In the eleventh century Lü Ta-Lin reported that '... with toad-grease and a Khun-Wu knife, jade can be worked like wax'. Of course many literary writers in all centuries failed to realise that the abrasive contained in the grease was the important factor, and thought the grease itself must exert some softening effect on the stone. The most ancient abrasive was quartz sand; corundum (i.e. emery; oxides of aluminium and iron) was a discovery of the twelfth century A.D., while crushed garnets (i.e. silicates of calcium and iron) came into use a couple of centuries earlier.

Precious stones, including the diamond

Gems have always been prized both for aesthetic and supposedly medical reasons, and since it was impossible to classify them rationally before modern times, their history is primarily of archaeological concern. Moreover, the subject is bedevilled by uncertainties in the identification of various terms used both in East and West. We do not know of any medieval Chinese treatise devoted solely to precious stones, although, of course, there is a mass of references to them in lapidaries, antiquarian books and encyclopaedias.

Unlike asbestos with its chain-like arrangements of molecules, or the layer-like structure of mica, the diamond is fully three-dimensional with a cross-linked lattice construction. There can be little doubt that the main ancient source of this remarkable crystalline carbon, the hardest of 'stones', was India. Among the earliest references is one in the first century B.C., and the Greek Dioscorides knew of it in the first century A.D., while it is mentioned as an Indian export at this time. The earliest Chinese reference was thought to be in a report of the third century A.D. about the Tunhuang region presenting diamonds to the emperor; there it is said they are derived from gold, come from India, can be used to carve jade, and although scoured many times, do not dissolve. However, this is not the earliest mention; there is a passage of A.D. 114.

The folklore connected with the collection of diamonds is similar in East and West. A method later used in China involved people with grass-sandals walking up and down over diamond-containing sands, after which the sandals were burnt and the gems collected from the ashes. Presumably it was the extreme hardness of the diamond which gave rise to the notion of its indestructibility. Surprisingly, the Chinese were not acquainted with cut and polished diamonds until the Portuguese brought some to Macao in the sixteenth century.

The touchstone

The touchstone deserves notice because of its connection with an interesting technique. It was usually coarse-grained or flinty velvet-black jasper (basanite), i.e. silica with concealed crystalline structure, associated with quartz, and coloured by iron and other oxides. The use of it was in assaying alloys of gold and silver. Medieval metallurgists had a set of 24 'needles', or little bars of gold, each of a known standard marked from 1 carat up to 24 carats. Taking the bar of gold to be assayed, they rubbed it on the touchstone, and then rubbed one or other of their standard bars alongside the streak it had left, choosing first the standard probably nearest in composition to the unknown. In ancient times the colour of the streak was the only guide, but after the introduction of mineral acids, a little aqua regia (a mixture of concentrated hydrochloric and nitric acids in a ratio of 4:1) was applied and the action on the two streaks noticed. If any differences were seen, other standard bars were used. The method depends on the reflective power and colour which cannot be detected from the metal surfaces themselves but are visible only when the metal is finely divided on the dark abrasive surface of the stone.

The touchstone of Greece is evidently very old, since there are references to it in the poets back to the sixth century B.C., and it was minutely described later. The earliest Chinese reference so far found seems to be in the *Ko Ku Yao Lun* (Handbook of Archaeology) of A.D. 1387, where the 'gold-testing stone' is said to come from Szechuan. It was used also for testing silver. However, a reference of about 1730 mentions that the stone was said to be the colour of a new ink block, and to be greatly prized by the merchants who tested alloys with it. There are possible earlier references, but one cannot be certain that these were not to stones used for grinding. The exact date for the use of the touchstone in Chinese metallurgy therefore remains somewhat obscure.

THE SEARCH FOR MINERAL DEPOSITS

Geological prospecting

The methods used by ancient miners for finding the location of deposits of ores and minerals must have been primarily based on traditional geological lore, the observation of the 'lie of the land', the direction of strata and knowledge of what kinds of rock were likely to be associated with the mineral sought. The mining works of sixteenth-century Europe give hints of such information, and there must have been similar traditions in China. Today new methods of prospecting have been introduced depending on geophysics (as in the measurements of local variations of gravity and the results of underground explosions), as well as observation and chemical analysis of soil and plants growing in the area under investigation. Chemical analysis of soil samples was not possible before the rise of modern science, but the association of certain plants with certain minerals was, and it goes back in China a surprisingly long way. Indeed, prospecting by association of one rock or mineral with another was presumably part of the most ancient traditional Chinese miner's lore, since echoes of it find their way into Han or pre-Han texts. Thus in the *Kuan Tzu* book compiled in the late fourth century B.C. in part from older material, we read:

> Where there is cinnabar above, yellow gold will be found below. Where there is magnetite above, copper and gold will be found below ... Where there is haematite above, iron will be found below.

Geo-botanical and bio-geochemical prospecting

The Chinese not only recognised the association of ores and rocks; but as already mentioned they also noted certain connections between plants and ores. That the ore and plants might be separated by a few or many metres, or greater distances, was not felt by them as a difficulty; the idea of 'action at a distance' appears in many ancient texts, encouraged perhaps by their appreciation of the organic oneness of the universe.

The best known example of such an association is probably that in the *Yu-Yang Tsa Tsu* (Miscellany of the Yu-Yang Mountain (Cave)) written by Tuan Chhêng-Shih about A.D. 800. There we find the following:

> When in the mountains there is the *hsiai* plant (a kind of shallot), then below gold will be found. When in the mountains there is the *chiang* plant (ginger), then below copper and tin will be found. If the mountain has precious jade, the branches of the trees all around will be drooping.

This and later texts make it clear that not only the presence or absence of plants was looked for, but also their physiological condition. In general we should not be far wrong in thinking that this kind of empirical lore was growing steadily from the Han time onwards to the Sui. Naturally there was magic in it as well as true knowledge, for surely no plant could betray the presence of a mineral like magnesium silicate, no matter how much men might value it as jade. Nevertheless we now know from experiments on the cultivation of the lower plants on synthetic media, that the presence of metallic and other elements, even if only in traces, is necessary for success.

It is now realised that some metals accumulate in certain plants. For instance, gold is known to accumulate in the horsetails (which may contain as much as 113 g per tonne), and umbellate chickweed (*Holosteum umbellatum*), when growing on mercury-rich soils, shows droplets of mercury between its cells. In view of these facts it is remarkable to find that Chinese literature contains a long-standing assertion that metals could actually be obtained from certain plants, as we find in the *Kêng Hsin Yü Tshê* (Precious Secrets of the Realm of Kêng and Hsin) of A.D. 1421 which gives a number of examples. In at least one case practical use was made of such knowledge in the extraction of mercury.

One would think that miners in all civilisations must have acquired some practical knowledge of plant–mineral associations, but there seems to be no trace of medieval European ideas on the subject. Agricola says little about it and traces of geo-botanical prospecting do not appear until 1760. But it is evident that the medieval Chinese observations were the forerunners of what is now a vast and rapidly growing body of scientific theory and practice.

6

Physics

Though physics has sometimes been regarded as the fundamental science, it was a branch of study in which traditional Chinese culture was never very strong. This is in itself a striking fact and it may later on (in a further volume) appear of special significance when placed in the context of a general discussion on the factors in East Asian society which inhibited the rise there of modern science. Nevertheless there is no lack of Chinese material on physical subjects. In considering fundamental ideas in Chinese science (volume 1 of this abridgement) some physics was mentioned, and here we shall continue with a consideration of the work of the Mohist school in the fourth and third centuries B.C. It will then appear that the idea of atoms never became important in Chinese thinking. Just as Chinese mathematics was algebraic rather than geometrical, and Chinese philosophy organic rather than mechanical, so we shall find that Chinese physical thought (one can hardly speak of a developed science of physics) was dominated by the notion of waves rather than particles.

The most important Chinese contribution in this field lay in their development of knowledge about magnetism, but we shall find it more convenient to deal with that in a later volume, where its application in the magnetic compass and therefore in navigation may be more thoroughly discussed. The Chinese were far in advance of the West in this matter and if the Renaissance had been Chinese, rather than European, the whole sequence of discoveries might have been different, for the Chinese seem to have had no parallels to the medieval Western students of motion. Chinese literature appears to contain no discussion of the trajectory of missiles or falling bodies, which was strange, considering the interest of the Mohists in military technology. Nevertheless, this theoretical vacuum did not in the least hinder the development of engineering in China, which before A.D. 1500 was frequently much superior to anything which Europe could show.

WAVES AND PARTICLES

It must now be made clear, as a further preliminary, to what extent Chinese thinking about physics was dominated throughout by the idea of waves rather than of atoms. Atomism is one of the most familiar features of European and Indian theorising, yet although at various times some Chinese thinkers watered its seed, the idea never took root among them, presumably because it was not in harmony with the organic wholeness of the universe on which Chinese thought was based. First, though, let us glance at some of the passing appearances of atoms on the Chinese stage.

The fundamental idea of atoms could be expected to arise in all civilisations independently, since in every country there were men engaged in cutting up lengths of wood, and the question would inevitably arise as to what would happen if successive cuttings were to go on until the uncuttable was reached. Logically strictest in this sense were the Mohists, with their atomic definition of the geometrical point (see page 40). The word for this used in the *Mo Ching* is *tuan* (端). But they also considered instants of time in the atomic sense:

C (Canon)	The 'beginning' (*shih*) means (an instant of) time.
CS (Exposition)	Time sometimes has duration and sometimes not, for the 'beginning' point of time has no duration.

Since this is fourth century B.C. or earlier, it is clear the notion of atomic instants of time can hardly have come into China with Buddhism, which arrived in the first century A.D.; to believe the Mohists were influenced by Indian thought is much more difficult. When Indian atoms of time did become current in China they were termed *chha-na*, a usage which we find adopted by Hsü Yo probably in the Later Han.

The word *shih* (始) used by the Mohists derives from a graph ( ) which represents the birth of a child, while *tuan*, used for 'point', was originally ( ), a graph showing the first visible sprouts of a plant. There are other words such as *wei* meaning something very minute, and derived from a graph ( ) showing two hands holding something small. *Wei* is sometimes translated as 'atom', but if acceptable in literary versions, this can hardly be admitted as correct for scientific purposes. There is also *chi* which is a more biological term meaning 'germ', used in the third century B.C., in the sense 'all things come from germs' and return to 'germs', but it would be stretching the meaning too far to insist on an atomistic interpretation. Lastly, terms for the minutest of weights are referred to almost in an atomistic sense in Han medical literature, but to translate them as atoms would be to give them a connotation more definite than the original language would bear.

In contrast to these rare examples of atomistic thinking, we find the texts speak as one voice with regard to the wave-like progression of Yin and Yang forces, reciprocally rising and falling. Throughout Nature there is a tidal flow of the two elementary influences. Our oldest example of this is from the *Kuei Ku Tzu* (Book of the Devil Valley Master) some of which goes back to the fourth century B.C.

> The Yang returns cyclically to its beginning; the Yin attains its maximum and gives place to the Yang.

but there are plenty of others. By the first century A.D. we find Wang Chhung saying:

> The Yang having reached its climax retreats in favour of Yin; the Yin having reached its climax retreats in favour of Yang.

So dominant in Chinese thought was the idea of wave motion that it seems sometimes to have acted as a brake on the advance of scientific knowledge. Traditional natural philosophy in China conceived of the whole universe as undergoing slow pulsations of its fundamentally opposed but mutually necessary basic forces. As the radiating mutual influences of individual things pulsed too, it was entirely in accord with the grain of Chinese thinking to envisage intrinsic rhythms in natural objects. Yet this view was not altogether propitious for physics as, for instance, in the late Sung, about A.D. 1140, when a writer denied the reflection of the sun's light by the moon, on the ground that the periodical rise and fall of the Yin force was a much better explanation for the moon's phases. Yet this was only following a tradition which had been well established in the Han, when Wang Chhung, as we have seen, argued vigorously against the correct theory of eclipses, preferring the view that the sun and moon had intrinsic rhythms of brightness of their own.

To summarise, then, the Chinese physical universe in ancient and medieval times was a perfectly continuous whole. *Chhi* condensed into tangible matter was not in any significant sense composed of particles, but individual objects acted and reacted with all other objects in the world. Such mutual influences could be effective over very great distances, and worked in a wave-like or vibration-like manner dependent in the last resort on the rhythmic alternation at all levels of the two fundamental forces, the Yin and the Yang. Individual objects thus had their individual rhythms. And they came together like the sounds of the individual instruments of an orchestra, but acting spontaneously and merging into the general pattern of the harmony of the world.

Chinese natural philosophers tended to think of cyclical recurrences. In its simplest and oldest form this had to do with little more than the rhythm of the seasons and the rise and fall of the individual lives of men, but later generations developed more precise, if more symbolical, cyclical representations of natural phenomena. Of course, a cyclical approach was imposed from the start by the subject-matter of certain sciences, such as calendar-making in astronomy or the recognition of the water-cycle in meteorology. But ideas of circulation were also prominent in physiology and medicine, where a movement of *chhi* and pulsing blood was believed to take place around the body daily. We know now that this circulation, as envisaged by the physicians of the Han, was only sixty times slower than the circulation discovered by William Harvey in 1628 and which we recognise today.

The one parallel among ancient Western thinkers was with the Stoic philosophers of ancient Greece, who had a kind of wave theory. In times corresponding to the Chhin and Han, it was they who laid stress on propagation in a continuous medium of three dimensions, using the analogy of water waves and ripples. They also found it necessary that this medium should be under tension. These ideas, which found application in fields as far apart as physiology and theology, clearly parallel the forms of thought which came to birth in China, where there was a concept of a continuous medium or continuum, as we shall shortly see. What is particularly interesting and significant is that the Stoics adopted an organic outlook, thought of a universal spirit or pneuma, of 'sympathies' of things on other things even at great distances, and physical 'fields of force' within the hierarchy of organisms. Indeed it was no coincidence that both the Stoics and the Chinese discovered the true cause of the tides.

Of course the wave conceptions of medieval Chinese thought were never applied specifically and systematically to the interpretation of the world of physics, but then the Greeks and the Latins made no such attempt either. A full understanding of the experimental method of science and what it implied was needed first, and this did not come until the sixteenth and seventeenth centuries in Europe. Then Robert Hooke spoke much about 'vehement vibratory motions' and applied wave-theories to light, heat and sound. The Dutchman Christiaan Huygens later built a whole structure of optics on them. At first sight it would seem most improbable that the contact with China which Europe had then had for nearly two centuries provided any stimulus; no doubt the development of wave ideas was essentially a consequence of the study of vibration itself, as in springs, yet we should not overlook the personal acquaintance which Hooke himself had with Chinese visitors in London, one of whom might well have dropped a hint about the great importance accorded in China to the long-term vibrations of the Yin and Yang.

Atomic theory was one of the greatest currents of Greek thought. Founded in the fifth century B.C. by Leucippus and Democritus, it was then attacked by Plato, Aristotle and other philosophers, but reaffirmed by the Epicureans and immortalised in the first-century B.C. poem *De rerum natura* by Lucretius. It was banished throughout the early Christian centuries, but was resurrected in seventeenth-century Europe and formed one of the cardinal features of the growth of modern science. There was atomism too in India and among the Arabs, who derived their ideas on it mainly from the Indians rather than the Greeks.

And it is a striking and perhaps significant fact that the languages of all these civilisations which developed atomic theories were alphabetic. Just as an almost infinite variety of words may be formed by different combinations of the relatively small number of letters in an alphabet, so the idea was natural enough that a large number of bodies with different properties might be composed by the association in different ways of a very small number of different constituent particles. Lucretius, indeed, remarks on this. On the other hand, the Chinese written character is an organic whole, and minds accustomed to an ideographic language would perhaps hardly have been open to the idea of an atomic constitution of matter. All the same, there are 214 radicals into which the fundamental elements in written Chinese can be reduced ('atoms'), and an immense variety of words ('molecules') formed from their combinations. Moreover, the combinations of the components of the Symbolic Correlation groups of five elements (volume 1 of this abridgement, pages 153–157) were understood from very early times to produce all natural phenomena. Therefore, while it is plausible that there was a correlation between alphabetic languages and atomism, it should not be pressed too strongly.

As for the great debate between continuity and discontinuity as such, Chinese organic philosophy was bound to be on the side of continuity. This might be seen even in the realm of mathematics. The Greeks were so far from the idea of continuity that they could not imagine 'irrational' numbers like $\sqrt{2}$ to be true numbers at all, because they could never be reduced to a simple whole number. But, as we have seen, the Chinese, even if they realised the special nature of these, were neither puzzled by them nor much interested in them. Their universe was a continuous medium within which interactions occurred, not by the clash of atoms, but by radiating influences. It was a wave world, not a particle world. And thus to the Chinese, as also to the Stoics, one of the great halves of modern 'classical' physics is owing.

MASS, MENSURATION, STATICS AND HYDROSTATICS

Many Chinese scholars have been inclined to suppose that nothing of any consequence could be found about these subjects in ancient and medieval Chinese writings. But such an impression was too pessimistic. There was basic knowledge on the properties of levers and the strength of materials, there was theory as well as practice concerning the physical properties of matter, and some fundamental work on measurement. However, before discussing this, it may be of interest to read a remarkable passage in the *Huai Nan Tzu* book (about 120 B.C.): a lyrical exposition of measure and rule, as seen in the operations of Nature no less than those of man, and all the more striking in that it belongs to a civilisation which produced neither Greek deductive geometry nor the mathematical approach characteristic of the world of Galileo. It runs thus:

> As for the great rules for regulating and measuring the Yin and the Yang there are six measures. Heaven corresponds to the plumb-line, earth to the water-level, spring to the compasses, summer to the steelyard (balance with unequal arms), autumn to the carpenter's square, and winter to the balance. The plumb-line serves to align the ten thousand things, the water-level to level them, the compasses to round them, the steelyard to equalise them, the square to square them and the balance to weigh them.
>
> The plumb-line as a measure is erect and unswerving. Draw it out as you will, it has no end. Use it as long as you wish, it will never wear out. Set it as far away as you can, it does not disappear. In its virtue it accords with Heaven, and in its brilliance with the spirits. What is desired it obtains, what is disliked it destroys. From antiquity until today its straightness has remained unchangeable. Vast and profound is its virtue, broad and great so that it can encompass (all things). This is why the Rulers of Old used it as the prime standard of things.
>
> The water-level as a measure is even and without slope, flat and without declivity. Broad and great it is, so that it can encompass (all things), wide and all-embracing so that it can harmonise (them). It is soft and not hard, blunt and not sharp, flowing and not stagnant, shifting and not choked. It issues forth and penetrates (everywhere) according to a (particular) principle. It is widespread and profound without being dissipated, and levels and equalises without any error. Thus the ten thousand things are kept in equilibrium, the people form no dangerous plots, and hatreds or resentments never arise. This is why the Rulers of Old used it as the equaliser of things.

The text continues, extolling in turn the compasses, the steelyard, the carpenter's square and the balance. Thus was faith affirmed in order, precise, clear, numerical, unvarying and repeatable, not vague and chaotic, not wholly composed of excess and defect. If there were no laws in Nature like the laws of men, there was no disorder either, but pattern above pattern, recognisable and measurable, level beyond level, from wisps of *chhi* to stars. And measures of the world and the principles of its measurement did not change; before even the Rulers of Old, they were.

The Mohists and measurement

By way of background let us recall previous references to this subject. The mathematics section showed us how far back in Chinese history it is possible to trace the use of powers of ten for units of measurement, and the astronomical section revealed an early attempt to develop something for measuring gnomon shadow lengths equivalent to the standard platinum metre of modern times. Here is a Mohist proposition dating from about Aristotle's time; that is, the late fourth century B.C.:

C A thing can be 'very' or 'not very'. The reason is given under 'following a standard'.

CS (People in different places) use very long or very short standards, but the 'very long' and the 'very short' should not be longer or shorter than the long and the short standard respectively. A 'standard's' standardisation may be true or false. All (individual) standards should conform to the (accepted) standard.

The writer was doubtless thinking of the confusion of standards in different parts of the country, each feudal State having at least one set of measures of its own. Other Mohist quotations show concern with establishing a standard, and using units of measurement neither too large nor too small for practical use.

The Mohists, the lever and the balance

It should again be emphasised that we have only surviving fragments of rather garbled texts to work on, so that one can hardly judge the extent of physics in the Warring States period without much guesswork. Nevertheless a few quotations will help to show Mohist thoughts on these subjects.

Force and weight

C Force is that which causes shaped things (i.e. solid bodies) to move.

CS Weight (heaviness) is a force. The fall of a thing, or the lifting of something else, is motion due to heaviness.

Balance of forces: consideration of pulley and balance

C A suspending force acts in the opposite direction to the force which pulls (downwards). The reason is given under 'beating against'.

CS For suspension force is necessary, but free fall occurs without the application of force (by us). A suspending force is not necessarily confined to the actual point at which it is applied (as in the case of a beam or a bridge). (Note how) cord is used for drilling (i.e. in a rotary bow- or pump-drill). (Consider a beam suspended on a rope. The side where) the distance from the point of suspension (the fulcrum) is greater, and/or the weight heavier, will go down; (the side where) the distance from the point of suspension is shorter, and/or the weight lighter, will go up; so that the more the upper side gains the more the lower side will lose. When the rope is at a right angle to the beam, the weights are the same (on both sides) and a mutual balance is struck.

(Consider two weights suspended by a rope over a pulley.) The more loss from the 'upper side' (by reduction of the amount of 'weight hanging'), the more gain there will be on the 'lower side'. If the 'upper' weight is entirely taken away, the 'lower' (side) will fall altogether.

In the last example we have an experiment using an ancestor to Atwood's machine, a device developed by George Atwood in A.D. 1780 for studying the relations between force, acceleration and mass.

Combination of forces

C A force made up of several forces, can act against one force. Sometimes there is a reaction and sometimes not. The reason is given under 'parallelism' (*chü*).

The fact that the exposition (CS) is missing and the brevity of the proposition make it difficult to be certain what is concerned. If *chü* (矩) is the right character, then the Mohists were making some attempt to work out a 'parallelogram of forces', and whether or not there would be a reaction would depend upon whether the structure under consideration were in equilibrium or not. However, should *chü* (拒) ('pushing') be the character intended it could be concerned with the transmission of energy by impact, but this seems less likely. In practical engineering in ancient and medieval China there must have been many occasions for empirical knowledge of the combination of forces, though the theoretical work necessary to resolve them into their separate components could not be undertaken.

Lever and balance

C balance can lose its equilibrium. The reason is given under 'gaining'.

CS If a weight is added to a balance, that side will fall.
As for the steelyard, let a quantity of material and a weight be balanced, the distance between the fulcrum and the point where the material is suspended being shorter than the distance between the fulcrum and the point where the weight is suspended. This will then be the longer. If now to both sides the same weight is added, the weight must go down (because the distance between the fulcrum and the point where the weight hangs is greater than that between the fulcrum and the material).

The most important thing about this excerpt is that it shows that the Mohists must already have been in possession of the whole theory of equilibrium as stated by Archimedes half a century later. That it was widely understood in the Han is clear from a passage in the *Huai Nan Tzu* (120 B.C.):

> Therefore if one has the benefit of 'position', a very small grasp can support a very large thing ... So a beam of 10 *wei* can support a house 1,000 *chün* in weight; a hinge only 12½ centimetres in length can control the opening or closing of a large gate.

It is generally agreed that the steelyard is of much later date in the ancient West, and the device never became dominant in Europe. It is curious that in China the opposite development occurred, and the steelyard seems to have been much more prevalent, at least from the Han onwards, than the simple balance with equal arms. Moreover, Chinese steelyards were often provided with more than one point of suspension so that weighting in a series of different ranges could be carried out. Nevertheless for accurately weighing smaller quantities the equal-armed balance held its own in China, especially with the Sung alchemists, and pharmacists from the Han onwards.

Tension, fracture and continuity

Let us begin with two Mohist propositions.

Strength of materials

C (Suppose a weight) to be supported (by a beam) which does not break. The reason is given under 'bearing'.

CS From a horizontal piece of wood a weight may be suspended, but the wood will not be broken because the centre of the wood can bear the weight. But (in the like circumstances) a hand-twisted rope may break, if its centre is not able to bear the weight.

Tension, breakage and continuity

C (It is upon) evenness, or continuity, that breaking or non-breaking depend. The reason is given under 'evenness or continuity'.

CS Let a small weight hang on a hair. Even if it is very light, the hair will break. This is because the hair is not (truly) even, or continuous. If it were it would not break.

The first of these is an example of attention to those problems of strain and load, bending and fracture, which must have occupied the minds of all ancient engineers, architects and practical physicists, but which were not tackled theoretically until the Renaissance. The second is of even wider interest. A parallel passage exists in the *Lieh Tzu* book where, although the *Mo Ching* is quoted, the purpose of its argument is exactly the opposite. In the Canon the Mohist writer seems to be affirming that the reason why a fibre breaks under tension is that it is formed of elements unequally strong, or unequally cohesive, so that a breaking-plane must occur somewhere. This is an essentially atomic or particulate point of view and fits in with the Mohist definitions of geometrical points and indivisible instants of time. But the writer of the *Lieh Tzu* book, on the contrary, is supporting continuity – as one would expect him to do, in view of what we have seen in the discussion of waves and particles. Continuity is, he says,

> the greatest principle in the world. The connected-togetherness of all shapes and things (in the world) is due to continuity. Now a hair (might be thought to have) continuity. But 'let a small weight hang on a hair. Even if it is very light, the hair will break. This is because the hair is not (truly) even, or continuous.' But with real continuity there cannot be any fracture, or any separation. Many people do not believe this, but I will prove it by examples.

Our Taoist, Master Lieh, is therefore in the great tradition. It is not in his power to prove his point by arguments which might have influenced Aristotle, but he embarks upon a series of parables and legends in the Taoist manner which shows us clearly what his conception of the natural world was. Remember Chan Ho, he says, who could bring up enormous fishes out of the abyss with a fishing-line made of a single silk fibre, such was his mental concentration and projection. Other stories follow, one of which teaches that music assures continuity between human beings, and between man and the rest of Nature. This is highly significant, for in the present context music means acoustics, and acoustics means wave transmission in a continuous medium. The theme recurs in powerful form in many other books of the Warring States and Chhin and Han periods.

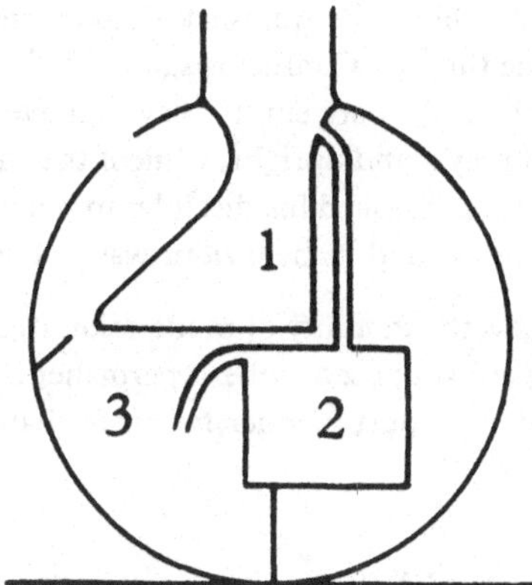

Fig. 157. Tentative reconstruction of an 'Advisory Vessel'.

Some may feel that we are straying far from physics, but this is not really so. In the ancient conceptions of the physical world, where sometimes *ching* (essence) can almost be translated as 'radiant energy', waves and cycles were supreme. There was no room for discontinuity or atomic particles. And so it was throughout centuries of indigenous Chinese thought.

Centres of gravity and the 'Advisory Vessels'

There seems to have been no theoretical treatment of the centre of gravity in China corresponding to the work of Heron of Alexandria on suspended objects of irregular shape. However, some trial-and-error methods must have been followed, notably in the suspension of chime-stones – L-shaped pieces of flat stone. Moreover, a remarkable application of knowledge about centres of gravity subsisted in the famous hydrostatic 'trick' vessels which altered their positions according to the amount of water they contained. For us it is very easy to imagine building a number of compartments into a bronze vessel with overflow channels into one another so arranged as to give a variety of effects (Fig. 157). In ancient China, however, it was regarded as a great marvel, and evidently went back to a respectable antiquity. The oldest description of such a vessel is in the *Hsün Tzu* book, and if, therefore, the invention did not come from Confucius' time, it was certainly known in the third century B.C. Part of the passage runs:

> Confucius inspected the temple of Duke Huan of Lu State and saw an inclining vessel. He asked the guardian of the temple what it was,

> and the guardian replied 'This is the Advisory Vessel which stands at the right hand side of the throne.' Confucius said: 'Ah, I have heard of these Advisory Vessels. If they are empty they lean over to one side, if they are half full they stand upright, while if they are full they fall over altogether.' Confucius asked his disciples to pour water into one of the vessels; they did so and its behaviour was just as he said.

The text then goes on to discuss the morality of moderation in all things, of which principle the vessels were supposed to be a permanent reminder to princes. The vessels persisted as a court wonder for more than a thousand years.

Specific gravity, buoyancy and density

The general idea of specific gravity must have existed from time immemorial. In the fourth century B.C., Mencius (Mêng Tzu) remarked that gold was heavier than feathers, otherwise how could it be said that a hook of gold was heavier than a cartload of feathers? But there seems to have been nothing equivalent to the treatise of Archimedes on floating bodies. Use by trial and error, of course, was made of his principle, as in the floating of arrows and vehicle wheels in water by the Chou and Han technicians, in order to determine their equilibrium.

Readers will doubtless know the so-called principle of Archimedes and the story connected with the philosopher in his bath pondering the problem raised by the possible adulteration with silver of King Hieron's golden crown. Archimedes noticed that the bath-water was displaced by his body; he weighed the crown and compared its displacement with that of equal weights of gold and silver. It has been suggested that the technicians of the Han were already familiar with the principle, and an analysis of a passage in the *Chou Li* (Record of the Institutions of the Chou Dynasty) makes it seem not unlikely, since at that time there were two terms for weighing, one for weighing in air (*chhüan*) and another for weighing in water (*chun*).

General notions of buoyancy must have been widespread among sailors in China, as elsewhere, from early times. From the third century A.D. there comes a famous story of weighing an elephant by observing the displacement of a boat loaded with different weights. The *Shen Tzu* (between the second and eighth centuries A.D.) says:

> Though a thing may be as heavy as the cauldron of Yen, or as much as 1,000 *chün* in weight; if it is placed upon a boat of Wu it will be transportable; this is the principle of floating.

This would have been no news to the founders of China's canal system, but as will be discussed in a later volume, the Chinese were the first to use water-tight compartments in ships, and in the Thang there was the monk Hui-Yuan with his floating and sinking water-clock bowls. Another monk, Huai-Ping, was responsible for a method of raising heavy objects from the bottom of a river by the use of buoyancy, analogous to the pontoons filled alternately with water and air used in modern salvage operations.

As for the specific gravity of liquids, the question rose particularly in connection with the assessments of strength of brine, and as salt was, at least from the Han onwards, a government monopoly or source of revenue, the procedures got a mention in literary works. From time immemorial salt-workers have used the swimming or sinking of eggs as a test of brine density, although the favourite test object in China was the lotus-seed.

China and the metric system

In current common speech and thought the metric system is primarily associated with the decimal ordering of coinage, weights and measures. But decimalisation is not of the essence of the metric system; the real significance of this is that it was the first great attempt to define terrestrial units of measurement in terms of an unvarying constant based on astronomy or on the size and shape of the earth. The metre was in fact defined as one ten-millionth part of one quarter of the earth's circumference at sea-level. It is said that the metric system took its origin in the need imposed by the development of scientific thought for unchanging, but at the same time conveniently related, units of physical measurement, and it is implied that this need was not satisfied until the last decade of the eighteenth century. This may be true enough for Europe, but an approach was made to such an unchanging unit in China in the first decade of that century. Moreover, as in the case of so many post-Renaissance scientific developments, there is an earlier pre-history of this celestial–terrestrial bond, and we can find already in the eighth century A.D. in China a large-scale attempt to establish it.

Though decimalisation is not the main issue here, it is worth recalling the predilection for decimalised measurement on the part of the ancient Chinese which we came across in describing Chinese mathematics (page 36). This goes well back into the Chou period, as foot-rules dating from the sixth century B.C. bear witness, and was adopted on a more considerable scale in 221 B.C. Nowhere else in the world was the decimalisation of weights and measures so early and so consistent. But the real progress of the science of measurement depended on fixing convenient length measures to comparatively unvarying natural reference standards, far beyond the range of all those whims which might from time to time affect princely law-givers. A great step forward was therefore made when the idea arose in China of fixing terrestrial length measures in terms of astronomical units. That this

Table 38. *Data of the Meridian Arc Survey of I-Hsing and Nankung Yüeh* (*A.D. 724 to 726*)

Station		The Meridian Line of +725				
No.	Place	Summer solstice shadow of 8-ft (2·4 m) gnomon (Ch. ft)	Equinoctial shadow (Ch. ft)	Polar altitude (Ch. degrees)	Approximate latitudes of stations from modern maps (Ch. degrees)	Distances in *li* and (in brackets) *pu*
1	Thieh-lê 鐵勒 (in the country of the Tölös horde of Turkic nomads beside Lake Baikal)	+5·13	+9·87	52·0°	*c.* 52·76°	6900
2	Wei-chou 蔚州 (Hêng-yeh Chün 横野軍) (northern Shansi)	+2·29	+6·44	40·0	40·38	1861 (214)
2*a*	Thai-yuan 太原	—	+6·0	—	38·22	
3	Hua-chou 滑州 (Pai-ma hsien 白馬縣)	+1·57	+5·56	35·3	36·07	198 (179)
4	Pien-chou 汴州 (Ku-thai piao 古臺表 (Khaifêng) near Chün-i 浚儀)	+1·53	+5·5	34·8	35·31	167 (281)
5	Yang-chhêng 陽城 (the central astronomical observatory)	+1·48	+5·43	34·7	34·93	526 (270)
6	Hsü-chou 許州 (Fu-kou piao 扶溝表)	+1·44	+5·37	34·3	34·65	160 (110)
7	Yü-chou 豫州 (Wu-chin piao 武津表 near Shang-tshai 上蔡)	+1·36	+5·28	33·8	34·09	1826 (196)
7*a*	Hsiang-chou 襄州	—	+4·8	—	32·46	5023
8	Lang-chou 郎州 (Wu-ling 武陵)	+0·77	+4·37	29·5	29·42	6112
9	Chiao-chou 交州 (Hu-fu 護府)	−0·33	+2·93	21·6	21·03	
	(capital of An-nan)			20·4		
10	Lin-i 林邑 (capital of Lin-I; near Hué)	−0·57 −0·91	+2·85	17·4	17·50	

Note: It has been supposed that the meridian line was as long as 13,000 *li* (*c.* 3,800 km) and included the most northerly station (no. 1). But the texts seem to indicate that the figures for this place were extrapolated and not derived from observations made at this time. There is reason for hinking that the distance 1–5 was an independent estimate dating from the Chên-Kuan reign-period could have occurred to scholars at all was due to the fact that the sun's shadow at summer solstice cast by a 2·5-m (8-foot) gnomon at the latitude of Yang-chhêng (the supposed centre of the Central Land) was about 0·5 m (1·5 feet). The relation with everyday measurement came about because of the long-standing idea that the shadow length increased 2·5 cm for every *li* north of the 'earth's centre' at Yang-chhêng, and decreased in the same proportion as one went south. After the Han, measurements made as far south as Indo-China soon disproved this numerical relation, but it was not until the Thang that a systematic effort was made to determine a great range of latitudes. This, when it came, had the effect of establishing a relationship between the lengths of terrestrial and celestial measures by finding the number of *li* which corresponded to 1° of terrestrial latitude, and thus fixing the length of the *li* in terms of the earth's circumference. The meridian line so set up takes its place between the lines of Eratosthenes (about 200 B.C.) and those of the astronomers to the Caliph al-Ma'mun (about A.D. 827).

The establishment in China of this basic metric measurement involved immense effort, most notably between A.D. 723 and 726 when the Astronomer-Royal, Nankung Yüeh, and the Tantric Buddhist monk I-Hsing, one of the most outstanding mathematicians and astronomers of his age, established at least eleven observing stations (Yang-chhêng was one) with a range of latitudes from 17·4° at Lin-I (not far from modern Hué in Vietnam) to 40° at Weichow (an old city in the neighbourhood of modern Ling-chhiu near the Great Wall). Locations of the stations are given in Table 38. The results obtained showed that the change was close to 10 cm for every thousand *li*, or four times the amount accepted by 'scholars of ancient times'. Admittedly these results were somewhat inaccurate by modern standards, for various reasons, but the exercise was noteworthy on a number of counts. The mathematics for analysing the evidence was very advanced for the time, and there was absolutely no inhibition about discrediting the beliefs of the scholars of past ages, an enlightened attitude which no contemporary European would have thought of taking. Above all, on account of its scale alone, this whole operation must surely be regarded as the most remarkable example of organised research carried out anywhere in the early Middle Ages.

THE STUDY OF MOTION (DYNAMICS)

The study of motion seems to have been, on the whole, conspicuously absent from Chinese thinking about physics. Nevertheless there are statements on the subject in the *Mo Ching*; indeed, some remarkable anticipations.

Movement in space (frames of reference)

C When an object is moving in space, we cannot say (in an absolute sense) whether it is coming near or going further away. The reason is given under 'spreading' (i.e. setting up coordinates by pacing).

CS Talking about space, one cannot have in mind only some special district. It is merely within a certain district that one can say that the first step (of a pacer) is nearer and his later steps further away. (The idea of space is like the idea of) duration. (You can select a certain point in time or space as the beginning, and reckon from it within a certain period or region, so that in this sense) it has boundaries, (but time and space are alike) without boundaries.

Ssuma Piao's commentary in the third century A.D. says 'The distance between Yen and Yüeh is limited, but that between the south and the north is infinite. Observing the limited from the viewpoint of the infinite, we find that Yen is not really separated from Yüeh. Space has no directions except that the place where you are yourself is the centre. Similarly circulating time (the course of the seasons) has no end and no beginning; you can make any specific time start whenever you like, according to what you are doing.' Many centuries later in Europe Nicholas of Cusa (1401 to 1464) and Giordano Bruno (1548 to 1600) spoke in the same way using almost the same words.

Movement and duration

C Movement in space requires duration. The reason is given under 'earlier and later'.

CS In movement the motion must first be from what is nearer, and afterwards to what is further. The near and the far constitute space. The earlier and the later constitute duration. A person who moves in space requires duration.

Motion

C Motion is due to a kind of looseness (i.e. to the absence of an opposing force).

CS There is motion (if a force is) allowed to work at the edge, just as a door pivot is free when the bolt is not fixed.

This may be the remains of an attempt to discuss circular motion.

Forces and motion

C The cessation of motion, is due to the (opposing force) of a 'supporting pillar'.

CS If there is no (opposing force) of a 'supporting pillar' the motion

> will never stop. This is as true as that an ox is not a horse. Like an arrow passing through between two pillars (without anything standing in its way, and following a straight-line motion without changing its direction).
>
> If there is (some kind of) 'supporting pillar' (some other force interfering with the motion), and nevertheless the motion does not stop (it may still be called motion but it will not be a straight-line motion because there will have been a deflection). This is a case of something being 'a horse and yet not a horse'. It is like people passing over a bridge (i.e. they have to climb up to the top of the arch and down again, although they continue in motion).

To see these Mohist definitions, as well as later Chinese thoughts, in their correct perspective, it is necessary to glance at the development of studies of motion in Europe. The Greek ideas of motion were crystallised in Aristotle (fourth century B.C.) who believed that everything had its natural place and that 'natural' motion occurred in a straight line towards that natural place. 'Violent' motion was motion caused by some external force. Celestial motions, on the contrary, were naturally circular. He believed a vacuum was impossible, so every body in motion moved through air or some other resisting medium. This medium was necessary for motion to continue, for Aristotle supposed that air rushed in behind a moving body to prevent a vacuum and so to help the body on its way. Opposition to this last view was strong from the sixth century A.D. onwards and the idea that from the outset a moving body had a certain 'impetus' was revived (it had first been suggested in the second century B.C.). This led to a whole movement of thought about dynamics among philosophers in the Middle Ages.

Further significant development came in the Renaissance when Galileo considered the motion of bodies in free fall, and ignored the distinction between 'natural' and 'violent' motion. He discussed the path of a projectile and concluded that it remains uniform except as influenced by friction or air resistance. In 1687 Newton published his laws of motion, the first of which states that 'every body continues in a state of rest, or uniform motion in a right (straight) line, unless it is compelled to change that state by forces impressed on it'. Two other philosophers of the time had said nearly the same thing and Galileo had used the principle for projectiles. What is astonishing is that we now find something extremely like it in the *Mo Ching* of the fourth or third century B.C. and quoted above under 'Motion' and 'Forces and motion'. The Mohist technical term 'supporting pillar' is to be understood only as that force which in Newton's first law changes the otherwise permanent state of motion of the moving body.

The other Mohist definitions given illustrate the relativistic and dialectical qualities of this school of thinkers, while another definition (not quoted) about movement along slopes was not taken up in the West until Pappus did so in the fourth century A.D. and Galileo analysed the whole problem in the sixteenth. Yet another concerned with moving spheres was not considered again until the thirteenth and the fifteenth centuries in Europe.

After these brilliant Mohist insights, it seems almost incredible that through the subsequent millennia of Chinese history there are no recorded discussions of the motions of bodies, whether impelled or free falling. This lack of evidence seems well established, but it appears to have had no restraining effects upon practical technology. So far as vehicles, projectiles and engines of all kinds were concerned, Chinese mechanical practice was ahead of the European, not retarded, down to the very time when Western medieval philosophers were preparing the way for the contributions of Galileo and Newton. For example, the early development of gunpowder weapons, from the first invention of the formula to the appearance of the metal-barrel gun and cannon, all took place in China before these last were known in Europe at all. And that was only one instance of a basic projectile-mindedness that characterised Chinese tactical attitudes all through the centuries.

The reason for the lack of later Chinese contributions to the theory of motion is a puzzle; indeed it is extraordinary when one considers two factors, action-at-a-distance and the relative values of states of motion and of rest. As regards the first, Chinese thinking in terms of the continuity and oneness of Nature meant that they never had any problem when it came to considering action-at-a-distance. This is why, early on, they could give a correct explanation of the tides, and why they could readily consider a magnetic field in connection with the compass, an idea which was later to exert a revolutionary force on European scientific thinking. Yet in the West the inability to conceive of action-at-a-distance was one of the factors which long delayed an appreciation of the true nature of gravity. The Chinese, however, in spite of having no such inhibition, never developed a theory of gravitation, so important for the development of a science of motion. The second factor concerns the relative value of the states of rest and motion. The Greeks and philosophers of the European Middle Ages seem to have considered a state of rest as intrinsically 'superior' to a state of motion, possibly because this attitude was engendered by social distinctions, where the philosopher who sat and thought was considered socially superior to the practical man. But the Chinese held no such opinion. No doubt in their culture lords and scholars sat still, while the slaves and the people moved about their work – yet certain healthier factors, hard to define but connected with the general industriousness prevalent in China throughout the ages, influenced Chinese thought in such a way that motion was regarded as, if anything, superior to rest. The sage, like Heaven, should never rest. Yet in spite of this, dynamics remained a study begun by the Mohists and then left undeveloped.

Perhaps the reason for this lack of later development was that Chinese thinking was so averse to atomic conceptions. Admittedly Aristotle, who can be said to have laid the foundations of dynamics, was no supporter of atomic theories, but nevertheless he participated in the same European tradition that was prepared to consider what happened to individual portions of matter in motion. Yet it was the Mohists, and only they, with their distinct leanings towards atomism in their geometrical definition of points and instants, and their notion of imperfect continuity as the reason why fibres broke in tension, who conceived of the motion of bodies in an abstract way.

SURFACE PHENOMENA

The transition from the study of motion to that of heat should have come about through the study of friction. But in spite of the fairly high level of practical technique, Chinese literature seems to lack theoretical discourses on the subject. The production of fire by rubbing pieces of wood was of course anciently known, and is referred to from time to time in books such as the *Huai Nan Tzu*; there is also a note on the friction of wheels in the *Chou Li*. But, as in Europe, the physics of rubbing solids had to wait until the Renaissance.

However, in the Thang period some interest was taken in the fit of smooth surfaces, by the prince Li Kao and others. Then, in the thirteenth century A.D., Chou Mi noted some properties of what we should now call monomolecular films, when he wrote that:

> 'Bear alum' (presumably bear-fat mixed with alum) can disperse dust. It can be tested as follows. A vessel of clean water is scattered over with dust. When a grain of this alum is then put in, the collected dust opens (becomes clear).

In another Sung book, the *Yü Huan Chi Wên* (Things Seen and Heard on My Official Travels), Chang Shih-Nan wrote:

> For testing the quality of lacquer or tung-oil, they make a ring of a small sliver of bamboo, and dip it into the liquid. If upon being withdrawn a film is formed across the ring, the lacquer or tung-oil is good; if they have been adulterated no ring-film will be formed.

The ring-test must have been used as a quality test for centuries; in 1878 it was introduced as a method for precisely determining surface-tension by finding the force necessary to break the surface.

HEAT AND COMBUSTION

In the West the understanding of heat was one of the last achievements in the elaboration of the structure of physics as we know it. In antiquity and the Middle Ages the necessary conceptions and definitions were completely lacking; in spite of the rich store of practical information about expansion and contraction, change of state from solid to liquid and liquid to solid, transitions from liquid to vapour, with many similar such processes, accumulated in trades and industries. In these circumstances it is perhaps not surprising that little can be found on heat as such in ancient and medieval Chinese writings. But like their Western counterparts these do yield many interesting examples of the observation and use of thermal phenomena in the field of technology.

In the sixteenth century Li Shih-Chen in his *Pên Tshao Kang Mu* (The Great Pharmacopoeia) gives the Chinese view as it had then crystallised. Fire has *chhi (pneuma)* but not *chih* (actual matter). Unlike all the other elements, which are unitary, fire is of two kinds, Yin and Yang, and there is an elaborate classification (Table 39). When we remember that even at the beginning of the nineteenth century in Europe it was by no means generally conceded that heat was a form of motion of matter, and indeed was more widely thought of as a weightless fluid known as 'caloric', then the efforts of Li Shih-Chen to classify different forms of flame and combustion do not seem as archaic as might appear at first sight. Looking at the table we can see that he may have been trying to differentiate between the essential heat production of the body and the heat connected with muscular movement, foreshadowing our categories of basal metabolic rate and total heat-output. The 'meditation fire' notion was probably derived from the yogis' feat of maintaining a high body-temperature under continued exposure to cold. 'Dragon fire' is not very obvious, perhaps lightning which may strike and melt without burning. In his time there was little point in separating the different forms of heat and light produced by friction and impact, but it is curious that he did not know where to classify the 'cool' flames of natural gas, a source of heat which the Chinese perhaps of all peoples were the first to make use of on an industrial scale, and which had been known for seventeen hundred years before Li Shih-Chen. Lastly he did not realise that oils are 'inextinguishable' because they continue to burn when floating.

Spontaneous combustion of substances at lower external temperatures was a matter of interest in China for many centuries, and in the third century A.D. there was a classic case of this when a calamitous fire occurred in the arsenal of the Chin emperor Wu. It was traced to the spontaneous combustion of oiled cloths caused by heat generated from chemical reactions in the material. Other examples make it seem likely that some Chinese observations were made of what we should now call the successive boiling

Table 39. *Li Shih-Chen's classification of the varieties of fire*

	Fire	
	Yang	Yin
Heavenly	(1) *Thai yang chen huo* (heat of the sun)	(1) *Lung huo* (dragon fire)
	(2) *Hsing ching fei huo* (light of stars and meteors)	(2) *Lei huo* (lightning)
Earthly	(1) *Tsuan mu chih huo* (fire and heat produced by drilling wood)	(1) *Shih yü chih huo* (fire from burning petroleum)
	(2) *Chi shih chih huo* (sparks produced by striking stone)	(2) *Shui chung chih huo* (ignis fatuus; burning methane?)
	(3) *Ka chin chih huo* (sparks produced by tapping metal)	
Human	(1) *Ping ting chün huo* (general metabolic heat)	(1) *Ming mên hsiang huo* (heat of the viscera and generative organs)
		(2) *San mei chih huo* ('samādhi' or meditation heat)
Unclassified	(1) *Hsiao chhiu huo* (= *han huo*; cold heat; flame of natural gas)	
	(2) *Tsê chung chih yang yen* (phenomena like ignis fatuus; burning methane?)	
	(3) *yeh wai chih kuei lin* (see marsh lights p. 347 below)	
	(4) *chin yin chih ching chhi* (glitter of gold, silver and gems)	

phases of liquids. Indeed the stages of boiling of a kettle in making tea for the Japanese tea-ceremony were long known, so that the tea was made only with 'aged hot water' – water at the third stage of boiling. Thus the tea-kettle which accompanies the Taoist sage or naturalist in so many pictures by Chinese artists acquires new meaning for us.

A curious passage about steam occurs in the *Huai Nan Wan Pi Shu* (The Ten Thousand Infallible Arts of the Prince of Huai-Nan). It is at least a text of the Han period and may conceivably go back to the second century B.C. The eighth recipe in the book runs as follows:

> To make a sound like thunder in a copper vessel. Put boiling water into such a vessel and then sink it into a well. It will make a noise which can be heard tens of *li* away.

Assuming the vessel was almost full of steam when the lid was tightly closed, then plunging it into cold water would cause a vacuum by sudden cooling, and if the vessel were thin walled there would be a sudden implosion with a consequent report. This, of course, was the principle used in the seventeenth century A.D. in the earliest steam engines (having, of course, a cylinder with walls strong enough to withstand collapse), and the story is an astonishing manifestation of this power some eighteen hundred years earlier, a power which the Chinese used only for military and wonder-working purposes.

Lastly a word about fire-making. Ancient Chinese literature naturally contains references to the well-known prehistoric and primitive methods of igniting tinder by the heat generated by rubbing pieces of wood together and by sparks from flint and steel, but less familiar is the priority which China seems to have in the invention of the sulphur match. In the *Chhing I Lu* (Records of the Unworldly and the Strange), written about A.D. 950, Thao Ku tells us:

> If there occurs an emergency at night it may take some time to make a light to light a lamp. But an ingenious man devised a system of impregnating little sticks of pinewood with sulphur and storing them ready for use. At the slightest touch of fire they burst into flame. One gets a little flame like an ear of corn.

An enlarged note on the same subject appeared in 1366, with a note that the invention was not made by the people of Hangchow, as generally assumed, but by the impoverished court ladies of the Northern Chhi in A.D. 577 at the conquest of the empire by the Sui. However, although sulphur matches were certainly on sale in the markets of Hangchow when Marco Polo was there (about 1270), no positive evidence for sulphur matches in Europe exists before 1530.

Digression on luminescence

The study of thermal changes, with their accompanying phenomena of burning, was for many centuries bedevilled by a crowd of miscellaneous observations most of which had nothing to do with heat. The effects seen are classified today as luminescence. The emission of light from substances only while they are exposed to different kinds of radiation is known as fluorescence; if it persists after the radiation source is cut off we may call it phosphorescence. A delayed phosphorescence may appear on heating (thermo-luminescence). Lights accompanying electric discharges are termed electroluminescence, and there are emissions of light brought about by friction, especially the crushing or rubbing of crystals (tribo-luminescence and piezo-luminescence). Finally there is chemi-luminescence, where light is given off during certain chemical reactions, especially notable where phosphorus is involved, and bio-luminescence when the light is emitted from living organisms (although this is really a form of chemi-luminescence). What did the Chinese have to say about these?

To begin with let us return to Li Shih-Chen's classification of 1596. He was baffled by marsh-lights or *ignis fatuus*, but so are modern investigators, and we are still not sure whether to ascribe the famous 'will-o'-the wisp' (or 'jack-o'-lantern') to burning methane, some specific chemical reactions, or to electrical discharges. As might be expected there are many references to this phenomenon in Chinese literature besides that of Li Shih-Chen. There were references, too, to the accumulated observations on the bio-luminescence of putrefying substances due, as we now know, to luminous bacteria and fungi. As the living origin of this light was not known, its appearances were naturally confused with the marsh-lights, and the idea grew up that the *ignis fatuus* was derived from old blood. The 'five-element' chapters in the dynastic histories contain many records of observations of this kind; for example in the *Sung Shu* we find the following entry:

> In the time of the emperor Min of the (Liu) Sung, on a *ping-wu* day in the fifth month of the second year of the Thai-Shih reign period (A.D. 466), a Taoist, Shêng-Tao, from a mountain temple near Huang-chhêng in Lin-i district of southern Lang-hsieh, (reported that) a pillar in an apartment next to the Hall of Salvation was spontaneously shining brightly in the dark. The wood had lost its natural properties. Some people said that when wood goes rotten it shines of itself.

Here the chief point of interest is that the association with rotting was clearly understood. And not only were records made of marsh-lights and glowing bacteria and fungi, but also the 'phosphorescence' of sea-water was noted. Bio-luminescence quite obviously from larger animals had of course long been known in China as in most other cultures; such was the light from fireflies and glow-worms.

The luminescence of inorganic materials was also recognised. There was the 'night-shining jewel' (volume I of this abridgement, page 71), probably a variety of fluorspar which displayed tribo- or piezo-luminescence. Perhaps the most famous of these was the 'Bononian stone' or *lapis solaris* which so much interested scientists in the seventeenth century. This mineral was a native barium sulphate rich in sulphur which glowed in the dark after being heated, and perhaps baryta deposits in China may have provided some 'night-shining jewels' of this kind. There is also a remarkable story which indicates that artificial phosphors may have been known in the Sung. It occurs in a book of miscellaneous notes by the monk Wên-Jung, written in the eleventh century A.D. and entitled *Hsiang Shan Yeh Lu* (Rustic Notes from Hsiang-shan):

> The Provincial Legate Hsü Chih-O was fond of collecting curios ... He also got hold of an extraordinary painting ... On the painting there was an ox which during the day appeared to be eating grass outside a pen, but at night seemed to be lying down inside it. None of the officials could offer any explanation of this phenomenon. Only the monk Lu Tsan-Ning, however, said that (he understood it). According to him the 'southern dwarf island-barbarians', when the tide is out and there remains only a little dampness on the shore, collect certain remaining drops of liquid from a special kind of oysters and use this to mix a coloured paint or ink which appears only at night and not by day...

Now the interest of the matter lies in the fact that in 1768 John Canton did in fact describe a phosphor made from oyster shells – an impure calcium sulphide made by heating the carbonate with sulphur. This became known as Canton's phosphorus. By adding the sulphides of arsenic, antimony or mercury, phosphors with blue or green luminescence can be obtained. Lu Tsan-Ning was a learned naturalist much respected by his contemporaries, and it would seem not unlikely that he and other alchemists of the early Sung may have prepared such luminescent substances, while the book of Wên-Jung contains many other accounts about the effects of different kinds of phosphorescent phenomena.

Not only were there observations of marsh-lights, biological and chemical luminescence, as well as artificially produced phosphorescence, but sparks of static electricity and the electro-luminescence known as *ignis lambens* were also noted. The *Po Wu Chih* (Record of the Investigation of Things) dated about A.D. 290 is quite specific:

> In places where fights have been fought and people have been slain, the blood of men and horses changes after a number of years into

will-o'-the-wisps. These lights stick to the ground and to shrubs and trees like dew. As a rule they are invisible, but wayfarers come into contact with them sometimes; then they cling to their bodies and become luminous. When wiped away with the hand, they divide into innumerable other lights, giving out a soft crackling noise, as of peas being roasted. If the person stands still a good while, they disappear, but he may then suddenly become bewildered as if he had lost his reason, and not recover before the next day.

Nowadays it happens that when people are combing their hair, or when dressing and undressing, such lights follow the comb, or appear at the buttons when they are done up or undone, accompanied likewise by a crackling sound.

LIGHT (OPTICS)

If Chinese optics never equalled the highest level attained by the Islamic students of light such as Ibn al-Haitham, who benefited from the availability of Greek geometry, it nevertheless began at least as early as the optics of the Greeks. Once again the importance of the Mohists must not be underestimated. The Taoists might talk about the wonders and beauties of Nature, the naturalists might bring forward their generalised explanations of her phenomena, the Logicians might argue about the best way of discussing, but only the Mohists actually took mirrors and light-sources and looked carefully to see what happened.

Before speaking of light-sources in the abstract, a few words should be said of the lights themselves. Presumably the walls of the feudal Princes of the Chou flared with torches made of various types of combustible material – bamboo, pine resin, etc. For smaller rooms throughout historical times oil was burnt with wicks in lamps or cruses, at first of pottery and bronze. From the beginning of history, China probably never lacked vegetable oils, and animal oil was also sometimes used. The 120 iron lamps in the assembly hall of Shih Hu, emperor of the Later Chou (about A.D. 340) were long famous.

The small simple flat pan or dome-shaped cruse served far and wide in ancient civilisations to hold oil and wick. But an ingenious medieval Chinese development introduced a reservoir of cold water underneath, which cooled the oil and so reduced evaporation. In 1190 Lu Yu wrote in the *Lao Hsüeh An Pi Chi* (Notes from the Hall of Learned Old Age):

> In the collected works of Sung Wên An Kung, there is a poem on 'economic lamps'. One can find these things in Han-chia; they are actually made of two layers. At one side there is a small hole into which you put cold water, changing it every evening. The flame of an ordinary lamp as it burns quickly dries up the oil, but these lamps are different for they save half the oil . . .

This industry must have started in the Thang, perhaps early in the ninth century A.D. It was thus contemporary with the cooling reservoirs of Chinese distillation apparatus, which had probably originated by A.D. 500. It was also an interesting anticipation of water-jackets and chemical condensers, foreshadowing that great development of such devices which came about in the thirteenth century and which we shall discuss in a later volume.

The word *chu*, which originally also meant torch, came early to be applied to candles, using wicks of fibre or bamboo. The problem is, exactly when? The oldest mention of candles made specifically of beeswax may be about 40 B.C., but the oldest occurrence of the surer term *la chu* is in A.D. 322. Nevertheless other references make it seem likely that beeswax candles were being made already in the Warring States period, so that the Mohist natural philosophers may well have used them in their experiments. In China candles were always made by dipping, never by moulding.

Mohist optics

Let us now examine the propositions on optics contained in the *Mo Ching* (fourth century B.C.), truncated and fragmentary though they are.

Shadow formation

C A shadow never moves (of itself). (If it does move it is owing to the moving of the source of light or of the object which casts the shadow.) The reason is given under 'charging action'.

CS When light arrives the shadow disappears. But if it were not interfered with, it would last for ever.

Umbra and penumbra

C When there are two shadows (it is because there are two sources of light). The reason is given under 'doubling'.

CS Two (rays of) light grip (i.e. converge) to one light-point. And so you get one shadow from each light-point.

This clearly indicates that the Mohists realised that light travels in straight lines.

Size of shadow dependent on position of object and of source of light

C As for the size of the shadow – the reason is given under 'whether slanting like a steering-oar' (i.e. not perpendicular to the direction of the light-rays), or upright (i.e. perpendicular to the direction of the light-rays); whether far or near.

CS If the post is slanting, like a steering-oar (not perpendicular to the rays of the sun or other light-source), its shadow is short and intense. If the post is upright (perpendicular to the rays of the sun or other light-source), its shadow will be long and weak. If the source of light is smaller than the post, the shadow will be larger than the post. The further (from the source of light) the post, the shorter and darker will be its shadow; the nearer (to the source of light) the post, the longer and lighter will be its shadow.

Here the experimentalist must have had a fixed light-source and a fixed screen, with the post able to move between them.

Pinhole

C The 'collecting place' (*khu*, 庫) (or, the 'wall', *chang*, 牆) is the place where the 'change' (i.e. where the inversion of the image) starts.

CS It is an empty (round) hole, like the sun and moon depicted on the imperial flags.

We do not know which is the right word in the first part of this proposition. *Khu* might refer to the whole enclosed space of the camera obscura (page 356). Another proposition traces in detail the formation of the inverted image obtained in a camera obscura and brings in the concept of the focal point, the point of intersection of the rays from the object outside to the image formed inside.

The Mohists also considered mirrors. Plane mirrors and combinations of such mirrors, together with the phenomenon of lateral inversion of the image so that left becomes right, follow in other propositions, and convex and concave mirrors are also considered. The proposition about concave mirrors is particularly interesting:

C With a concave mirror, the image may be smaller and inverted or larger and upright. The reason is given under 'outwards from the centre area' (i.e. away from the centre of curvature); and 'inwards from the centre area' (i.e. from the focal point towards the surface of the mirror).

CS (Take first) (an object in) the region between the mirror and the focal point. The nearer the object is to the focal point (and therefore the further away from the mirror) the weaker the intensity of the light will be (if the object is a light-source), but the larger the image will be. The further away the object is from the focal point (and therefore the nearer to the mirror), the stronger the intensity of light will be (if the object is a light-source) but the smaller the image will be. In both cases the image will be upright. From the very edge of the centre region (i.e. almost at the focal point), and going towards the mirror, all the images will be larger than the object, and upright.

(Take next) (an object in) the region outside the centre of curvature and away from the mirror. The nearer the object is to the centre of curvature the stronger the intensity of light will be

> (if the object is a light-source), and the larger the image will be. The further the object is from the centre of curvature, the weaker the intensity of light will be (if the object is a light-source), and the smaller the image will be. In both cases the image will be inverted.
>
> (Take lastly) (an object in) the region at the centre (i.e. the region between the focal point and the centre of curvature). Here the image is larger than the object (and inverted).

This is a striking passage. The Canon distinguishes between what we now call the real and inverted image, while the Exposition records a series of meticulous observations, the last one of which must have involved very careful procedure, and although the Mohists seem to have had no special technical terms for focal point and centre of curvature, there appears to be no doubt that they could differentiate between them. Refraction was also discussed, the proposition on this dealing with the apparent bending of a stick in a vessel of water.

Once again, to place these contributions in perspective, it is necessary to glance very briefly at the parallel origins of this science in Greece. The oldest and most widely accepted theory of light and vision was the Pythagorean belief that visual rays were emitted by the eye, ran in straight lines to the object, and by touching it gave the sensation of light. Next in importance was the view held by the Epicurean philosophers in which objects themselves were supposed to give off images in all directions, some of which penetrated the eyes and gave rise to sight. The only work contemporary with the Mohists was that of Euclid. This consists of 58 theorems, treated like his geometrical ones, and based on four definitions all of which were known to the Mohists, but what he wrote on mirrors is unknown. The oldest Greek work on this subject is by Heron of Alexandria, about A.D. 100, so what we have from the Mohists is rather earlier than anything from Greece. However, the *Optics* of Ptolemy, written early in the second century A.D. (about the time of Chang Hêng) goes much beyond any systematic exposition which has remained in Chinese literature, and deals not only with mirrors but also with refraction, the results of which are applied to the problem of finding positions of celestial objects refracted from their true positions by the earth's atmosphere.

The fundamentally wrong Greek conception of light and vision was held until the correct view prevailed in the revolutionary work of Ibn al-Haitham (Alhazen) between A.D. 965 and 1039. Yet the Chinese seem never to have held any theory about the eye emitting rays, and had more in common, therefore, with the Epicurean minority.

Mirrors and burning mirrors. A development parallel to, and probably much older than, the optical work of the Mohists, was the use of burning mirrors for igniting tinder by sunlight. References to these are very common in ancient Chinese writings; indeed mirrors of bronze must go back extremely far into the Chinese Bronze Age, for one of the earliest mentions is dated to 672 B.C. The oldest existing dated mirrors, however, are from A.D. 6 and A.D. 10, while there are many from Later Han times. Among Han writings, the *Chou Li* refers to two kinds of officials, the 'Directors of Fire Ceremonies' who obtained 'new fire' by means of a fire-drill, and the 'Directors of Sun Fire', who made ceremonial use of burning mirrors. There are other references, and these are contemporary with descriptions of such mirrors in European (Latin) literature; perhaps they indicate the spread in both directions of a technique originally Mesopotamian or Egyptian.

The Han mirrors give evidence of considerable metallurgical knowledge, for the bronze never contains enough tin to make it excessively brittle, while it also has enough added lead to improve the casting properties. The resulting 'speculum metal' is truly white, reflects without tinning or silvering, resists scratching and corrosion well, and is admirably suited to its particular purpose.

Concave mirrors were used not only with the sun but also with the moon. Kao Yu, of the Later Han, describes very clearly the use of both:

> The burning mirror ... is heated by being made to face the sun at noon time; in this position cause it to play upon mugwort tinder and this will take fire. The ... Yin mirror ... is like a large clam(-shell). It is also polished and held under the moonlight, at full moon; water collects upon it, which can be received in drops upon the bronze plate...

What was happening here was just that dew was being collected. That this was so much prized was a kind of philosophical superstition, though there was a good deal of confusion between this and the belief, which modern research has shown to be true, that certain marine animals waxed and waned in correspondence with the moon's phases. Perhaps the confusion arose because of the concave shape of the shells of certain molluscs; as Kao Yu said, the mirrors were like large clam-shells.

Some of the scientific accuracy of the Mohists still persisted in the Han, as appears from the *Huai Nan Tzu* (Book of (the Prince of) Huai-Nan) of about 120 B.C., where it is made clear that the tinder must be at precisely the right position with respect to the mirror if it is to catch fire. Thus they knew of the focal point of a concave mirror. It is clear, too, that they had plane as well as curved mirrors, all of which appear to have been of

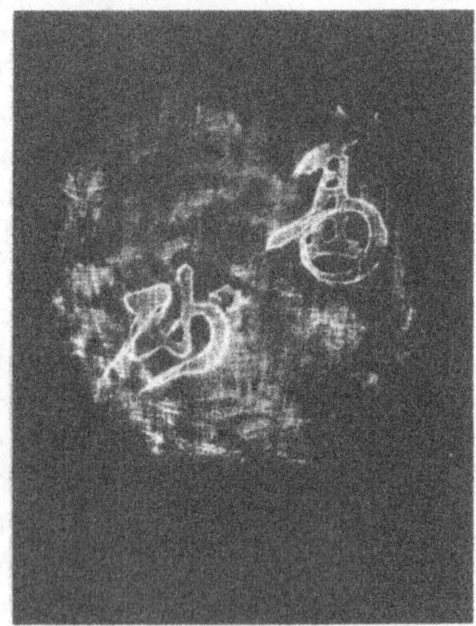

Fig. 158. A 'magic mirror' of Japanese provenance (from H. Dember, *Ostasiatische Zeitung*, 1933, vol. 9 (19), p. 203). Although the polished face (not seen here) appears to the eye perfectly smooth, the characters executed in relief on the back (left) are clearly visible in the reflection (right). They read *Takasago*, the name of a *nō* play.

high quality, some with their surfaces coated with a layer of tin. The *Huai Nan Wan Pi Shu* (Ten Thousand Infallible Arts of (the Prince of) Huai-Nan), a book of Taoist alchemical and technical recipes of the second century B.C., gives some details of these processes and also provides evidence of an interest in multiple reflections.

> A large mirror being hung up (above a large trough filled with water), one can see, even though seated, four 'neighbours'.

The Taoist Than Chhiao of the tenth century A.D. was clear that even though an image of a distant mirror may be so small as to be on the limit of visibility, the rays of light go to the mirror all the same. And a century later Shen Kua wrote of how 'the ancients' made mirrors either plane or concave (if large) and convex (if small) because the whole face could not be seen in a small flat or concave mirror whereas it could in a small convex one. He believed the art of mirror-making to have been superior in past times, a hint that perhaps the Mohists were in closer contact with the technicians of their time than scholars were in his own day, though Shen Kua's strictures on the abilities of the twelfth-century technicians seem rather too severe. It is clear that he made pinhole experiments and saw that the collection of rays at a pinhole was similar to the collection of rays at the focus of a concave mirror.

Fig. 159. Diagrammatic cross-section of a 'magic mirror' (greatly exaggerated to show the nature of the relief).

Mirrors of unequal curvature

Shen Kua also wrote of another kind of mirror (Fig. 158).

> There exist certain 'light-penetration mirrors' which have about twenty characters inscribed on them in ancient style which cannot be interpreted. If such a mirror is exposed to the sunshine, although the characters are all on the back, they 'pass through' and are reflected on the wall of a house, where they can be read most distinctly ...

There were three mirrors of this kind in Shen Kua's family and he had seen others treasured in other families. Since ordinary mirrors, however thin, do not 'let light through', Shen Kua remarked that ancient mirror-makers did indeed seem to have some special art.

Only in the later part of the nineteenth century were these mirrors subjected to a careful investigation and not until 1932 was a definitive explanation prepared. The secret was that the surfaces of the mirrors were of more than one curvature (Fig. 159). The design was cut in the back of the mirror, and the strains imposed gave a mirror which differed imperceptibly in thickness from place to place. The convex curvature on the front (reflecting) surface of the mirror varied, the thicker portions of the mirror having a rather flatter curvature than the thinner ones. It has proved possible to make such mirrors in the West, and they were still being produced in Japan at the time when modern physical investigations started. However, such mirrors must have been made in China as far back as the fifth century A.D., and probably already in Han times. Finally, such mirrors with a false back could be made to throw designs or characters on the screen different from those apparently inscribed on the back, thus inducing even greater wonder.

Camera obscura

It seems that in the Thang and the Sung there was much interest in experiments with light passed through pinholes into darkened rooms. Shen Kua has an important passage on it in his *Mêng Chhi Pi Than* (Dream Pool Essays) of A.D. 1086:

> The burning mirror reflects objects so as to form inverted images. This is because there is a focal point in the middle (i.e. between the object and the mirror). The mathematicians call investigations about such things Ko Shu. It is like the pattern made by an oar moved by someone on a boat against a rowlock (as fulcrum). We can see it happening in the following example. When a bird flies in the air, its shadow moves along the ground in the same direction. But if its image is collected (like a belt being tightened) through a small hole in a window, then the shadow moves in the direction opposite to that of the bird ... Take another example. The image of a pagoda, passing through the hole or small window, is inverted after being 'collected'. This is the same principle as the burning mirror. Such a mirror has a concave surface, and reflects a finger to give an upright image if the object is very near, but if the finger moves further and further away it reaches a point where the image disappears and after that the image appears inverted. Thus the point where the image disappears is like the pinhole of the window...

This is certainly interesting. The reference to the mathematicians suggests, with other such references, that active study had been going on for a long time previously, while the oar analogy is striking – indeed the space swept out on each side of the rowlock is even now a good way to describe the light-cones on each side of the focus. The identification of three separate radiation effects (pinhole, focal point and burning point) is excellent.

The camera obscura is mentioned in the *Yu-Yang Tsa Tsu* (Miscellany of the Yu-Yang Mountain Cave) which appeared early in the ninth century A.D., though its explanation of the optics is wrong. The famous Arabic physicist Ibn al-Haitham in the tenth century used it to observe solar eclipses, but it seems that, like Shen Kua, he had predecessors in his study of its optics. In brief, both Arabs and Chinese seem to have been interested in the camera obscura from the eighth century onwards, yet its origin is still unknown.

Lenses and burning-lenses

Rock-crystal and glass. Until recently there has been great uncertainty about glass-making in China, for the earliest literary references are obscure. Rock-crystal, a pure crystalline form of quartz (silicon oxide), was known there in ancient times, but it could only be worked like jade since temperatures high enough to fuse it and make vessels of it have been attainable only in modern times. The first pharmaceutical natural history to mention it was the *Pên Tshao Shih I* (Omissions from Previous Pharmacopoeias) of A.D. 725, but it seems to have been known earlier than this. It had a variety of names, just as it appeared in a variety of forms, more often coloured or translucent rather than clear and transparent, and some kinds such as 'tea-crystal' or black crystal were used in the Ming for making dark spectacles, just as we use dark sunglasses today. In the West, acquaintance with rock-crystal is ancient, a famous worked piece from Babylonia probably being of the ninth century B.C., while Greek and Roman authors mention it and many rock-crystal balls have been found at Pompeii and other sites.

Glass, in its commonest and earliest form, is essentially a silicate of sodium or potassium (or both) and calcium, produced by fusing sand (silica) with limestone (calcium carbonate) or other alkaline ingredients. Another form, developed at a later stage in history, is lead-silicate glass, in which some of the alkaline elements are replaced by lead. The oldest piece of glass which can be positively dated was made somewhere in Mesopotamia in the third millennium B.C., and by the time of the Romans the industry was a very great one, centred on Alexandria. Some of its products certainly reached China.

Chinese glass technology. The earliest known Chinese glass is in the form of opaque beads and dates from the Chou. It is of lead-silicate type, found only as rare and isolated examples in Western Asia from 700 B.C. onwards. But Chinese glass exhibits a further feature unknown in other ancient glasses; it contains relatively large amounts of barium as the oxide (sometimes almost 20 %) and very large amounts of lead (PbO), as much as 70 %. The designs of some of the oldest objects of Chinese glass were like those of Europe; an opaque bead was drilled and then layers of glass of different colours were inserted. These 'eye-beads' were common in Europe from about 480 B.C., and reached China by 300 B.C. But most Chinese glass was made into objects of distinctively Chinese character such as dragons and cicadas, or as *pi* emblems (page 160). It seems that during the time of the Warring States, and the Chhin and Han periods, glass was used by the poorer families as a standard substitute or cheap imitation of jade for funerary purposes.

The consistent use of lead, and still more the use of barium, in Chinese glass poses the question of how far this was intentional. On balance it appears to have been adopted on purpose in order to give glass an added brilliance. And as one moves from the Han to Thang periods, Chinese glass changes from the lead-barium type to a lead-sodium-calcium type and thence to ordinary soft glass without lead in it at all. However, it is not yet possible to say when the earliest glass was made in China, though there are indications that it may go back as early as the mid-sixth century B.C.

With this archaeological evidence in mind it is possible to see some of the literary references in a new light, but in doing so it is necessary to bear in mind that a considerable amount of glass was imported into China. Two terms were used for glass; *liu-li* which seems, by and large, to have been used for opaque glass, and *po-li* meaning more or less transparent glass. A third, earlier term, *pi-liu-li*, from which *liu-li* seems to have been derived, was originally used for foreign glass and this has sometimes been taken to indicate that the first glass to be seen in China was imported, and that Chinese glass manufacture was a late development. We now know this was not so, although we have the puzzling situation that the earliest recognisable term for glass was a word of foreign origin. Perhaps the Han court and the scholars were more interested in rarities from abroad than in the imitations of jade produced for the popular market by Taoist artisans at home, so that the original term for glass was lost.

This pattern of the double origin of glass, indigenous and imported, continued throughout the centuries, and it is impossible to sort out the separate strands. The earliest literary reference that may have something to do with glass-making is the legend of the female semi-creator, Nü-Kua, who 'fused minerals of five colours in order to repair the blue heavens'. It appears in a number of early sources, including the *Lieh Tzu* book compiled between the fourth century B.C. and the third century A.D., and in view of what is now known about glass-making in pre-Han times, it seems likely to be a garbled reference to glass. However, the first really suggestive literary evidence of glass-making occurs in the Later Han, in the *Lun Hêng* (Discourses weighed in the Balance) of Wang Chhung, while from the third century A.D. onwards references become more numerous.

In brief, archaeological and literary allusions indicate that almost from the middle of the first millennium B.C. onwards there was an indigenous Chinese glass industry, the roots of which lay in ancient Mesopotamia, running parallel with a considerable trade in imported glass wares of particular kinds and some special raw materials. Chinese glass manufacture seems to have been an obscure industry, often distinctly localised, so that here and there it had to be revived from time to time. But the chief question from our point of view is whether the Chinese could have had burning-glasses or other lenses from the Han onwards, and the balance of evidence seems clearly to be that they could.

Burning-glasses and the optical properties of lenses. In the *Lung Hêng* Wang Chhung, writing in A.D. 83, has three passages about an instrument for bringing the sun's rays to a focus. In the first he simply says that, at the height of summer, people 'liquefy and transform five minerals, casting therewith an instrument with which they can catch fire (from heaven)'. The second adds the information that this was done by Taoist technicians, but the third, contrasting natural and artificial products, is the most interesting of all.

> The Tribute of Yü (chapter of the *Shu Ching* (Historical Classic)) speaks of bluish jade and *lang-kan* (possibly ruby, agate or coral). These were the products of the earth, and genuine like jade and pearls. But now the Taoists melt and fuse five kinds of minerals and make 'jade' of five colours out of them. The lustre of these is not at all different from that of true jade. Similarly pearls from fishy oysters are like the bluish jade of the Tribute of Yü, all true and genuine (natural products). But by following proper timing (i.e. when to begin heating and how long to go on) pearls can be made from chemicals, just as brilliant as genuine ones...
>
> Now by means of a burning mirror one catches fire from heaven. Yet of five mineral substances liquefied and transmuted on a *ping-wu* day in the fifth month, an instrument is cast, which, when brightly polished and held up against the sun, brings down fire too, in precisely the same manner as when fire is caught in the proper way.

Can there be much doubt that this is an account of the making of glass burning-lenses? At one time the passages were taken to refer to casting bronze, but it is not obvious why, if this were the case, such care should be taken to specify five minerals. Bronze would require only two ores, perhaps even only one, with the possible addition of a flux. Glass needs silica, limestone, an alkaline carbonate, and perhaps lead oxide or barium, together with colouring matter. Of course the text does not clearly tell us that lenses were being made, yet the fact that the passage on burning-glasses follows immediately on the discussion of pearls is very suspicious, since the term 'fire-pearl' in later centuries undoubtedly meant a burning-lens. The whole quotation also fits in perfectly with what we have seen above concerning the substitution of glass for jade and bronze objects in tombs.

At the end of the third century A.D. there is a curious story of using ice as a burning-lens. In a section on conjuring tricks, the *Po Wu Chih* (Notes on the Investigation of Things) says:

> A piece of ice is cut into the shape of a round ball and held facing the sun. Mugwort tinder is held to receive the bright beam from the ice, and thus fire is produced. There has been much talk of getting fire by the use of pearls, but this (ice) method has not been employed.

Although ice can be used in this way, it seems more likely that the authors were really talking about a lens of rock-crystal or glass; indeed, there was a persistent theory in China that ice turned into rock-crystal after thousands of years. The first widespread mention of 'fire-pearls', however, comes in the Thang, when the largest is said to be the size of a hen's egg, 'white (transparent), and emitting light to a distance of several feet'. It is also said that the material 'looks like rock-crystal'.

It is at the period between the Thang and the Sung that one of the most interesting of all Chinese references to lenses is found. The *Hua Shu* (Book of Transformations), attributed to Than Chhiao, and datable at about A.D. 940, has a passage concerning four optical instruments. In the past this reference has been taken to refer to types of mirrors, but there are only three types of mirror – plane, convex and concave – whereas there are four types of lens – plano-concave with one side flat and one concave, bi-concave with both sides concave, plano-convex and bi-convex. If then we read the passage with lenses in mind we shall find it makes everything abundantly clear:

> I have always by me four lenses. The first is called *kuei* (the 'sceptre', a diverging bi-concave lens). The second is called *chu* (the 'pearl', bi-convex), the third is called *chih* (the 'whetstone', plano-concave). The fourth is called *yu* (the 'bowl', plano-convex).
> With *kuei* the object is larger (than the image).
> With *chu* the object is smaller (than the image).
> With *chih* the image appears upright.
> With *yu* the image appears inverted.
> When one looks at shapes or human forms through such instruments, one realises that there is no such thing as (absolute) largeness or smallness, beauty or ugliness...

Here there can be no doubt about the identification of *chu*, which was the old bi-convex burning-glass. *Chih* as plano-concave is strongly suggested by the shape of the traditional Chinese whetstone, which is not a wheel but a plate of stone held upright in supports, its upper surface becoming concave by the continual grinding. *Yu* could well be a solid glass hemi-

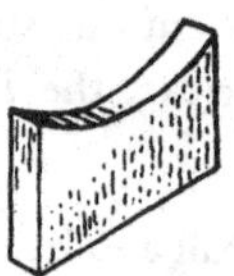

sphere and *kuei* takes the remaining place. One only wishes that a fuller account of Than Chhiao's experiments had come down to us.

The later development of optics in China has been inadequately investigated, but there can be no doubt that one must envisage a much more widespread use of glass and rock-crystal lenses in medieval times than has generally been supposed. When the Jesuits came in the early seventeenth century they brought much new optical knowledge, including accounts of Galileo's discoveries with the telescope, and published several tractates on the subject. But it looks as if there had been an indigenous development also, for two outstanding 'optick artists' are known to have been working in Suchow just at that time, Po Yü and Sun Yün-Chhiu; they made magnifying glasses, magic lanterns, searchlights, telescopes, simple and compound microscopes, kaleidoscopes, and many different kinds of mirrors. Indeed Po Yü may have been one of the several independent inventors who played with combinations of lenses and so stumbled upon the telescope (Lippershey, Digges, della Porta, etc.). What is certain is that he was the first to use the telescope as a gunnery sighting device for artillerists; this was in 1635. Then in the early nineteenth century came modern treatises such as the *Ching Ching Ling Chhih* of Chêng Fu-Kuang, and the *Kuang Lun* of Chang Fu-Hsi.

Eye-glasses and spectacles. It has sometimes been said that the invention of spectacles was Chinese, but careful study has shown that the literary evidence for this was a corrupt text. Spectacles probably arrived in China, as did cotton, overland from the north as well as through maritime contacts in the south, and this seems to have been very soon after their first invention in Europe about A.D. 1286. The Sung people, however, did have two techniques which may be considered introductory to spectacle lenses; one was the magnifying glass, and the other dark glasses as eye-protection. Liu Chhi wrote, in his *Hsia Jih Chi* (Records of Leisure Hours) some time before his death in A.D. 1117, that his contemporary, Shih Khang, and other judges, used various magnifying lenses of rock-crystal for deciphering illegible documents in legal cases. The judges also used dark glasses made of smoky quartz, not, as we do, to protect our eyes from the sun, but to disguise from litigants their reactions to the evidence. However, blank spectacles with slits were used from early times as snow-glasses by Tibetans and Mongolians, and the Chinese made use of these too.

Shadow-play and zoetrope. A celebrated incident in Chinese history concerns the evocation, by the magician Shao Ong, of a moving image of one of the dead concubines of the emperor Wu of the Han, in 121 B.C. The story is told both in the *Shih Chi* (Historical Records) and the *Chhien Han Shu*

(History of the Former Han Dynasty). The latter text runs as follows:

> (After the death of Li Fu Jen) the emperor could not stop thinking of her. Shao Ong, a magician from Chhi, said that he could cause her spirit to appear. So after certain offerings of wine and meat had been set forth, and when certain lamps and candles had been disposed about a curtain, the emperor took his place behind another (diaphanous) curtain. After a time he saw indeed at a distance a beautiful girl sitting down and walking back and forth. But he could not approach her . . .

This was always remembered in later times, and there are quite circumstantial accounts of the same kind of illusion being produced in the Thang. It is probable that the Han feat was simply one of those shadow-plays which have been traditional for centuries in many Asian countries. But bearing in mind Chinese optical knowledge in Thang times, it is just possible that someone had the idea of placing one or more lenses at the pinhole of a closed chamber. This was, of course, the new element in the invention of the magic lantern in 1630. Yet it seems more probable that the Thang continued what Shao Ong had started.

Another ancestor of the cinema was a variety of zoetrope, which may well have originated in China. This took the form of a light canopy hung over a lamp, and carrying vanes at the top arranged so that the ascending convection current of air would cause it to rotate. On the sides of the cylinder forming the lower part of the canopy there would be thin panes of paper or mica having pictures painted on them which, if the cylinder rotated fast enough, would give an impression of the movement of animals or men. Such a device certainly embodied the principle of a rapid succession of separate images, which may also have been the basis of representations of gods and goddesses with attributes held by many limbs. Hence the 'eleven-armed' or 'thousand-armed' Kuan-Yin (Fig. 160).

The twelfth-century Sung scholars Fan Chhêng-Ta and Chiang Khuei wrote poems describing how the 'horse riding lamp' showed shadow-horses prancing round after the lamp is lit. One of the Jesuit fathers Gabriel Magalhaens was delighted with the thing. Modern writers have described toys of a similar kind made in contemporary China, and in 1900 in his *Yen-ching Sui Shih Chi* (Annual Customs and Festivals of Peking), Tun Li-Chhen wrote:

> Pacing-horse lamps are wheels cut out of paper, so that when they are blown on by (the warm air rising from) a candle (fastened below the wheel), the carts and horses (painted on it) move and run round without stopping. . .

Fig. 160. Image of Kuan-Yin of the four Cardinal Points and the Thousand Arms, in the Chieh-thung Ssu temple, Chiangsu. Flashing action and infinite vigour symbolised by the multiplication of limbs and attributes. (Photograph, Boerschmann.)

It is usually claimed that the zoetrope was invented in the nineteenth century in the West, though it had in fact been described by John Bate as early as 1634, but it seems that neither Western writers nor Chinese authors like Tun Li-Chhen made sufficient allowance for the ancient mechanicians of genius, such as Ting Huan of the Han. His helicopter top was the object which awakened Sir George Cayley, the father of modern aerodynamics; and his gimbal suspension is still attributed to Jerome Cardan. There has been too much self-assurance on one side and too much diffidence on the other; only the balance of history could ultimately show that Europe had as much leeway to make up in the tenth century as China had at the beginning of the twentieth.

SOUND (ACOUSTICS)

The term acoustics may be defined either broadly to cover the nature of sound in general, as a branch of physics; or more narrowly to denote our knowledge of sound as applied to the properties of halls and buildings. It is taken here in the broad sense, describing not only the achievements of the Chinese in this field, but also their attitudes to acoustic phenomena in ancient and medieval times. The whole subject is of particular interest in the history of science because it was one of the earliest fields, both in East and West, where careful measurements were applied to Nature.

In acoustics, as in so many branches of science, the Chinese approach was rather different from the European. Where ancient Greece analysed, ancient China looked into relationships. Thus a Greek author like Plutarch of the first century A.D. could ask:

> why the narrower of two pipes of the same length should speak (sharper and the wider) flatter? Why, if you raise the pipe, all its notes will be sharp; and flat again if you stoop it? And why, when clapt to another, it will sound the flatter; and sharper again when taken from it? ...

But Tung Chung-Shu, for example, in the second century B.C. – one of the most scientific and philosophical minds of his age – was confronted by the much more striking experience of sympathetic resonance, whereby one note struck on one instrument causes another instrument to sound the same note. He accepted it simply as being 'nothing miraculous', since it accorded so well with the typically Chinese view of the world as a single organic whole.

> Try tuning musical instruments ... The *kung* note or the *shang* note struck upon one lute will be answered by the *kung* or *shang* notes from the other stringed instruments. They sound by themselves. This is nothing miraculous, but the Five Notes being in relation; they

are what they are according to the Numbers (whereby the world is constructed).

In China we have to deal with two distinct currents, the literary tradition of the scholars, and the oral tradition of the craftsmen who were expert in acoustics and music. From what follows it will be seen that the latter must have done a great deal of experimentation, asking questions quite parallel to those asked by the Greeks – but the details were only rarely recorded. Nevertheless, Chinese interest in sound, though it followed a different course from that of the Greeks, was by no means fruitless.

The relationship of sound with flavour and colour

Few peoples ancient or modern have proved themselves more sensitive than the Chinese to the timbre of musical sounds. For instance, there are sixteen different 'touches' in playing on the silk strings of the classical lute (*ku chhin*), as well as other methods of striking and plucking them. To take one example only, the vibrato on a string could take more than ten forms, so suggesting the infinite subtlety with which any given note could be played. Indeed, even today an expert *chhin* player will himself remain intently listening long after a note has become inaudible to other listeners. As Taoist thought put it 'The greatest music has the most tenuous notes'; and indeed the playing of the *chhin* mainly depends on exploiting the production of different timbres at the same pitch. This technique was developed to perfection by the Later Sung (seventh century A.D.).

What then did the early Chinese believe sound and the subtleties of musical notes to be? The Greeks had various ideas: the Pythagoreans thought of it as the reality of number *par excellence*, others derived a relationship between speed and sound, or even suggested that sound was speed itself. In ancient China, on the contrary, no such analysis or abstraction was made. Sound was regarded as one form of an activity of which flavour and colour were others. The background of this acoustic thinking was largely determined by an idea which stemmed from the vapours of the cooking pot, with its fragrant steam, for which the word was *chhi*. This is a word the use of which in acoustics has no English equivalent; it can mean 'vapour, air, breath, vital principle, to present food, to pray, beg or ask', but the common context is that of sacrifice to ancestors. They are prayed in the *Shih Ching* (Book of Odes) of the ninth to fifth centuries B.C. to return and invigorate their descendants and their crops:

> Sonorous are the bells and drums. Brightly sound the stone-chimes and flutes.
> They bring down with them blessings – rich, rich the growth of grain!
> They bring down with them blessings – abundance, abundance!

The ancestors are tempted to return to earth not only by the prayers of their descendants chanting liturgical phrases, but by the sounding of musical instruments and the delicious emanations which rise up from the magnificent bronze cooking-vessels. When they arrive their eyes are also feasted with the sight of an assembly dressed in ceremonial clothing, furs and emblems all conforming to traditional themes in colour.

From the earliest historical periods the Chinese were concerned with a synthesis of sound, colour and flavour, responding to the synthesis of Nature manifested in thunders, rainbows and spicy herbs. One *chhi* rises up from earth to heaven like steam from cooking-pots symbolising the prayers of the living; another descends from heaven to earth, like rain and dew, symbolising the invigorating influence of the ancestors. Their intermingling produces wind, wherewith heaven makes music, and brings into being not only rainbows, which are heaven's colours, but the flowers of the changing year and with them the flavouring herbs in due season. All were signs and symbols of the great climatic processes on which the life of the ancient Chinese people depended, balancing ever between flood and drought. Such was the environment which brought forth their philosophy of the whole of Nature as an organism. A purely analytic treatment of sound would hardly have been consistent with this.

Chhi *and its relation to acoustics*

Chhi, then, had two main sources. It could go up from earth to the ancestors, and it could come down from heaven with the ancestors to earth. But there was a third, very important source, and that was the breath of man himself although, with increasing sophistication, *chhi* was thought of as something more rarefied than steam or breath. It becomes an emanation, a spirit. Thus we read in the *Tso Chuan* (Master Tsochhiu's Enlargement of the Spring and Autumn Annals), compiled between 430 and 250 B.C.: 'There are the six *chhi* of heaven. Their incorporation produces the five flavours; their blossoming makes the five colours; they proclaim themselves in the five notes'. Sometimes *chhi* is used in a general way for the emanation which goes up to and comes down from heaven, and sometimes for a particular form of its descent. Elsewhere in the *Tso Chuan* it is stated that the *chhi* themselves make the five flavours. In a later volume we shall see how important were these 'sapidities' in pharmacology and medicine. It is quite hard to know what is implied in 'descent', but the term 'six channels' is sometimes used in place of the six *chhi*. This suggests a connection which is important for early Chinese acoustic theories, for if *chhi* is something which can be canalised or piped off, the obvious instrument for the purpose would be a bamboo tube, such as was used in China for irrigation. Consequently it is not surprising to find early references to the shaman-musician piping off his own *chhi* through bamboo tubes in an attempt to alter the processes of Nature – of heaven's *chhi* – by sympathetic magic. Should we not see a rather late echo of this practice in the story of Tsou Yen (fourth century B.C.) blowing on his pitch-pipes for the benefit of the crops?

All this should make it clear that the acoustics of the early Chinese was highly 'pneumatic'. Yet it must not be forgotten that they thought of *chhi* as something between what we should now call matter in the form of a rarefied gas, and radiant energy.

Channels for chhi; *the military diviner and his humming tubes.* The use of hollow tubes, bones or branches as speaking trumpets for disguising or amplifying the voice of the shaman or witch doctor is widespread among primitive peoples. That it should occur in China is not remarkable. What is remarkable is that the Chinese should have attempted in this, as in so many other of their activities, to reduce the practice to a clearly regulated and classified system.

In a lost 'Book of War', quoted in the Thang by Chang Shou-Chieh, reference is made to five different states of morale, all of which can be known by a skilful diviner. It is claimed that every man has within his body his own *chhi* and that, in any army where men are massed together, there is a collective *chhi* which floats above it, and can be seen as a coloured cloud. The diviner uses his *chhi* to react with the outside world when he blows through his humming tube, and since one *chhi* will then react on another *chhi* by a kind of mysterious resonance, so the diviner can use the collective *chhi* of an enemy army to discover the enemy's morale. As the Thang scholar Ssuma Chên commented:

> Above every enemy in battle array there exists a vapour-cloud. If the *chhi* is strong, the sound (note) is strong. If the note is strong, his host is unyielding. The pitch-pipe (or humming-tube) is (the instrument) by which one canalises (or communicates with) the *chhi*, and thus may foreknow good or evil fortune.

There is a certain reasonable basis for this strange belief. It is probable that in the tense moments before a battle, the shaman might from anxiety or excitement fail to emit his small jet of breath properly so that fluctuating, feeble or dead notes would result. This, then, would provide the basis for divination, since the variation in the sounds, though arising from the state of morale of the shaman's own side, could be attributed, because of 'resonance', to the *chhi* of the enemy.

Later a more detailed appraisal of enemy morale was attempted by a five-fold division of sound, the terms for which were *kung*, *shang*, *chio*, *chih* and *yü*. In works of the fourth century B.C., these are the names of the five notes of the pentatonic or five-note scale (a major scale with the fourth and seventh omitted), but earlier they had another meaning as well. This alternative did not use the terms to refer to the pitch of the notes but rather referred to their quality or timbre.

Classification of sound by timbre

We have so far seen that early Chinese ideas of sound were based on the idea of *chhi*, and this view persisted until comparatively recent times. But their ideas on sounds did progress, from the early stage where they boded good or ill, to an exact appreciation of how one sound differed from another in timbre, volume and pitch.

Today we regard timbre as that which distinguishes one note from another by overtones, so giving the note a certain quality. Thus the timbre of the same note on a flute or a clarinet differs because the strengths of the overtones of the note differ for each instrument. The ancient Chinese music-masters would not have expressed themselves in this way for the different elements which make up a sound were probably not thought of in isolation; nevertheless, timbre was very important for them.

Material sources of sound. The Chinese grouped instruments into 'the eight (sources of) sound' but the complete classification was only arrived at gradually. Thus, although the *Yo Chi* (Record of Ritual Music and Dance) compiled from Chou sources (late sixth century A.D.) lists eight instruments – bells, drums, pipes, flutes, ringing-stones, feather (wands or dresses), shields, and axes – yet these contain only four 'sources of sound', namely metal, skin, bamboo and stone (Fig. 161). Indeed, the book goes on to state that musical instruments comprise four sources of sound – metal, stone, silk and bamboo. From this and other references it seems clear there was a period in which the 'eight sources' were not yet classified and settled. References to the eight winds, on the other hand, are frequent in the far earlier *Tso Chuan*, and these also affected sound classification.

Winds and dances. The annual climatic cycle on which life in China depended is specifically related to mimes and music in the *Yo Chi*, which states that former kings arranged these as a form of sympathetic magic to ensure that the seasons would arrive on time. The dances in the ritual mimes fell into two classes, warlike and peaceful, the latter including beast dances and rain-making dances, and there are many accounts in early texts of grand performances of music, with bells, stone-chimes, etc. followed by a great wind or storm with thunder and rain.

The winds were eight in number, one from each compass point and one from each intermediate point. Thus the Later Han commentator Chêng

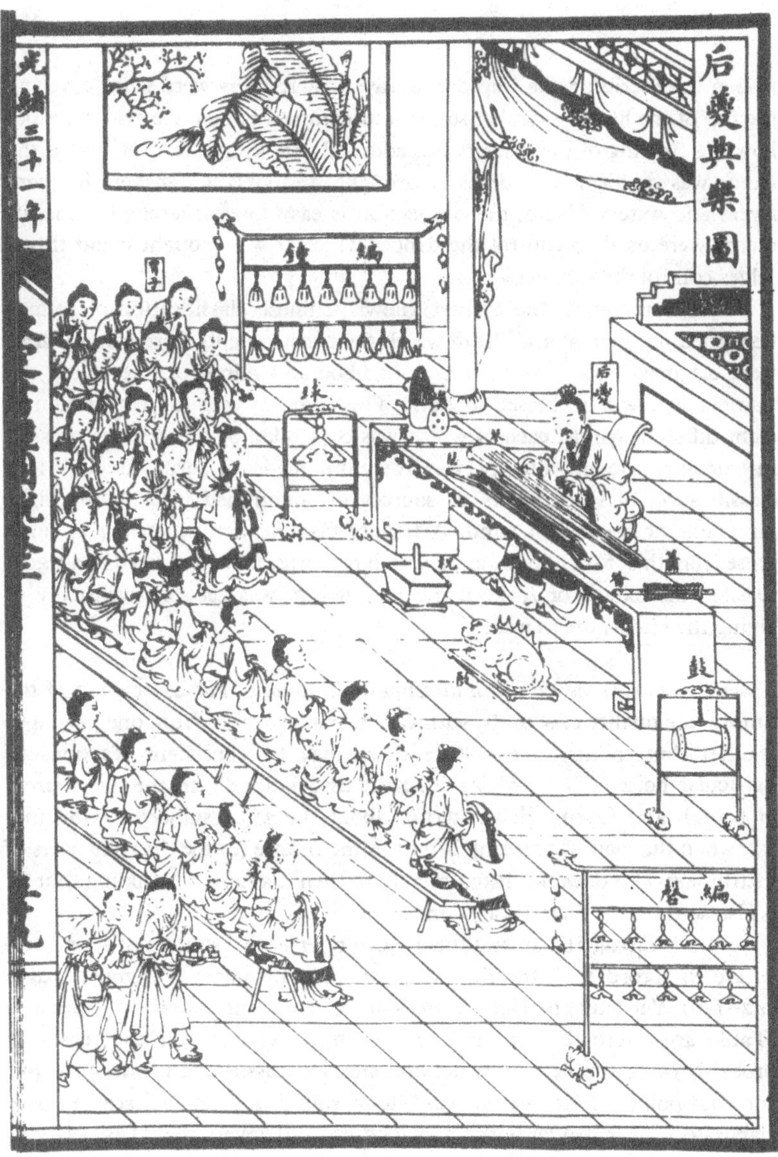

Fig. 161. A late Chhing representation of the instruction of musicians by the legendary music-master Hou Khuei. Before him, on the table, are the zithers *chhin* and *sê*; to his right the reed-organ (*shêng*), the globular flute (*hsüan*) and the transversely blown straight flute (*chhih*); to his left the pan-pipes (*hsiao*) and the vertically blown bamboo flute open at both ends (*yo*). In the background the stand of bells and the great chime-stones; in the foreground the stand of chime-stones and the standing-drum. In front of Khuei's table, on the ground, are the percussion tub (*chu*) and the tiger-box (*yü*). Tea is being served. From *Shu Ching Thu Shuo*.

Hsüan in describing the 'cap dance' says that feathers were worn covering the top of the head to symbolise the four compass points. He also describes how the clothing of the dancers was adorned with kingfisher feathers; as the dance was performed in times of drought, and since kingfishers frequent rivers and watery places, the connection is easy to see. Percussion instruments were used to control the dances which it was thought could themselves control the weather.

To summarise, the earliest known Chinese classification of sounds depended on the materials from which their instruments were made. These were originally four – stone, metal, bamboo and skin or leather – but the number was later increased to eight. This fitted in with the recognition of eight different winds, each of which was controlled by a particular musical instrument, and an attempt was therefore made to form an eight-fold classification of instruments or sources of sound tied in with the eight directions or sources of wind. A further classification according to pitch arose from this, for an instrument which had originally been used as a signal to start the dance, or to keep the beat going, was capable eventually of giving the pitch-note as well.

Timbre and its connection with directions and seasons. The association of the timbre, the quality of sounds, with various moods varies from one person to another, but in any one culture there may be consistent reactions to particular noises. The *Yo Chi* describes associations with the five sources of sound, and Chêng Hsüan writes about the same subject, saying that '... when the man of breeding listens to the timbre (of the different sorts of instruments) he does not listen merely to their clanging or tinkling, but he is also sensitive to their associations'.

Both Chêng Hsüan and the *Yo Chi* refer to five sources of sound and so are in accord with the theory of the Five Elements (volume 1, pages 142–159). Yet later the eight sources are given in most works of reference (Table 40). Here the seasons seem strangely erratic, but the reason is probably quite simple. An earlier and simpler classification in fours – four compass-points, four seasons and four sound sources – became more elaborate as the calendar with its agricultural and governmental associations developed, and it is this which the table echoes.

Classification of sound by pitch

To describe the different pitches of sound, the Romans used 'cutting' or 'sharpened' for those at one end of the range, and 'heavy' for those at the other; English uses a consistent metaphor based on a scale or ladder on which the range goes from 'low' to 'high'. The Chinese used a metaphor which is not surprising for a people whose economy was so bound up with thought of an analogy of a court and its officials in which the notes are

Table 40. *Traditional list of the eight sources of sound*

Source of sound	Compass-point	Season	Instrument
1. Stone	North-west	Autumn–Winter	Ringing-stone
2. Metal	West	Autumn	Bell
3. Silk	South	Summer	Lute or zither
4. Bamboo	East	Spring	Flute and pipe
5. Wood	South-east	Spring–Summer	Tiger-box (*yü*)*
6. Skin	North	Winter	Drum
7. Gourd	North-east	Winter–Spring	Reed-organ (*shêng*)
8. Earth	South-west	Summer–Autumn	Globular flute

*The tiger box is a hollowed block of wood shaped to resemble a tiger with a serrated back; it gives a rasping noise when brushed or struck smartly with a stick split at its end into twelve leaves.

hydraulic engineering, namely clear (*chhing*) and muddy (*cho*). But there is no such simplicity when it comes to the words for the original five notes of the Chinese scale (page 367), although they seem, possibly, to have referred once to the positions of the instruments used in controlling the music and dancing in a mime. Indeed several literary references suggest that rather than a ladder stretching from low to high, as in the West, the Chinese arranged either side of the chief or *kung* note. As the commentator of the *Huai Nan Tzu* (The Book of (the Prince of) Huai-Nan) succinctly put it: 'The *kung* note is in the middle; therefore it acts as lord.'

The development of musical acoustics

The pentatonic scale. With the recognition of pitch intervals and the naming of notes, accurate measurement, observation and test become possible, and the science of acoustics has been born. One cannot say precisely when the Chinese first gave names to their notes, but the *Tso Chuan*, in passages which seem to be from the fourth century B.C., contains references to the fact that there were five notes in the scale. On the other hand, the notes are not named. By 300 B.C., however, no doubt can remain: the names *kung*, *shang*, *chio*, *chih* and *yü* were being used to distinguish different notes on stringed instruments, as the *Erh Ya* (Literary Expositor) encyclopaedia makes clear. A century and a half later, we find Tung Chung-Shu naming notes when referring to sympathetic resonance (page 364).

There was doubtless a differentiation of pitches before the fourth century B.C., and these may well have been different terms for use with different instruments, a flute-player teaching a pupil tunes by naming the finger-holes on the flute, and a lute-player naming the different strings.

Indeed, according to Ssuma Chhien in about 100 B.C. there was some form of notation for stringed instruments as early as the sixth century B.C., for Duke Ling is said to have commanded his music-master Chüan to write down an important tune. Still later (about 150 B.C.) there comes the work of Tung Chung-Shu, already mentioned, who referred to instruments being tuned to certain notes such as *kung* and *shang*. It is worth noting also that, at this time, the Chinese were tuning even their drums and noting sympathetic resonance when they were struck, while in Europe as late as the sixteenth century A.D. drums were still untuned, being described, for instance, as 'rumbling tubs'. Certainly by 120 B.C. the *Huai Nan Tzu* book is in a position to give an explicit statement not only mentioning the five named notes, but also stating that in combination with the twelve absolute pitches sixty 'mode-keys' can be formed.

The heptatonic scale and later elaborations. A pentatonic scale was thus in use in China in the fourth century B.C., but there is also a tradition that heptatonic (seven note) music was invented by the Duke of Chou. According to Chêng Hsüan, writing in the second century A.D., the five notes (using modern notation) were CDEGA, and the seven notes were CDEF GAB (if we pitch the *kung* note on middle C). Whether, in fact, truly heptatonic music was played in Chou times cannot be known for certain, yet more than a hint is contained in many references to a 'New Music'.

The twelve-note series and the set of standard bells. The evolution of the Chinese theory of sound passes from the formation of scales in relative pitch to that of a series of notes of fixed or absolute pitch. This Chinese gamut of twelve notes is intimately associated with the history of Chinese bells, which had a double use in their early orchestras – for giving the pitch and for starting the music. By the time of the Chou period at the latest, things had advanced beyond the stage of striking lumps of ringing-stone or slabs of bronze which gave out the desired note by chance, and the Chinese were producing bells which they were able to tune with accuracy. Their range of twelve notes, known as the twelve *lü*, did not form what today we should call a scale, but were simply a series of fundamental notes from which scales could be constructed.

It is not possible to say when the process of standardisation of bells was first completed, but the earliest literary reference to the full set of twelve bells may be that in the *Kuo Yü* (Discourses on the (ancient feudal) States), where they are mentioned in a discussion said to have taken place in the year 521 B.C., though a reference in the *Yüeh Ling* (Monthly Ordinances of the Chou Dynasty) may be earlier. Like bronze mirrors, bells were regarded in Chou times as instruments of high magical potency, and they were divided into two groups, Yin and Yang. The *Kuo Yü* lists them as in

Table 41. *Classification of bells in the* Kuo Yü

Yang bells		Yin bells	
Huang-chung	'yellow bell'	*Ta-lü*	'great regulator'
Ta-tshou	'great budding'	*Chia-chung*	'compressed bell'
Ku-hsien	'old and purified'	*Chung-lü*	'mean regulator'
Jui-pin	'luxuriant'	*Lin-chung*	'forest bell'
I-tsê	'equalising rule'	*Nan-lü*	'southern regulator'
Wu-yi	'tireless'	*Ying-chung*	'resonating bell'

Table 41. The names of these bells, once established, became the names of the twelve notes which formed the classical Chinese gamut or series of notes.

The introduction of the arithmetical cycle. In tracing the evolution of this series, three stages have been mentioned so far. First, there was the primitive stage preserved in the *Chou Li* (Records of the Institutions of the Chou Dynasty), in which the notes had names, though some of them differed from those ultimately adopted. Secondly we have the twelve bells listed in the *Kuo Yü*, also divided into sixes, and here all the names agree with those of the ultimate orthodox gamut of notes. We cannot, however, say anything positive about the intervals between these notes or the pitches (frequencies) of the notes themselves. And thirdly, from evidence in the *Lü Shih Chhun Chhiu* (Master Lü's Spring and Autumn Annals) which appeared about 240 B.C., it is clear that a new stage had been reached, for although the pitches still remain unknown, we can see how the series of notes was obtained.

The easiest way to appreciate the elegant simplicity of the Chinese mothod is to glance first at the Greek method – the so-called Pythagorean scale – as this is the one from which our musical scales were originally derived. The Greeks used the lyre and cithara, both stringed instruments, as the guides to their scales, and the framework of their system was the octave, that interval to which the outer strings of the lyre were tuned. The two remaining strings were tuned to a fourth and fifth. This was done by ear until, in the sixth century B.C., the discovery attributed to Pythagoras was made, concerning the lengths of strings, and how lengths of a half, two-thirds and three-quarters of a freely vibrating string were needed to give the octave, the fifth and the fourth. A musical scale based on such mathematical relationships was then devised by the Pythagoreans, although another Greek school still favoured tuning by ear.

The Chinese gamut of pitches, on the other hand, requires only the simplest mathematics and does not use the octave as the starting point.

Indeed it does not even include an octave at all. The Chinese simply took a fundamental note and then multiplied by two-thirds and three-quarters alternately, thus embarking on a process which evolves an unending spiral of notes. Our oldest source for any actual lengths is the *Shih Chi* (Historical Records) of Ssuma Chhien, written about 90 B.C., and the manner in which these are expressed – by use of a decimal system in conjunction with a system based on thirds – has a distinctly Babylonian flavour. Indeed, certain common features in the Chinese and Pythagorean methods of drawing up scales make a common origin seem likely; the Babylonians, with their highly developed stringed instruments, may well have been this source.

The search for accuracy in tuning

The discovery that the intervals between musical notes are determined by mathematics put the art of tuning on an entirely new basis. In China the musician or scholar with almost miraculous ability to detect small differences in tone was still revered, but the limitations of the ear were recognised nonetheless. A stringed instrument of some kind was needed for precise tuning.

Resonance phenomena and the use of measured strings. There is evidence that in Chou times the Chinese had such an instrument. Commenting on the method of tuning bells in the *Kuo Yü*, Wei Chao (third century A.D.) states that 'a board seven feet (2·1 m) long (was used) having a string (or strings). They fixed them and so tuned.' Although his description is far from clear, he states that the Office of the Grand Revealed Music of the Han had a *chün* or 'tuner', and the great length of the instrument to which he refers is interesting in itself. A long string would emit a good loud tone suitable for producing a sympathetic tone in a bell, and its great length would also make possible the accurate division of the string.

Tuning bells by resonance was certainly done, and there is an interesting note on this in the *Chin Hou Lüeh Chi* (Brief Records set down after the (Western) Chin (Dynasty)) of about A.D. 317.

> Using the standard pitches they gave them their summons, and all (the bells) responded though they had not been struck. The notes and the sympathetic tones agreed and became one.

The cosmic tide in buried tubes. Tuning bells by resonance was, in a sense, straightforward enough, but how to verify that the tubes of flutes and similar instruments were of the correct length presented a great problem. However, as we have seen, bamboo tubes were used for canalising *chhi*, and one of the great manifestations of *chhi* was wind. Moreover the winds of the eight directions were summoned each by its appropriate magical dance, led off by a note from an instrument made from one of the eight sources of sound. There was, therefore, a clear relationship between notes, winds and directions. Probably no one was ever so simple as to hope that if bamboo tubes were pointed in the right direction the appropriate wind would blow through them and sound the right note. But some ancient nature-philosophers set out to trap the *chhi* another way, combining the *chhi* which rose up from the earth with the *chhi* which descended from heaven to produce the different types of wind which blew at different seasons of the year. Thus arose the strange practice of *chhui hui* (the blowing of the ashes).

This technique is referred to by the philosopher Tshai Yuan-Ting (A.D. 1135 to 1198), an expert in acoustics and music. In 1180 he wrote:

> The (pitch-pipes) are blown in order to examine their tones, and set forth (in the ground) in order to observe (the coming of) the *chhi*. Both (these techniques) seek to (determine the correctness of the) Huang-chung tube by testing whether its tone is high or low, and whether its *chhi* (arrives) early or late. Such were the ideas of the ancients concerning the making (of the pitch-pipes)...

and goes on to explain that if one has no standard note available, pipes of different lengths are cut and that with the correct tone is chosen before being buried, for the choice to be verified. In other words, Tshai Yuan-Ting refers to an experimental approach for obtaining a standard pipe from which others may be tuned. 'Later generations', he says, 'have sought (to construct accurate pitch-pipes) only by measuring with the foot-rule'.

The practice of *chhui hui* was to place the pitch-pipes in a circle, each one pointing in the direction of a compass point. They were held in stands so that one end was low, the other high, and the upper end stuffed with ashes. The single-roomed building in which this took place was sealed against draughts. Then when the *chhi* for a given month arrived, the ashes of the appropriate pipe were supposed to be blown out. Later, modifications like burying the tubes in the ground with only their upper ends protruding, were tried, for the procedure was clearly not entirely satisfactory, and in due course a sceptical attitude developed to the whole idea. Yet semi-magical though this technique was, it led finally to the construction of standard pitch-pipes of prescribed lengths and, from the Han onwards, these were used as the orthodox method for giving pitch to other instruments.

Tuning by means of vessels containing water. This was another method for tuning. The Chinese, like the Greeks in Alexandria, noticed that different notes sounded when vessels were filled with varying amounts of water, and in China this phenomenon was fully exploited. One of the earliest accounts is given about A.D. 320 by Kan Pao, using the *chhun* (Fig. 162).

> Water is filled in (to the *chhun*) to a height of one foot above the ground, and a container is filled with water and put underneath. The *mang* (an apparatus on which strings were set) is placed between them. If the *mang* is shaken by hand, a tremendous noise like thunder is produced.

In Thang times bowls of water were used for tuning and a set of vessels set aside for this became a customary method. Perhaps it was partly derived from the use of water for testing the measurements of standard vessels of capacity.

The manufacture and tuning of bells. The tradition of bell-making in China is very old, and the part which bells played there in music and in tuning instruments was of great importance. The art of the Chinese bell-founders, therefore, deserves careful study, and the way the bell developed is a significant part of this.

The small hand-bell *to* (Fig. 163) was probably the ancestor of all Chinese bells. In the earliest examples the diameter of the barrel was greater than its length and the *to* was normally held with mouth uppermost. When the mouth began to point downwards, the hanging bells called *chung* (Fig. 164) came into being. The bells were struck with a stick or hammer to make them sound, for bells with clappers (generic name *ling*) only came later. Such developments, including a clapper, were accomplished in China by the fifth century B.C., but to turn the bell into a musical instrument was a different matter: it had to be properly tuned, and this is regarded, even today, as a highly intricate matter. That is because a bell's pitch and tone-quality depends upon the nature and proportions of the metals used, the precise inner and outer contours of the bell, the distance between them, and the amount of metal needed to fill them. The temperature of pouring the metal and the rate at which it is allowed to cool are other factors. Tuning has to be achieved by removing small amounts of metal so as to thin it slightly in certain places. This corrects the fundamental note of the bell and brings several partial tones into a pleasing consonant relationship with it.

It is clear from the description in the *Chou Li*, even if they represent the views of Han scholars rather than the traditions of actual foundry-men, that the Chinese in early times were at least aware of the many factors involved in the tuning of bells.

> Thinness and thickness, that is what produces vibration and throbbing (respectively); purity or impurity (of the metal), that is what (causes the sound) to proceed outwards (i.e. from the vibrating walls

Fig. 162. The *chhun*, a bronze bell of elliptical section, wide at the mouth and narrowing towards the round base, suspended by a loop, usually, as here, in the form of a tiger. Like the *ling*, it had a tongue, but hung from a cross-bar as the *chhun* was open upwards. *Chhun* filled with various amounts of water were sometimes used for tuning purposes.

> of the bells themselves); the open or closed (form of the mouth), that is what (causes the sound) to proceed upwards...

These observations seem reasonable, and elsewhere in the *Chou Li* there is evidence that the foundry-men had their own technical terms and much specialised knowledge. This comes in the section on the twelve different types of sound from a bell:

> The sound (produced in) the upper part (of the bell) is rumbling (*kun*).
> The sound (produced in) the straight part (of the bell) is slow (*huan*).
> The sound (produced in) the lower part (of the bell) is spreading (*ssu*).
> The sound (produced by) the parts which curve outward is scattered (*san*).

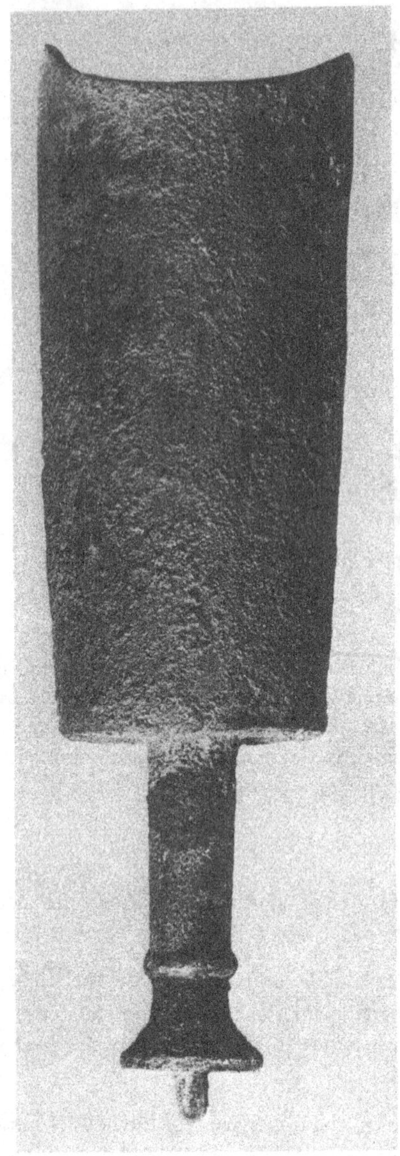

Fig. 163. Clapperless upward-facing hand-bell (*to*). Chou period. Winkworth Collection (photograph, Koop). Height 43 cm.

Fig. 164. Clapperless downward-facing hand-bell (*chung*). Chou period, perhaps as early as the sixth century B.C. It bears an inscription saying 'The Elder of Hsing in Ting (district) has made this bell, named Mysterious Harmony, with the note Jui-pin, for use'. Victoria and Albert Museum (photograph, Koop). Height 56 cm.

> The sound (produced by) the parts which curve inward is hoarded (*lien*)...
> The sound (produced by) thin (walls) is a staccato shaking (*chen*).
> The sound (produced by) thick (walls) is like stone (*shih*).

There seems little doubt that this text contains an analysis of the factors which interested the Chinese over two thousand years ago in the production of suitable partials from bells, especially when we remember that even now timbre is not an exact science: it can still only be described by using a metaphor, e.g. 'rich', 'thin', etc.

Pitch-pipes, millet-grains and measurement

While other early civilisations concerned themselves with standards of length, capacity and weight when drawing up their systems of measurement, the Chinese were apparently unique in including measures of pitch (*lü*), and that not merely on a par with, but as the basis of, the other three. The *Kuo Yü* describes this as follows:

> For this reason the ancient kings took as their standard the *chung* vessel, (and decreed that) the 'size' of its pitch should not exceed that (produced by the string) of the *chün* (seven-foot tuner) and that its weight should not exceed a stone (*tan*) (120 catties). The measures of pitch, length, capacity and weight originate in this (standard vessel).

The standard measuring vessel, the *chung*, we have met already – the word can also mean a wine bowl, a grain measure and, more usually, a bell. We can indeed think of the simple grain scoop as having evolved into a bell, and the simple bell into a standardised measure of fixed dimensions, capacity and weight, as well as musical pitch. When pipes became the standard pitch-givers, it was natural that they should inherit the measuring functions which had at one time belonged to the bells. Consequently we read of the number of grains of millet which the Huang-chung pipe ought properly to contain. It has sometimes been supposed that precise numbers of millet-grains governed the length and capacity of the Huang-chung tube and thus checked its pitch. Though this may have been so in and after the Han, it is quite contrary to the earlier doctrine that the *lü* are the basis of all other measurements; for it is the *lü* which gave the lengths to the string tuner, which in turn gave the standard pitches. The standard measure, the *chung*, had to emit the Huang-chung note. Nevertheless, the use of millet-grains in the reverse role of checking measuring instruments focused thought on the relationship between length and diameter in pitch-pipes, and is highly relevant in the construction of acoustic theory.

The chief point of interest in the use of cereal grains is that it indicates an increasing awareness of the need for accuracy. The old measures based on the human body, such as the foot, or the inch measured from the pulse in the wrist to the base of the thumb, were obviously not sufficiently accurate for measurements designed to achieve exact pitch.

The recognition of sound as vibration

With the arrival of millet-grain counting as a measure of volumes we enter a phase in Chinese acoustic development which can properly be regarded as scientific. In the West, in the Roman Empire contemporary with the Han, Vitruvius (about 27 B.C.) gives a good deal of acoustical information, including something on the nature of sound. '(The voice)', he says, 'is moved in an infinite number of undulating circles similar to those generated in standing water if a stone is cast into it, when we see innumerable rings spread forth from the centre and travel out as far as they possibly can...' He speaks, too, of sound being somewhat of the nature of a blow on the taut membrane of a drum.

The Chinese also thought of sound in terms deriving from the observation of waves in a liquid, though distinct statements of this analogy are rare before the eighth century A.D. However, the following striking passage from the *Chhun Chhiu Fan Lu* (String of Pearls on the *Spring and Autumn Annals*) of Tung Chung-Shu (second century B.C.) shows him boldly applying a concept of radiating waves to all substances, however viscous they are, including the aetheric *chhi*.

> Man's (activity) brings about the growth of the ten thousand things below, and unites him with Heaven and Earth above. Thus it is that in accordance with his good government or disorderly, the *chhi* of movement or rest, of compliance or contrariness, act either to diminish or increase the transformations of the Yin and Yang, and to agitate everything within the Four Seas. Even in the case of things difficult to understand, such as the spiritual, it cannot be said to be otherwise. Thus then, if (something) is thrown on to (hard) ground, it is (itself) broken or injured, and causes no movement in the latter; if thrown into soft mire, it causes movement within a limited distance; if thrown into water, it causes a movement over a greater distance. Thus we may see that the softer a thing is, the more readily does it undergo movement and agitation...

Two words which distinctly indicate a mental connection between waves in water and air are *chhing* and *cho*. Their ordinary meaning is 'clear' and 'turbid' respectively, but in texts on acoustics they are used as technical terms. Chêng Hsüan says *chhing* means the six upper notes, while *cho* means the six lower notes. If a small stone is dropped into water it produces a relatively high-pitched sound, and generates small ripples close together;

moreover, being small it does not much disturb the bed of the lake or stream so that the water remains clear. If, on the other hand, a large stone is dropped in water, it produces a relatively loud deep sound, sending out large widely-spaced ripples, and it does disturb the river bed so that the water becomes turbid. Whether this theory of the origin of the terms *chhing* and *cho* be true, it certainly seems to fit the contexts where these words appear not entirely divorced from the other associations of sound such as timbre and volume.

Chinese thinking was correlative; it linked things together, and in its musical thought, the linking factor was *chhi*. 'The *chhi* of the Earth ascends above; the *chhi* of Heaven descends from the height' says the *Yo Chi* (Music Record) of the first century B.C.; Yin and Yang rub together as a consequence, so heaven and earth shake together and their drumming is thunder. 'Thus it is that music is a bringing together of Heaven and Earth.' Thirteen or fourteen centuries later there is naturally a more sophisticated approach. Chang Tsai, writing about A.D. 1060 in his *Chêng Mêng* (Right Teaching for Youth) on sound, says:

> The formation of sound is due to the friction (literally mutual grinding) between (two) material things, or *chhi* (or between material things and *chhi*). The grinding between two *chhi* gives rise to noises such as echoes in a valley or the sounds of thunder. The grinding between two material things gives sounds such as the striking of drumsticks on the drum. The grinding of a material thing on *chhi* gives sounds such as the swishing of feathered fans or flying arrows. The grinding of *chhi* on a material thing gives sounds such as the blowing of the reeds of a mouth-organ.

This extract shows both the strength and the weakness of the traditional Chinese approach to such problems. One must admire the ability to classify and distinguish, but making distinctions is not the same as being able to analyse a complex phenomenon into its components. Yet a little earlier, during the Southern Tang (A.D. 938 to 975) Than Chhiao or some other Taoist writing in the *Hua Shu* (Book of the Transformations in Nature) did better. He says:

> *Chhi* follows sound and sound follows *chhi*. When *chhi* is in motion sound comes forth, and when sound comes forth *chhi* is shaken.

This is an important contribution. The advent of sound transforms still air into *chhi* (air in a state of agitation) and air in agitation produces sound. The use in the text of the transitive verb *chen* 'to shake' is particularly significant, for it embodies so clearly the idea of vibration. It has occurred before in the description of the sounds produced by bells (page 380) as a 'staccato shaking'. One can well understand that the Chinese should have realised that sound is produced by the 'shaking' of the air if in fact they gave such close attention to the timbre of bells.

Than Chhiao goes on to point out that if sound is produced by disturbance of the *chhi*, all *chhi* everywhere will be in a state of disturbance and therefore capable of being heard wherever a hollow or resonant chamber exists to receive it. The mechanism he visualised comes from another quotation, and it is clear he thought of the world as an empty void with a potentiality for energy, for out of nothing it can produce power as in lightning. This lightning can, he says, produce *chhi* or atmospheric agitation, and atmospheric agitation can produce sound. Thus, within limits, he gives quite an accurate account of how the noise of thunder is generated.

Although Than Chhiao did not discuss hearing, the medieval Chinese did not overlook the question. Another Taoist, Thien Thung-Hsiu, in the *Kuan Yin Tzu* (Book of Master Kuan Yin), about A.D. 742, wrote:

> It is like striking a drum with a drumstick. The shape of the drum is possessed in my person (in the form of the ear). The sound of the drum is a matter of my responding to it.

A very interesting comment, for it seems he believed that sounds strike the inner ear like drumsticks on a drum, i.e. they exert a pressure, and what is more, it is the human response which enables one to describe this as sound.

From a somewhat garbled account in the Ming book *Hsiang Yen Lu* (Rustic Notes from Hsiang-shan) there is some reason to believe that at least by the fourth century A.D. there was also study of the speed of travel of echoes, or perhaps the time-interval between the visible action giving rise to a sound and its arrival at the observer's ear.

The detection of vibrations. The *Mo Tzu* book, in its discussion on fortifications, written perhaps as early as the fourth century B.C. by Chhin Ku-Li, mentions the use of hollow objects as resonators (geophones) for determining the presence and direction of tunnelling and mining by the enemy besieging a city. The text points out that if it seems that sapping and mining are taking place, excavations must be made.

> Within the city shafts are to be dug five paces distant from one another, to a depth of fifteen feet (4·6 m) below the level of the base of the city wall, until one reaches a three foot (0·9 m) depth of water. Then large pottery jars are to be prepared each of a size sufficient to hold more than 40 *tou* (i.e. more than 200 litres); their orifices are closed by a membrane of fresh skin, and they are sunk in the shafts. If men with good hearing are then set on watch to listen carefully,

they will be able to hear clearly in which direction the enemy is digging.

This dates from about 370 B.C. In Europe, the use of hollow shields as listening posts was made at the siege of Barca by the Persians (late sixth century B.C.). It is hard to believe that this technique did not arise out of independent empirical observations made in both East and West.

One of the most curious later applications of Chinese vibration detectors was the use of an instrument by the fishermen of northern Fukien for obtaining audible warning of the approach of shoals of fish. When the fishermen expect a shoal, they take a piece of bamboo about 5 cm in diameter and 1·5 m long, plunge it into the water to a depth of 1 m or so, and apply their ears to the upper hole of the bamboo near the boat. Western observers present have heard sounds like a confused distant rumble when the shoal was said by the fishermen to be some 1·5 km away, an estimate which was confirmed by the catch in due course.

The free reed. Something has been said earlier about the shamans trying to canalise *chhi* through pipes. The metallurgists were, however, also interested in this (perhaps indeed they were sometimes the same people), and hence in due time the process was bound to be mechanised. Chinese developments of special bellows and pumps will be discussed in the next volume of this abridgement; here it is only necessary to point out that there is a close connection between valves – or more specifically the valve flap or 'clack' – in a pump and reeds in musical instruments. The beating reed is exactly like the valve in that it can completely close the aperture, but the free reed is able to vibrate within the aperture.

The 'mouth-organ' (*shêng*) goes back far into the Chou, and the generally accepted view is that the principle of the free reed came to the West from China. The *shêng* is therefore the ancestor of the harmonica or reed-organ group of instruments (harmonium, concertina, accordion, etc.) and there is concrete evidence that it was transmitted through Russia in the nineteenth century.

On the other hand, the piston bellows was first applied to organs in Alexandria, and Vitruvius gives a minute description of them towards the end of the first century B.C. In the thirteenth century A.D., a reed-organ with bellows invented by the Arabs arrived in China and created so much interest it was reconstructed to play the Chinese scale. This Muslim invention preceded the European invention of the reed-organ in about A.D. 1460, while as reconstructed by the Chinese with free reeds ('apricot-leaves') it anticipated the European harmonium by no less than five and a half centuries.

The evolution of Equal Temperament

With such understanding of the nature of sound produced by vibrating strings and columns of air, it was to be expected that the Chinese should make some achievements in musical intonation. That European music owes them a debt in this respect is not usually appreciated.

In Western European music the Greek system of tuning and developing scales was for long the fundamental scheme of reference. Usually attributed to Pythagoras, it used the lengths of vibrating strings to give the basic reference notes; these were the octave and the fifth above a given note. A scale was then built up by upward and downward steps of octaves and fifths from a chosen note. Such a scheme, known as 'just intonation', had limitations, for the intervals between the notes in any scale depended on the note chosen as the starting point. It was not therefore possible to move from one scale to another: if an instrument was tuned, say, in C, then it could not play a scale of F, because the intervals between the notes in C required a different tuning from those in F. The Chinese had similar problems in their five-note (pentatonic) scales which were constructed by upward and downward steps of fifths only. However, to achieve harmonious music with choirs of men and women or boys and girls singing together, they also took the octave into account, although the just intonation they arrived at differed slightly from the Greek.

In Western Europe the tuning known as Equal Temperament arose in the seventeenth century. Here adjustments were made by slightly sharpening or flattening the tuned notes so that the musical intervals in every scale were the same. It was thus possible to move from one scale or key to another at will, a scheme which received its greatest publicity in the *Wohltemperierte Clavier* of J.S. Bach (1685 to 1750). Now in China in 1584, the distinguished mathematical and musicological scholar Chu Tsai-Yü, a Manchu prince, managed, after some thirty years of careful study and experiment, to produce an Equal Temperament method of tuning (Fig. 165). He arrived at his result mathematically by dividing the notes in the octave equally throughout, thus gaining a perfectly tempered scale. For example, for string and pipe lengths this meant dividing successively by the twelfth root of two ($\sqrt[12]{2}$), and for pipe diameters as well by the twenty-fourth root of two ($\sqrt[24]{2}$). His tempered scale was identical to that used in Europe at a later date.

A study of the appearance of Equal Temperament in Western Europe shows that it was unknown in 1577 but stated by the Flemish physicist Simon Stevin a little before 1620. Stevin's system was calculated on a similar basis to that adopted by Chu Tsai-Yü. It may be that this was an independent discovery, but if so it was the second remarkable invention of

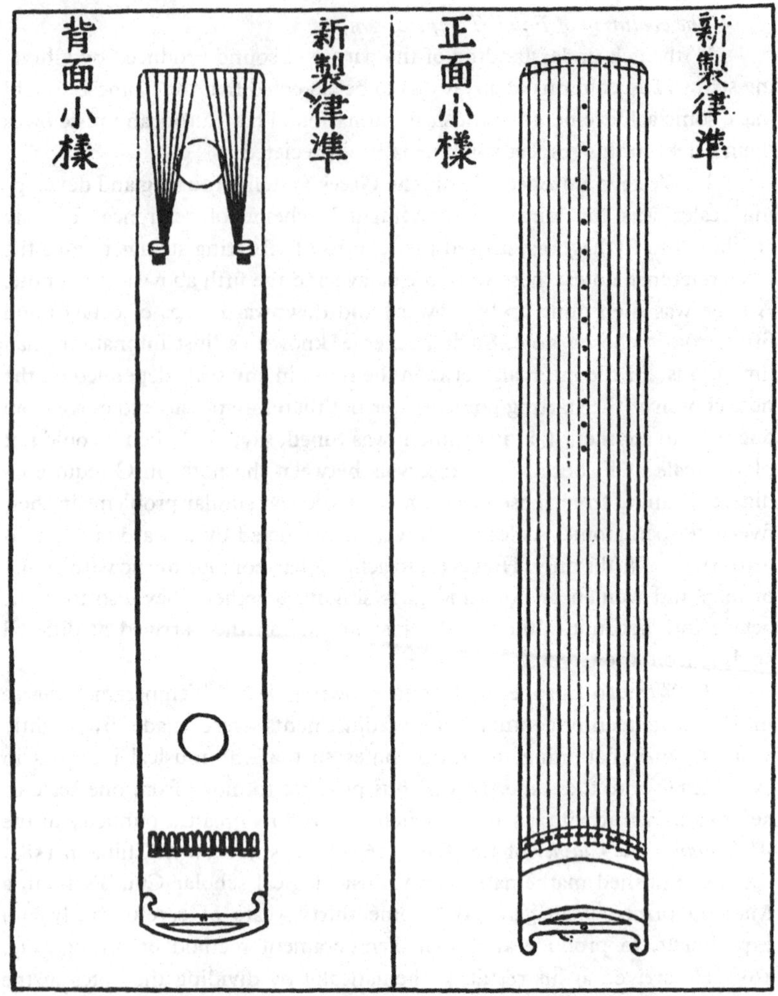

Fig. 165. Chu Tsai-Yü's tuning instrument (*chun*). From his *Lü Hsüeh Hsin Shuo* (A.D. 1584). The legends at the top read 'Sketch of the New *Chun*'; on the right the front, on the left the back of the instrument.

Stevin's which had previously appeared in China, the first being his sailing-carriage (see a later volume). In any case Chu Tsai-Yü's discovery may justly be regarded as a crowning achievement of China's two millennia of acoustic experiment and research.

Thus here on the borderline between science and art China and Europe came into the modern world hand in hand, even though Chu and Stevin probably never knew anything about each other.

Table of Chinese dynasties

Period	Chinese	Dynasty	Subdivision	Dates
	夏	HSIA kingdom (legendary?)		c. −2000 to c. −1520
	商	SHANG (YIN) kingdom		c. −1520 to c. −1030
	周	CHOU dynasty (Feudal Age)	Early Chou period	c. −1030 to c. −722
			Chhun Chhiu period 春秋	−722 to −480
			Warring States (Chan Kuo) period 戰國	−480 to − 221
First Unification	秦	CHHIN dynasty		−221 to −207
	漢	HAN dynasty	Chhien Han (Earlier or Western)	−202 to +9
			Hsin interregnum	+9 to + 23
			Hou Han (Later or Eastern)	+25 to +220
	三國	SAN KUO (Three Kingdoms period)		+221 to +265
First Partition	蜀		SHU (HAN)	+221 to +264
	魏		WEI	+220 to +265
	吳		WU	+222 to +280
Second Unification	晉	CHIN dynasty: Western		+265 to +317
		Eastern		+317 to +420
	劉宋	(Liu) SUNG dynasty		+420 to +479
Second Partition		Northern and Southern Dynasties (Nan Pei chhao)		
	齊		CHHI dynasty	+479 to +502
	梁		LIANG dynasty	+502 to +557
	陳		CHHEN dynasty	+557 to +589
	魏		Northern (Thopa) WEI dynasty	+386 to +535
			Western (Thopa) WEI dynasty	+535 to +556
			Eastern (Thopa) WEI dynasty	+534 to +550
	北齊		Northern CHHI dynasty	+550 to +577
	北周		Northern CHOU (Hsienpi) dynasty	+557 to +581
Third Unification	隋	SUI dynasty		+581 to +618
	唐	THANG dynasty		+618 to +906
Third Partition	五代	WU TAI (Five Dynasty period) (Later Liang, Later Thang (Turkic), Later Chin (Turkic), Later Han (Turkic) and Later Chou)		+907 to +960
	遼		LIAO (Chhitan Tartar) dynasty	+907 to +1124
			West LIAO dynasty (Qarā-Khiṭāi)	+1124 to +1211
	西夏	Hsi Hsia (Tangut Tibetan) state		+986 to +1227
Fourth Unification	宋	Northern SUNG dynasty		+960 to +1126
	宋	Southern SUNG dynasty		+1127 to +1279
	金		CHIN (Jurchen Tartar) dynasty	+1115 to +1234
	元	YUAN (Mongol) dynasty		+1260 to +1368
	明	MING dynasty		+1368 to +1644
	清	CHHING (Manchu) dynasty		+1644 to +1911
	民國	Republic		+1912 to +1949
		People's Republic		+1949

NOTE. When no modifying term in brackets is given, the dynasty was purely Chinese. During the Eastern Chin period there were no less than eighteen independent States (Hunnish, Tibetan, Hsienpi, Turkic, etc.) in the north. The term 'Liu chhao' (Six Dynasties) is often used by historians of literature. It refers to the south and covers the period from the beginning of the third to the end of the sixth centuries A.D., including (San Kuo) Wu, Chin, (Liu) Sung, Chhi, Liang and Chhen. The minus sign (−) indicates B.C., and the plus sign (+) is used for A.D.

BIBLIOGRAPHY

Balazs, E., *Chinese Civilisation and Bureaucracy; Variations on a Theme*, Yale University Press, New Haven, 1964; reprint in reduced format, 1968.

Beazley, C.R., *The Dawn of Modern Geography*, 3 vols., London, 1897 & 1901 (vols. 1 & 2), Oxford, 1906 (vol. 3).

Brown, Lloyd A., *The Story of Maps*, Little Brown, Boston, 1949.

Budge, E.A. Wallis, *Cuneiform Texts from Babylonian Tablets, etc. in the British Museum*, 41 vols., 1896–1931.

Bushell, S.W., 'The Early History of Tibet', *Journal of the Royal Asiatic Society*, 1880, N.S., **12**, 435.

Cameron, N., *Barbarians and Mandarins; Thirteen Centuries of Western Travellers in China*, Walker and Weatherhill, New York and Tokyo, 1970.

Cammann, S., 'The Evolution of Magic Squares in China', *Journal of the American Oriental Society*, 1960, **80**(2), 116.

Cammann, S., 'Old Chinese Magic Squares', *Sinalogica*, 1962, VII(1), 14.

Carter, T.F., *The Invention of Printing in China and its Spread Westward*, Columbia University Press, New York, 1925; revised editions 1931 and 1955.

Chang Kuang-Chih, *The Archaeology of Ancient China*, Yale University Press, New Haven and London, 1963.

Chhen Hêng-Chê (Sophia H. Chen Zen) (ed.), *Symposium on Chinese Culture*, China Institute of Pacific Relations, Shanghai, 1931.

Clark, D.H. and Stephenson, F.R., *The Historical Supernovae*, Pergamon, Oxford and New York, 1977.

Eberhard, W., *Conquerors and Rulers; Social Forces in Mediaeval China*, Brill, Leiden, 1952, 2nd edn; revised, 1965.

Eichhorn, W., *Chinese Civilisation; an Introduction*, Faber and Faber, London, 1969.

d'Elia, P.M., *Galileo in China*, Harvard University Press, Cambridge, Mass., 1960.

Fêng Yuan-Chün, *A Short History of Classical Chinese Literature*, tr. Yang Hsien-Yi and Gladys Yang, Foreign Languages Press, Peking, 1958; reprinted 1959.

Feuchtwang, S.D.R., *An Anthropological Analysis of Chinese Geomancy*, Vientiane, Laos, 1974.

Fitzgerald, C.P., *China; a Short Cultural History*, Cresset Press, London, 1935.

Goodrich, L. Carrington, *Short History of the Chinese People*, Harper, New York, 1943.

Graham, A.C., *Later Mohist Logic, Ethics and Science*, Chinese University Press, Hong Kong, and School of Oriental and African Studies, London, 1978.

Guitel, G., *Histoire comparée des numérations écrites*, Flammarion, Paris, 1975.

Hartner, W., 'The Obliquity of the Ecliptic according to the *Hon Han Shu* and Ptolemy.' Communication at the 23rd International Congress of Orientalists, Cambridge, 1954.

Herrmann, A., *Historical and Commercial Atlas of China*, Harvard-Yenching Institute, Cambridge, Mass., 1935; 2nd edition, *An Historical Atlas of China*, ed. N. Ginsburg, with preface by P. Wheatley, Edinburgh University Press, Edinburgh, and Aldine, Chicago, 1966.

Hirth, F. and Rockhill, W.W. (trs.), *Chau Ju-Kua; His work on the Chinese and Arab Trade in the 12th and 13th centuries, entitled 'Chu-Fan-Chi'*, Imperial Academy of Science, St Petersburg, 1911.

Ho Ping-Ti, *The Cradle of the East, an Enquiry into the Indigenous Origins of Techniques and Ideas of Neolithic and Early Historic China, 5000 B.C. to 1000 B.C.*, Chinese University, Hong Kong, and University of Chicago Press, Chicago, 1975.

Ho Ping-Yü (Ho Peng Yoke), 'Chin Chiu-shao', *Dictionary of Scientific Biography*, 1971, vol. 3, p. 249, Scribner, New York, 1971; and 'Chu Shih-chieh', vol. 3, p. 265; and 'Li Chih', vol. 8, p. 313, Scribner, New York, 1973.

Hsi Tse-tsung *et al.*, 'Heliocentric Theory in China – in Commemoration of the Quincentenary of the Birth of Nicholaus Copernicus', *Scientica Sinica*, 1973, 16, 364.

Hucker, C.O., *China, a Critical Bibliography*, University of Arizona Press, Tucson, 1962; reprinted 1964 and 1966.

Hudson, G.F., *Europe and China; A Survey of their Relations from the Earliest Times to 1800*, Arnold, London, 1931.

Hughes, E.R., *Chinese Philosophy in Classical Times*, Dent, London, 1942 (Everyman's Library, no. 973).

Jacobs, N., *The Origin of Modern Capitalism and Eastern Asia*, University Press, Hong Kong, 1958.

Kaltenmark, O., *Chinese Literature*, tr. from the French (1948) by A.M. Geoghegan, Walker, New York, 1964.

Karlgren, B., *Sound and Symbol in Chinese*, Oxford, 1923, reprinted 1946.

Kroeber, A.L., *Configurations of Chinese Growth*, University of California Press, Berkeley and Los Angeles, 1944.

Lam Lay Yong, *A Critical Study of the Yang hui suan fa. A Thirteenth-Century Chinese Mathematical Treatise*, Singapore, 1977.

Lattimore, O., *Inner Asian Frontiers of China*, Oxford University Press, London and New York, 1940.

Lattimore, O. and Lattimore, E. (eds.), *Silks, Spices and Empire; Asia seen through the Eyes of its Discoverers*, Delacorte, New York, 1968.

Laufer, B., *Sino-Iranica; Chinese Contributions to the History of Civilisation in Ancient Iran*, Field Museum of Natural History (Chicago) Publications – Anthropological Series, 1919, vol. 15, no. 3 (Pub. no. 201) (rev. and crit. Chang Hung-Chao, Memoirs of the Chinese Geological Survey, 1925 (ser. B), no. 5).

Laufer, B., *Chinese Pottery of the Han Dynasty*, Brill, Leiden, 1909; reprinted Tientsin, 1940.

Libbrecht, U., *Chinese Mathematics in the Thirteenth Century. The Shu-shu chiu-chang of Ch'in Chiu-shao* (MIT East Asian Science Series, vol. 1), MIT Press, Cambridge, Mass., 1973.

Loewe, M., *Imperial China; the Historical Background to the Modern Age*, Allen and Unwin, London, 1966.

Loewe, M., *Everyday Life in Early Imperial China; during the Han Period, 202 B.C. to A.D. 220*, Batsford, London, and Putnam, New York, 1968.

Maeyama, Y., 'On the Astronomical Data of Ancient China (ca. −100 − +200): A Numerical Analysis', *Archives internationales d'histoire des sciences*, 1975, **25**, 247; and 1976, **26**, 27, 'Quantitative study of the *Hsing ching*, of unprecedented astronomical sophistication'.

Moule, A.C., *Christians in China before the year 1550*, Society for the Promotion of Christian Knowledge, London, 1930.

Moule, A.C. and Pelliot, P. (tr. and annot.), *Marco Polo (A.D. 1254 to A.D. 1325); The Description of the World*, 2 vols., Routledge, London, 1938; reprinted AMS Press, New York, 1976.

Nakayama, S., *A History of Japanese Astronomy. Chinese Background and Western Impact*, Harvard University Press, Cambridge, Mass., 1969.

Needham, Joseph, 'The Translation of Old Chinese Scientific and Technical Texts', in *Aspects of Translation*, ed. A.H. Smith, Secker and Warburg, London, 1958, p. 65, Studies in Communication, no. 2; and *Babel*, 1958, **4** (no. 1), 8.

Needham, Joseph, Wang Ling and Price, D.J. de S., *Heavenly Clockwork; the Great Astronomical Clocks of Mediaeval China*, Cambridge University Press, Cambridge, 1960 (Antiquarian Horological Society Monographs, no. 1).

Needham, Joseph, *Time and Eastern Man* (Henry Myers Lecture, Royal Anthropological Institute, 1964), Royal Anthropological Institute, London, 1965.

Needham, Joseph and Lu Gwei-Djen, 'The Optick Artists of Chiangsu', *Proceedings of the Royal Microscopical Society* (Oxford Symposium Volume), 1967, **2**, 113.

Needham, Joseph, *The Grand Titration; Science and Society in China and the West* (Collected Addresses), Allen and Unwin, London, 1969.

Needham, Joseph, *Clerks and Craftsmen in China and the West* (Collected Lectures and Addresses), Cambridge University Press, Cambridge, 1970.

Phillips, G., 'The Seaports of India and Ceylon, described by Chinese Voyagers of the Fifteenth Century, ...', *Journal of the North China Branch of the Royal Asiatic Society*, 1885, **20**, 209.

Reichwein, A., *China and Europe; Intellectual and Artistic Contacts in the Eighteenth Century*, Kegan Paul, London, 1925, tr. from the German edition, Berlin, 1923.

de Reparaz-Ruiz, G., 'Historia de la Geographia de España' in *España, la Tierra, el Hombre, el Arte*, vol. 1, Barcelona, 1937.

Rufus, W.C., 'Astronomy in Korea', *Journal (Transactions) of the Korea Branch of the Royal Asiatic Society*, 1936, **26**, 1.

Sarton, George, *Introduction to the History of Science*, vol. 1, 1927; vol. 2, 1931 (2 parts); vol. 3, 1947 (2 parts); Williams and Wilkins, Baltimore (Carnegie Institution Publication, no. 376).

Schafer, E.H., *The Golden Peaches of Samarkand; a Study of Thang Exotics*, University of California Press, Berkeley and Los Angeles, 1963; Rev. J. Chmielewski, *Orientalische Literatur-Zeitung*, 1966, **61**, 497.

Schafer, E.H., *The Vermilion Bird; Thang Images of the South*, University of California Press, Berkeley and Los Angeles, 1967.

Schafer, E.H., *Pacing the Void; Thang Approaches to the Stars*, University of California Press, Berkeley and Los Angeles, 1977.

Sickman, L. and Soper, A., *The Art and Architecture of China*, Penguin (Pelican), London, 1956.

Singer, C., *A Short History of Scientific Ideas to 1900*, Oxford University Press, Oxford, 1959.

Singer, C., Holmyard, E.J. and Williams, T.I. (eds.), *A History of Technology*, 5 vols., Oxford University Press, Oxford, 1954–8.

Sivin, N., *Cosmos and Computation in Early Chinese Mathematical Astronomy*, Brill, Leiden, 1969.

Sivin, N., 'Copernicus in China', *Studia Copernicana*, 1973, **6**, 63.

Sivin, N., 'Shen Kua', *Dictionary of Scientific Biography*, vol. 12, p. 369, Scribner, New York, 1975.

Sivin, N., 'Wang Hsi-shan', *Dictionary of Scientific Biography*, vol. 14, p. 159, Scribner, New York, 1976.

Sivin, N. (ed.), *Science and Technology in East Asia*, Science History Publications, New York, 1977.

Sivin, N. and Graham, A.C., 'A Systematic Approach to the Mohist Optical Propositions', in *Chinese Science*, ed. N. Sivin and S. Nakayama, MIT Press, Cambridge, Mass., 1972.

Sivin, N. and Nakayama, S., *Chinese Science: Explorations of an Ancient Tradition* (MIT East Asian Science Series, vol. 2), MIT Press, Cambridge, Mass., 1972.

Stephenson, F.R. and Clark, D.H., *Applications of Early Astronomical Records*, Hilger, Bristol, 1978.

Teich, M. and Young, R., *Changing Perspectives in the History of Science*, Heinemann, London, 1973.

Têng Ssu-Yü and Biggerstaff, K., *An Annotated Bibliography of Selected Chinese Reference Works*, Harvard-Yenching Institute, Peiping, 1936.

Treistman, J.M., *The Prehistory of China; an Archaeological Exploration*, David and Charles, Newton Abbot, 1972.

Waley, A. (tr.), *The Way and its Power; a study of the 'Tao Tê Ching' and its Place in Chinese Thought*, Allen and Unwin, London, 1934.

Waley, A., *Three Ways of Thought in Ancient China*, Allen and Unwin, London, 1939.

White, W.C. and Millman, P.M., 'An Ancient Chinese Sun-Dial', *Journal of the Royal Astronomical Society of Canada*, 1938, **32**, 417.

von Wiethof, B., *An Introduction to Chinese History, from Ancient Times to 1912*, Thames and Hudson, London, 1975.

Willetts, W.Y., *Foundations of Chinese Art; from Neolithic Pottery to Modern Architecture*, Thames and Hudson, London, 1965, revised, abridged and rewritten version with many illustrations in colour.

Yabuuchi, K., 'Astronomical Tables (Calendars) in China from the Han to the Thang Dynasty'. English article in *Studies in the History of Science and Technology in Mediaeval China* (Chūgoku Chūsei Kagaku Gijutsushi no Kenkyū), Tokyo, 1963.

Yabuuchi, K., 'Astronomical Tables in China [the 'Calendars'] from the Wu Tai Period to the Chhing Dynasty', *Japanese Studies in the History of Science*, 1963, 2, 94.

Yang, Lien-Shêng, *Money and Credit in China; a Short History*, Harvard University Press, Cambridge, Mass., 1952.

Yule, Sir Henry, *Cathay and the Way Thither; being a Collection of Mediaeval Notices of China*, 1st edn 1866, Hakluyt Society Pubs, 2nd ser., London, 1913–15, revised by H. Cordier in 4 vols.

Yule, Sir Henry (ed.), *The Book of Ser Marco Polo the Venetian, concerning the Kingdoms and Marvels of the East*, tr. and ed. with notes, by H. Yule, 1st edn 1871, reprinted 1875; 3rd edn, 2 vols., ed. H. Cordier, Murray, London, 1903, reprinted 1921; 3rd edn also issued Scribner, New York, 1929.

INDEX

Printed in the United States
By Bookmasters